THE BRITISH ISLANDS

AND THEIR

VEGETATION

VOLUME II

For the convenience of readers this new edition is issued in two volumes. The list of contents for the whole work appears in each volume. The pagination is continuous and the index is at the end of Volume II.

The two volumes are not sold separately.

THE BRITISH ISLANDS

AND THEIR

VEGETATION

BY

A. G. TANSLEY, Kt,

M.A., F.R.S.

Sherardian Professor Emeritus of Botany, Oxford
Lately Chairman of the Nature Conservancy (Privy Council)
Honorary Fellow of Trinity College, Cambridge

With 162 Plates containing 418 photographs
and 179 Figures in the text

VOLUME II

VOLUME I contains Parts I—IV
VOLUME II contains Parts V—IX

Fourth Impression

CAMBRIDGE
AT THE UNIVERSITY PRESS
1965

PUBLISHED BY
THE SYNDICS OF THE CAMBRIDGE UNIVERSITY PRESS
Bentley House, 200 Euston Road, London, N.W.1
American Branch: 32 East 57th Street, New York, N.Y. 10022

First printed 1939
Reprinted with corrections 1949
and issued in two volumes
Reprinted 1953
1965

First printed in Great Britain at the University Press, Cambridge
Reprinted by offset-litho
by John Dickens & Co. Ltd, Northampton

CONTENTS

PART I

THE BRITISH ISLANDS AS ENVIRONMENT OF VEGETATION, *pp.* 1–146

CHAPTER I

PHYSICAL FEATURES AND GEOLOGICAL HISTORY, *pp.* 3–28

The continental shelf, 3. Endemic species, 4. Palaeogenic and Neogenic, 4. THE PALAEOGENIC REGION, 8. Scotland and northern England, 8. Wales, 10. South-western peninsula, 10. Ireland, 10. The great tectonic folds, 11. Caledonian folds, 11. Armorican folds, 13. Existing sculpture, 13. Palaeogenic soils, 14.

THE NEOGENIC REGION, 15. Permian and Trias, 16. The later Secondary rocks, 16. Jurassic, 18. Cretaceous, 18. Wealden, 18. Lower Greensand, 19. The Upper Cretaceous sea, 21. The Chalk, 21. Clay-with-Flints, 23. Eocene, 23. Oligocene, 23. Pliocene, 25. The Pleistocene ice age, 25. Glacial deposits, 25. References, 28.

CHAPTER II

CLIMATE, *pp.* 29–54

Climate and the distribution of vegetation, 29. Microclimates, 29. Heat and moisture, 30. Temperature and rainfall, 30. Ratio of precipitation to evaporation, 31. Saturation deficit, 33. The "Meyer ratio", 33. Diurnal variation of saturation deficit, 34. Seasonal variation, 35. Mist and fog, 35. Sunshine, 36. Frost, 39. Snow, 40. Wind, 41. Direction, 41. Velocity, 41. Wind at high altitudes, 43. "Arctic-alpine" climate, 43.

General characters of the climate, 43. Distribution of barometric pressures, 43. Passage of depressions, 46. Cyclonic rainfall, 46. Orographical rainfall, 47. Convectional rainfall, 47. Effects of the westerly winds, 47. Highest precipitations, 53. Lowest precipitations, 54. References, 54.

CHAPTER III

REGIONAL CLIMATES, *pp.* 55–77

Maritime climates, 55. Extreme Atlantic climate, 57. West coast of England and Wales, 59. South coast climate, 60. East coast climate, 61. Northern inland climate, 63. Scottish mountain climate, 63. English midland climate, 66. Irish inland stations, 67.

Seasonal distribution of rainfall, 67. Deviations from monthly and annual means, 68.

CHAPTER IV

SOIL, *pp.* 78–100

CHAPTER V

DISTRIBUTION OF ROCKS AND THE SOILS THEY PRODUCE, *pp.* 101–126

CHAPTER VI

THE BIOTIC FACTOR, *pp.* 127–146

PART II

HISTORY AND EXISTING DISTRIBUTION OF VEGETATION, *pp.* 147–210

CHAPTER VII

PRE-HISTORY, *pp.* 149–170

CHAPTER VIII

THE HISTORICAL PERIOD, *pp.* 171–193

PART IV

THE WOODLANDS, *pp.* 241–484

CHAPTER XII

NATURE AND STATUS OF THE BRITISH WOODLANDS, *pp.* 243–266

DOMINANT AND OTHER TREES. THE MORE IMPORTANT SHRUBS

CHAPTER XIII

OAKWOOD. INTRODUCTORY, *pp.* 267–290

PEDUNCULATE OAKWOOD (QUERCETUM ROBORIS)

CHAPTER XVII

MIXED OAKWOOD ON SANDY SOILS, *pp.* 350–357

(QUERCETUM ROBORIS ET SESSILIFLORAE OR QUERCETUM ERICETOSUM)

CHAPTER XVIII

BEECHWOOD, *pp.* 358–385

INTRODUCTORY. BEECHWOOD ON CALCAREOUS SOIL (FAGETUM CALCICOLUM). CHALK SCRUB AND YEW WOOD. SERAL ASHWOOD

CHAPTER XIX

BEECHWOOD ON LOAM (FAGETUM RUBOSUM), *pp.* 386–407

CHAPTER XX

BEECHWOOD ON SANDS AND PODSOLS (FAGETUM ARENICOLUM OR ERICETOSUM), *pp.* 408–426

SUMMARY OF BRITISH BEECHWOODS

CHAPTER XXI

ASHWOOD ON LIMESTONE (FRAXINETUM CALCICOLUM), *pp.* 427–443

CHAPTER XXII

PINE AND BIRCH WOODS, *pp.* 444–459

Volume II begins here

CHAPTER XXX

THE CUMBRIAN LAKES, *pp.* 596–621

CHAPTER XXXI

THE VEGETATION OF RIVERS, *pp.* 622–633

CHAPTER XXXII

MARSH AND FEN VEGETATION, *pp.* 634–647

CHAPTER XXXIII

THE EAST ANGLIAN FENS. NORTH-EAST
IRISH FENS, *pp.* 648–672

SUMMARY OF THE LATER HYDROSERES

CHAPTER XXXVII

THE HEATH FORMATION (*continued*), *pp.* 743–772

PART VIII

MOUNTAIN VEGETATION, *pp.* 773–813

CHAPTER XXXVIII

THE UPLAND AND MOUNTAIN HABITATS. MONTANE AND ARCTIC-ALPINE VEGETATION, *pp.* 775–796

CHAPTER XXXIX

ARCTIC-ALPINE VEGETATION (*continued*), *pp.* 797–813

PART IX

MARITIME AND SUBMARITIME VEGETATION, *pp.* 815–903

CHAPTER XL

INTRODUCTORY.   THE SALT MARSH FORMATION, *pp.* 817–843

CHAPTER XLI

THE FORESHORE COMMUNITIES.   COASTAL SAND DUNE VEGETATION, *pp.* 844–867

CHAPTER XLII

SHINGLE BEACHES AND THEIR VEGETATION, *pp.* 868–894

CHAPTER XLIII

SUBMARITIME VEGETATION, *pp.* 895–902

FIGURES IN THE TEXT

PLATES

Part V

THE GRASSLANDS

Chapter XXV

NATURE AND STATUS OF THE BRITISH GRASSLANDS

Grassland as biotic plagioclimax. While the majority of our deciduous woodlands are properly regarded as more or less modified climax communities, as seral woods clearly leading to climax, or as planted but roughly equivalent representatives of one of these, the vast bulk of our "permanent" and semi-natural grasslands, which occupy between them a far greater area of the country than any other type of vegetation (Chapter IX), are subclimax or (better) biotic plagioclimax vegetation, i.e. vegetation stabilised by pasturing (Chapter X). In other words, if pasturing were withdrawn their areas would be invaded and occupied, as they were originally occupied, by shrubs and trees.

Raunkiaer has pointed out, as already explained in Chapter XI, that the vegetation of the world can be classified in two distinct ways which give very different results. We may either divide it into regions according to the life forms of the dominant species of the climatic climax communities, such as tropical rainforest, temperate continental grassland, summer deciduous forest, northern coniferous forest, and so on; or we may calculate the percentages belonging to the different life forms represented in the total flora of an area, and divide our regions according to the life form or life forms which show a marked excess over their average percentage in the flora of the world. If we use this second method we find that the Mediterranean region, for example, in which the climax communities are dominated by evergreen sclerophyllous shrubs or small trees (*nanophanerophytes* or *microphanerophytes* in Raunkiaer's terminology) becomes a region of *therophytes* (annual plants), because the species of these are preponderant in number in the whole flora; and the temperate region of Central and Western Europe (including most of the British Isles) in which the climax communi-
The British ties are those of the deciduous summer forest, dominated by
Isles a *mesophanerophytes*, is a *hemicryptophyte* region because about
hemicrypto-
phyte region 50 per cent of all the native species are hemicryptophytes—
plants whose perennating buds are situated in or immediately in contact with the surface layer of the soil—against 27 per cent in the flora of the world as a whole.

Now in the grasslands of Britain the large majority of the species of flowering plants are hemicryptophytes. Thus in the chalk grassland of the South Downs 72 per cent of the whole number of species of flowering plants are hemicryptophytes, and 88 per cent of the thirty-three commonest species, making up the great bulk of the vegetation, belong to this life form (Tansley and Adamson, 1926, p. 31). Of our common meadow grasses the vast majority are hemicryptophytes. Thus the British grasslands are essentially hemicryptophytic communities, and we may conclude that the British climate is specially favourable to grassland because grassland is the most characteristic hemicryptophyte vegetation. If the dominant phanerophytic life forms, the trees and shrubs, are destroyed and prevented from returning by regular pasturing (or mowing) they are naturally replaced by hemicryptophytes. While the climate permits the establishment of forest on most soils it is also eminently suitable for grasses. Most of the grasses, and notably the meadow-grasses, require constant moisture during their main growing season (late spring and early summer) if they are to make good vegetative growth, and the predominantly moist climate and well distributed rainfall are excellently suited to their needs. This is most conspicuously seen in the more markedly Atlantic climate in the west of Great Britain and in Ireland (the "Emerald Isle") where the grasslands are particularly luxuriant. In years when rain falls at frequent intervals during the whole summer, as it often does, the meadow grasses remain green and continue to grow during the whole season. After a hot dry summer when the grasslands turn brown after the flowering period, the occurrence of heavy autumn rains leads to a resumption of growth in all but the coldest climates, and this is important for the grazing of stock. Deficiency of rain, low temperatures, and cold, drying winds during the primary growing season (April and May) are particularly inimical to grass vegetation and result in a poor hay crop and deficient pasture.

Turf or "sole" of grassland. The turf-forming grasses are specially well adapted to withstand and flourish under constant pasturing. They naturally form numerous lateral shoots ("tillers") from the base of the primary shoot axis and when the upward growth of these is prevented by the eating or mowing off of the tops, fresh buds are formed at their base, either in or just above the soil surface, and grow out laterally into new vegetative branches, till the whole surface of the soil is covered by a continuous felt-work of leafy shoots—the typical turf or "sole" of good pasture or a well-tended lawn. Flowering and the consequent expenditure of reserves on seed production is very largely prevented and the whole energy of the plants is devoted to vegetative growth. Fresh mineral food must of course be supplied to make up for what is removed by the stock or the mowing machine. In a pasture this is partly furnished by the dung of the grazing animals, often supplemented by extra supplies spread by hand in the winter, but since this is deficient in calcium and phosphorus, lime and phosphates should also be added. A mown lawn has to be constantly manured to replace the nitrogen lost, for example with sulphate of ammonia, and on many soils also with lime, phosphate and potash, to replace the losses involved in mowing. Otherwise the grasses languish, lose their competitive power, and are largely replaced by "weeds"—often deep-rooted herbaceous plants which draw their water and mineral supplies from lower levels of the soil, e.g. dandelion (*Taraxacum*), cat's ear (*Hypochaeris radicata*) and restharrow (*Ononis*), or by mosses which make very small demands.

Effects of overgrazing and undergrazing. In an "overgrazed" pasture the grasses are sometimes grazed down so close to the soil that many of their lateral buds are destroyed, and so much of the plant is eaten away as to impair its powers of vegetative regeneration. This result is seen in its extreme form in grassland heavily infested with rabbits. To keep a pasture or lawn "in good heart" an equilibrium must be established between the supply of mineral food and of water (in the form of frequent rain or from an artificial source such as a lawn "sprinkler") and the constant removal of aerial shoots. In an "undergrazed" pasture tall "rank" growth and flowering are not adequately prevented, so that fresh young shoots, which are by far the most nutritious for stock, are not produced, and no good turf is formed. Also the dead remains of the shoots of many grasses may form a mat or "mattress" which obstructs fresh growth. Then the grassland is invaded either by heath plants or by the seedlings of shrubs and trees, or by both together, and tends to "revert", as the agriculturist would say, to heath or woodland, or, as the ecologist would say, the normal succession to heath or woodland, interrupted by regular pasturing, reasserts itself. The earlier stages of this process can be seen in the pastures left derelict as the result of agricultural depression (cf. Chapter XXIV, p. 480).

"Rough grazings." Besides the large area of "permanent" pasture, which is fenced so that grazing can be regulated, and is often manured in addition, there is a great area of "rough grazings", mostly on open commons in the lowlands and on hillsides and mountain sides, over which sheep and sometimes cattle or ponies are allowed to range in summer. The vegetation of these is largely grass, but partly heath. The grassland is for the most part only lightly pastured and is open to invasion by heath plants and in the more protected places at lower altitudes by scrub and even by trees. The palaeogenic and partly mountainous country of Wales, Scotland and the north of England which is less suitable for agricultural crops naturally bears a much larger proportion of this type of vegetation than the neogenic lowlands which form the greater part of England.

The whole of this hillside vegetation, so far as it is dominated by grasses, was grouped under the general name of "hill pasture" by Robert Smith, the pioneer of ecological survey in Great Britain, in his papers on the Edinburgh District and on North Perthshire (1900). He distinguished different types on the different soils of various hill ranges. In the subsequent surveys of the southern Pennines (west Yorkshire) by W. G. Smith, Moss, and Rankin (1903) "heath pasture", with grasses dominant but heath plants associated, was distinguished from "natural pasture" mostly on limestones, without heath plants and with calcicolous associates. F. J. Lewis (1904), in his work on the northern

Types of rough grazing

Pennines, used the term "grass heath" instead of "heath pasture". In *Types of British Vegetation* (1911) the comprehensive term "siliceous grassland" was used for this kind of vegetation, because it is developed on soil derived from rocks containing a large proportion of silica and deficient in basic minerals; while "limestone grassland" was used as roughly equivalent to the "natural pasture" of the earlier surveyors, because it is mainly developed on limestones, though also on other basic rocks. In the present work the vegetation is described as communities named from the dominant plants in each case, and the communities are grouped according to soil reaction, which on the whole corresponds very well with their natural affinities. Thus the soil of the "grass heath" or "siliceous grassland" is always acid in reaction, that of the "natural pasture" or "limestone grassland" and related types is primarily alkaline because the soil is derived from basic rocks. Owing however to strong leaching under the heavy rainfall of many of the districts in which limestone grassland occurs, and the related accumulation of acid humus, the surface layers even of limestone soil may show an acid reaction and characteristic heath plants are often present.

Siliceous grassland [side note]

Invasion by heath and peat plants. Most of these grassland communities are based on an acid soil and many are developed in immediate proximity to peat moors, so that they are subject to invasion by heath and moor plants, and all transitions are found between the heath and moor communities proper and the grassland communities. The drier siliceous grasslands are invaded by heath plants (ling, bilberry, etc.) and easily pass into upland heath, while the moister may pass into communities dominated by various peat-forming plants which require a damp or wet soil. Grazing is the main and often the sole factor which maintains the dominance of the pasture grasses. This is notably the case with the drier and better drained grassland on the siliceous rocks, dominated mainly by the common bent (*Agrostis tenuis*) and the sheep's fescue (*Festuca ovina*), often with the red fescue (*F. rubra*), which form the basis of most of these hill pastures. This "grass heath" was perhaps all at one time covered with forest and is subject to invasion by heath and sometimes by scrub and trees. The bent-fescue community is also often invaded by the mat-grass (*Nardus stricta*) which has a rather wide range of edaphic habitats and is often dominant on the drier peats, giving rise to a community which has been called "grass moor". The purple moor-grass (*Molinia caerulea*) with an even wider range of habitats but a higher water requirement, is dominant on the peats which are wetter (the water completely stagnant) of many hill slopes of the north and west. While the *Nardus* grasslands are of little value for pasture, those dominated by *Molinia* are quite useful in the early summer (Chapter XXVI).

Bent-fescue community [side note]

Nardetum and Molinietum [side note]

Limestone grassland. Contrasting with the "siliceous grassland" there are in the north and west extensive areas of limestone and other basic rocks (igneous or metamorphic) which bear a distinctive grassland community, generally dominated by *Festuca ovina* or *F. rubra* or both and often referred to by the earlier surveyors of vegetation as "natural pasture", because they afford particularly good grazing for sheep and are as a whole less accessible to colonisation by woody plants. Much of this area too was probably formerly occupied by woodland or scrub, and owes its existing condition to grazing. With it must be classed the chalk and oolite grasslands of southern England (Chapter XXVII).

Arctic-alpine grassland. Closely related edaphically to the limestone grassland is the arctic-alpine grassland developed on basic rocks on the slopes of some of the higher mountains, mostly on southern aspects above 2000 ft. (610 m.). Like the limestone grassland this arctic-alpine community is usually dominated by the sheep's fescue (*Festuca ovina*), but usually the viviparous form, sometimes accompanied by an abundance of the alpine lady's mantle (*Alchemilla alpina*), and a mixture of arctic-alpine species with others from lower altitudes. The true arctic-alpine grassland is certainly a climax vegetation, not primarily dependent on grazing, and belongs of course to the arctic-alpine formation developed above the upper altitudinal limit of trees and shrubs (Chapter XXXVIII); but some of these grasslands, even at high levels, are grazed in summer by sheep.

Other grasslands. Less extensive types of grassland occur in a great variety of other situations. Some occupy man-made habitats like the grass "verges" of country roads, stabilised by cutting and sometimes by pasturing (see p. 561). Some are entirely natural seral types, such as the maritime *Glyceria maritima* and *Festuca rubra* communities of salt marsh (Chapter XL), which may be stabilised by pasturing, and the marram grass community of sand dunes (Chapter XLI): others are maritime rock and cliff communities constantly subjected to salt spray: others again submaritime cliff top (Chapter XLIII) and sand dune communities partly stabilised by grazing. Existing alluvial grasslands on the flood plains of rivers are probably always largely determined by grazing or cutting for hay; but other quite natural seral types exist on the borders of lakes and slow rivers and represent stages of the hydrosere (succession from fresh water), though these also may be stabilised by mowing or pasturing (Chapters XXXII, XXXIII).

Finally, streaks and patches of grassland may occur in heath and moor communities as the result of flushing by water relatively rich in mineral salts, along the lines of paths, or in places where sheep or cattle congregate as the result of manuring and trampling. At the sides of roads, too, the blowing of dust, especially when it contains lime, mineralises a wide verge. This encourages the growth of pasture grasses, and thus more intensive grazing, which leads to the development and maintenance of a good grass sward on each side of the road in place of the original moorland or heath vegetation (Stapledon, 1935). And there are almost endless other situations where the local conditions lead to development of communities dominated by grasses.

Common features of grassland habitats. The dominance of grasses in such a wide variety of situations is the most convincing evidence of the suitability of the British climate to this form of vegetation. The question now arises: can we find any common condition of these grass habitats, apart from the grazing factor, which will differentiate them as a whole from the habitats of the other forms of vegetation which are dominant in the same general climate? It must of course be remembered that though the great majority of our native grasses are hemicryptophytes there are a great many herbs also belonging to this life form, and that the hemicryptophytic grasses themselves vary a great deal in their soil and water requirements: further, that there are a number of grasses which are rhizome geophytes (such for example as *Phragmites, Ammophila* and species of *Agropyron*) whose habit of growth is very different from that of the hemicryptophytic meadow grasses. But most of these rhizomic grasses do not form grassland in the ordinary sense, and it remains true that turf and meadow grasses in the wide sense, such as the species of *Poa, Festuca, Agrostis, Avena, Dactylis, Alopecurus, Lolium, Cynosurus, Anthoxanthum,* etc. do have habit characters and habitat requirements sufficiently similar to be correlated with corresponding habitats.

The general nature of the grassland habitats has been best expressed by Smith and Crampton (1914). In regard to climate they point out that in Britain the "cold temperate moist summers and open winters favour grassland from sea level to the highest elevations"; but that the moist cold temperate climate also favours leaching and consequent exhaustion of soluble mineral salts in the surface soils of quickly drained places, leading to soil acidity and accumulation of raw humus or peat which are the great physical enemies of grassland, apart from the "grass moors" dominated by *Nardus, Molinia* and *Deschampsia flexuosa*. Hence grassland establishes itself on finely divided thin residual soil (correlated with the shallow and fibrous roots of most grasses) developed from a soluble or smooth-weathering rock basis, such as we get on chalk downs and other limestone rocks, and on dolerite and other hard basic rocks; or alternatively where there is periodical flushing of sloping surfaces with waters containing alkaline salts in solution. Grassland developed in the former type of habitat is considered by Smith and Crampton as "stable", in the latter "migratory". These last, they say, "are mainly found on alluvial rain-washed or spring-flushed surfaces along the rivers and coastal belts and on the flanks of mountains". Such grasslands "depend on periodic flushing, flooding and renewal of surface fertility...". In the absence of this dynamic factor constantly renewing fertility of the soil, leaching impoverishes the surface layers of the well-drained soils and heath tends to come in, while on the

Climatic factors

Edaphic factors

badly drained soils marsh and fen, or where soluble salts are deficient bog or "moss" are developed.

As we pass west and north in the British Isles the climatic factors hostile to forest are greatly intensified and lead to the increase of leaching and accumulation of raw humus, so that the heath, bog or "moss" vegetation becomes the most serious competitor of grassland: in the midlands and the east and south-east, with higher summer temperatures and lower rainfall, it is forest which naturally supervenes when grassland is left ungrazed. Superposed on these climatic and edaphic factors is the biotic factor of grazing, and this, as explained earlier in the chapter, everywhere favours grassland against the other plant formations, and may override the climatic and edaphic factors. Even the most strongly leached soils, in Wales, for example, remain grassland as long as they are grazed and not waterlogged.

Habit forms. Smith and Crampton divide grassland into five main types according to the habit of the dominant grasses. Of these the *turf-forming* type, characteristically the grazing type, is by far the most important, and next to this the related *meadow* type, consisting mainly of taller growing grasses, traditionally cut for hay, and not forming so close a turf as the first type, partly because of their freer natural habit and largely because their upward growing shoots are not constantly removed. The general vegetative economy of these two types has already been described. The *tussock* type is characteristically formed of coarse hard grasses whose dead basal parts decay with difficulty, accumulating soil and humus between the masses of shoot bases, so that a "tussock" is formed and gradually increases in size. The dominant peat grass *Nardus* belongs to this type, which is also represented by various species of sedge. The *stooled* meadow type (e.g. *Deschampsia caespitosa, Molinia* and various sedges) is an exaggeration of the tussock type. Here the grasses and sedges accumulate the silt deposited by frequent gentle flooding of the flat belts of land on the borders of lakes and slow rivers which are the typical habitats of the stooled meadow type. The tussock and stooled types, owing to the slow decay and constant accumulation of the dead shoots are much less suitable for grazing. Smith and Crampton's fifth type, the *lair* type, is quite specialised and local, originating from the congregation of cattle or sheep and the consequent trampling and manuring of the ground with highly nutritious organic substances.

The chief grass communities. The enormous variety in detail of the habitats and the composition of the British natural and semi-natural grasslands will be sufficiently clear from what has been already written, and in attempting to classify the various communities dominated by grasses it is obviously impossible to take account of every local variation. Still there are a number of important communities that are sufficiently plain, and may be summarised as follows. The first three may be considered together as "acidic" grasslands and are dealt with in more detail in Chapter XXVI.

Acidic grassland. (Chapter XXVI).

(1) ***Agrostis-Festuca* grassland.** This is dominated by species of bent (mainly *Agrostis tenuis*, but also *A. alba*, *A. canina* and hybrids) with sheep's and red fescue (*Festuca ovina* and *rubra*) and is developed on siliceous and sandy soils, well drained and distinctly (but not extremely) acid in reaction. Of the "natural" grasslands used for grazing ("rough grazings") it is by far the most widespread, occurring on many of the sandy heaths and commons of the lowlands and very extensively on the sides of hills and mountains composed of "siliceous" rocks, approximately up to the limit of former woodland, which it has replaced. The accompanying dicotyledonous plants are rather few, except for some woodland relicts (see Chapter XXVI). Among the few leguminous plants bird's foot trefoil (*Lotus corniculatus*) is the most widespread and is valuable as a pasture plant.

Subordinate types of grassland on more acid sandy and siliceous soil, with more tendency to peat formation, often replacing heath as the result of grazing, trampling or burning, are sometimes dominated by *Holcus mollis* (often a woodland relict), and especially by *Deschampsia flexuosa*.

(2) ***Nardus* grassland.** This is dominated by the mat grass or "white bent" (*Nardus stricta*) and is developed over siliceous soils of medium dampness where the conditions favour the accumulation of acid peaty humus, and on disintegrated peat washed down from the eroding edges of elevated peat moors. The sod is commonly 8–12 in. (20–30 cm.) thick, resting on an impermeable subsoil such as shale or heavy boulder clay from which it is easily separable. It is of much less value for grazing than the bent-fescue type, for though sheep nibble round the *Nardus* plants they fail to keep down the tough growth of the rootstocks and basal leaf sheaths, which accumulate a mat of dead material forming tussocks or stools. *Nardus* grassland is floristically very much poorer than the bent-fescue type, which it often invades when grazing is inadequate.

(3) ***Molinia* grassland.** Dominated by the purple moor-grass or "flying bent" (*Molinia caerulea*) this occurs on peaty soil which is kept much more continually wet than that of the Nardetum. There must, however, always be a certain slow movement and consequent aeration of the water to allow *Molinia* to become dominant. The flora, like that of the Nardetum, is usually relatively poor, but is marked by more species characteristic of boggy places. *Molinia* grassland has a certain grazing value for sheep in early summer, though it is much better if cattle are employed to keep it succulent, but *Molinia* itself is of no value for winter keep, though sheep may find green leafage of other species growing among and protected by the *Molinia* plants.

Grassland dominated by *Molinia* is also formed on fens under certain circumstances (Chapters XXXII, XXXIII) and in other places where it can obtain a constant moving water supply, such as in sandy heath where flowing underground water is available close to the surface.

Basic grassland. (4) Grassland on chalk, limestone and other basic rocks (Chapter XXVII). This is a very characteristic type, at the opposite extreme from the acid types, and corresponds with the "stable" grassland of Smith and Crampton (1914). It is no doubt "stable" in the sense that it does not change as a result of allogenic factors, but it is open to invasion by woody plants (chalk and limestone scrub, ash and beech) except at the higher elevations and in the most exposed places. The soil is shallow, finely divided, basic in reaction and very well drained, so that the vegetation is liable to suffer from drought. The most widespread and characteristic grass is *Festuca ovina*, often overwhelmingly dominant. On the Chalk *F. rubra* is almost as abundant, either mixed with *F. ovina* or itself dominant. And in parts of the south of England other grasses (such as *Bromus erectus* and *Brachypodium pinnatum*) may assume dominance. Associated with the sheep's fescue or other dominant, are a number of other grasses and herbs, some of which are commonly known as "calcicole" because they occur especially or almost exclusively on these basic soils.

Three main kinds of rock on which this grassland occurs may be distinguished: (1) the chalk and oolitic limestones of the south and east of England, (2) the harder Palaeogenic limestones of the north and west, of which by far the most important and widespread is the Carboniferous or "Mountain" Limestone, and (3) the hills and "bosses" of basic igneous or metamorphic rock specially characteristic of parts of Scotland, though by no means confined to that country. All three have characteristic rounded contours owing to smooth weathering, though this feature is most marked in the case of the chalk, which is the softest and most easily soluble in percolating rainwater carrying dissolved carbon dioxide, and which never develops crags and screes like the older and harder limestones. The flora and vegetation of the first and second are on the whole closely similar, though a number of species occur on the chalk which are rare or absent on the northern and western limestones. The vegetation of the third type, the basic igneous rocks, has not been closely studied, but it certainly has much in common with that of the harder limestones.

Calcareous grassland affords one of the best *natural* (i.e. unmanured) sheep pastures in the country. It has the reputation of being "dry, sweet and healthy", dry because of the porous soil, "sweet and healthy" because the dryness and the basic "buffer" prevents acidity developing and because the varied herbs and grasses of the community provide in the aggregate a ration that is rich in a variety of mineral nutrients.

(5) **Neutral grassland** (Chapter XXVIII). This collective term has been applied to grassland developed on soils which generally do not depart very widely from the neutral point, such as are derived from many lowland clays and loams (Tansley, 1911, p. 84). It is necessarily a comprehensive and

rather vague category, and the natural or semi-natural communities of which it is composed have not been subjected to close ecological investigation, though many agricultural studies of the effects of different regimes of grazing and manuring have been published. The majority of the grasslands on such soils are "permanent pasture" as opposed to rough grazing, i.e. it is enclosed and more or less intensively grazed, more rarely mown for hay. Most of it is derived from seed mixtures sown on ploughed land; but the grazing and manuring regime a has such decisive selective effect on the composition of the sod, some species dying out and others coming in, that the flora may be completely changed within an amazingly short time according to the treatment.

To this category belongs the great bulk of the grassland known to agriculturists as "permanent grass" and "long ley". It is mainly on village greens, and on some commons, on the edges of woods, and in alluvial meadows which are not too wet, that the more natural communities occur. Such land is practically always grazed (or mown)—otherwise it develops scrub and forest—but the grazing is rarely systematic. The soils on which these communities develop are often very fertile, and among the dominant grasses are the best meadow and pasture grasses, principally perennial rye-grass (*Lolium perenne*), cocksfoot (*Dactylis glomerata*), the rough stalked meadow grass (*Poa trivialis*), crested dogstail (*Cynosurus cristatus*), meadow fescue (*Festuca pratensis*), meadow foxtail (*Alopecurus pratensis*) as well as ubiquitous but less desirable grasses like common bent (*Agrostis tenuis*), sweet vernal grass (*Anthoxanthum odoratum*), and Yorkshire fog (*Holcus lanatus*). Among the grasses are many dicotyledonous herbs, which become more and more restricted in number when the land is enclosed and subjected to a proper grazing regime. The dominance of perennial rye-grass (*Lolium perenne*) and "wild" white clover (*Trifolium repens*) is the mark of the best pastures of this type. The less valuable have *Agrostis tenuis* in increasing quantity.

Footpaths and road verges. Everyone must have noticed the contrast between the vegetation of footpaths and that of the land they traverse, and graziers have found that horses and sheep often prefer to feed upon the footpath rather than upon other parts of the field. The most striking case is seen on footpaths crossing heath, where the edges of the path are inhabited by grass vegetation contrasting sharply with the dwarf shrubs of the heath itself; but footpaths crossing grass fields are marked by the darker green colour of their herbage, which is normally maintained throughout the winter, while the rest of the herbage may be more or less withered. The trampled grass verges bordering on highways show similar vegetation.

Bates (1935) and Davies (1938) have recently investigated the nature and causes of this phenomenon. Bates studied the vegetation of eight footpaths in Lancashire, Derbyshire and Norfolk and on very different soils. They included one traversing a hillside sheepwalk with a very acid soil (*p*H 4) and a high rainfall, and another through a larch plantation on the same soil: a third traversed a mixed wood, on a somewhat acid medium loam (*p*H 5): a fourth a neighbouring paddock where the soil was neutral owing to heavy liming. The fifth and sixth examples were on alluvium—one on silt land of excellent grazing value and the other on black riverside neutral soil. The remaining two traversed heath, one a chalk heath with neutral rather moist soil, the other a dry sandy heath (*p*H 5).

Bates found that the characteristic species were everywhere the same except on the sandy heath, where footpath and "surrounds" alike were dominated by *Agrostis* and *Festuca ovina*. The centre of a path, where much used, is of course bare, all plants being destroyed by the trampling. Adjoining this Bates describes an edging of *Poa pratensis*, and next a zone of *Lolium perenne*, followed by one of *Trifolium repens* (which is however absent in shaded situations such as woodland paths). The lateral boundaries are not sharply defined, the path vegetation merging gradually into the surrounding grassland or heath, for intermittent trampling occurs at the sides though it is much more concentrated in the centre. Of accompanying species *Cynosurus cristatus, Dactylis glomerata* and *Plantago* spp. were fairly constant.

Bates considered that this zonation was entirely due to the selective influence of trampling, *Poa pratensis* being the most resistant species, followed by *Lolium* and then *Trifolium repens*. This view he tested by planting ten species side by side in strips and treading a pathway across all the strips. In one experiment the path was also puddled when wet, and then all the species were killed except *Poa pratensis, Lolium perenne, Dactylis glomerata* and *Trifolium repens*. The growth of the three grasses was restricted, but they recovered and made fresh growth. The clover was also adversely affected but not destroyed. On the other hand *Agrostis tenuis, Alopecurus pratensis* and *Anthoxanthum* were completely destroyed by trampling; so were *Agrostis stolonifera, Agropyron repens* and *Festuca elatior* when the soil was puddled as well as trampled.

Davies (1938) insists on the point that trampling not only involves vertical pressure from above, but that there is a horizontal twisting action as the ball of the foot leaves the ground, and a similar action by the hooves of farm stock, which tends to break the surface, move, and therefore aerate the soil and 'earth up' the bases of the plants. He therefore holds that treading, unless it is too severe, benefits the majority of grassland plants. When the frequency of treading exceeds a certain point its selective action becomes effective, and beyond a certain degree of intensity all vegetation is destroyed. The two species *Lolium perenne* and *Trifolium repens* are well known to be the most valuable as pasture plants, and occur on the most fertile soils. Their occurrence or dominance on footpaths and grass verges would therefore indicate high soil fertility of these strips of grassland, and this may be partly due to the 'cultivating' action of the foot. The presence of ryegrass and white clover would in its turn attract stock, and the fertility would again be raised by the dropping of their dung along the path or verge.

Davies finds that *Poa annua*, not *P. pratensis*, is the prevalent grass where disturbance is great enough to kill other species. It "shows an amazing capacity to flower, set seed and re-establish itself, and though short-lived is not a typical therophyte, for when 'earthed up' it forms runners which live for more than a year and produce new tillers and roots at the nodes". *Poa annua* should therefore, Davies thinks, be classed as a hemicryptophyte along with the great majority of our meadow grasses.

Davies found a much greater variety of grasses on roadside verges than Bates found at the sides of footpaths, differing according to situation, soil and the natural or semi-natural vegetation of the surrounding land. He also criticises Bates' interpretation of the resistance of various species to differential intensity of treading in terms of leaf structure and life form.

In a later paper Bates (1938) replies that sheep tracks and grass verges cannot be directly compared with human footpaths because the effect of human treading is quite different from that of the hooves of sheep and cattle; and that it is difficult to see how the flat-soled human foot can have a 'cultivating' action on the soil. Its rotatory action rather grinds plants into the soil as well as bruising them. *Poa pratensis* (whose determination as the most resistant footpath grass has been confirmed by other authorities) is the characteristic grass of much trodden stable turf, while *Poa annua* is characteristic of unstable, shifting areas.

REFERENCES

BATES, G. H. The vegetation of footpaths, sidewalks, cart-tracks and gateways. *J. Ecol.* **23**, 470–87. 1935.

BATES, G. H. Life forms of pasture plants in relation to treading. *J. Ecol.* **26**, 452–4. 1938.

DAVIES, WILLIAM. Vegetation of grass verges and other excessively trodden habitats. *J. Ecol.* **26**, 38–49. 1938.

LEWIS, F. J. Geographical distribution of vegetation of the basins of the rivers Eden, Tees, Wear and Tyne. Parts I and II. *Geogr. J.* **23**. 1904.

MOSS, C. E. In *Types of British Vegetation*, pp. 131–6 (siliceous grassland); pp. 154–60 (limestone grassland). 1911.

SMITH, ROBERT. Botanical survey of Scotland. Part I. Edinburgh district. Part II. North Perthshire district. *Scot. Geogr. Mag.* 1900.

SMITH, W. G. and CRAMPTON, C. B. Grassland in Britain. *J. Agric. Sci.* **6**, 1–17. 1914.

SMITH, W. G. and MOSS, C. E. Geographical distribution of vegetation in Yorkshire. Part I. Leeds and Halifax district. *Geogr. J.* **22**. 1903.

SMITH, W. G. and RANKIN, W. M. Geographical distribution of vegetation in Yorkshire. Part II. Harrogate and Skipton district. *Geogr. J.* **22**. 1903.

STAPLEDON, R. G. "Permanent Grass", in *Farm Crops*, edited by W. G. R. Paterson. **3**, 74–136. 1925.

STAPLEDON, R. G. *The Land Now and Tomorrow*. London, 1935.

STAPLEDON, R. G. and HANLEY, J. A. *Grassland: its management and improvement.* Oxford, 1927.

TANSLEY, A. G. *Types of British Vegetation*, pp. 84–7 (neutral grassland), pp. 94–7 (grass heath), pp. 173–8 (chalk grassland). Cambridge, 1911.

TANSLEY, A. G. and ADAMSON, R. S. A preliminary survey of the chalk grasslands of the Sussex Downs. *J. Ecol.* **14**, 1–32. 1926.

Chapter XXVI

THE ACIDIC GRASSLANDS

I. *AGROSTIS-FESTUCA* (BENT-FESCUE) GRASSLAND

Grassland dominated by the bents, mainly *Agrostis tenuis* (= *A. vulgaris*), but also *A. stolonifera* (*alba*) or *A. canina*, together with sheep's and red fescue (*Festuca ovina* and *F. rubra*), all shallow-rooting grasses, probably covers a greater area in the British Islands than any other of the "natural" types of grassland. It is the typical community of the "grass-heath" or "siliceous grassland" of the earlier writers.[1] Its main develop-

"Siliceous grassland" ment is on grazed and well-drained hillsides and the lower slopes of mountains composed of "siliceous" rocks—grits, sandstones, mudstones, slates, and the less basic kinds of igneous and metamorphic rocks; and of one or other of these the great majority of the hills and mountains of the north and west are formed. Davies (1936) estimates that in Wales *Agrostis-Festuca* grassland occupies 5·5 per cent of the total "agricultural area", i.e. including rough grazings. In some siliceous hill regions the percentage area is certainly much higher.

But substantially the same "grazing community" is also developed on lowland sandy "heaths" and commons which are grazed or trampled, though here many of the associated species are different; and this last is the

"Grass heath" community to which the term "grass heath" is most appropriately applied. Commonly dominated, like the siliceous hill pasture, by bent and fescue, and with a similarly acid soil, grass heath contains a wide variety of more xerophilous species, in correspondence with the drier (sandy) soil and with the drier climate of the east and south-east of England.

Bent-fescue grassland is not determined by altitude since it may occur in the appropriate habitat at any elevation below the arctic-alpine zone. Characteristically it forms a lower belt on the hills and mountains up to about 1000 ft. (300 m.) or a little higher, above the enclosed pastures, and below the great areas of peat which commonly cap the plateaux and rounded summits. The bent-fescue grassland contrasts by its relatively brighter green (though it is not so vivid as limestone grassland) with the more neutral tints of the peat grasses and the darker colour of the ericaceous undershrubs.

---

[1] The earlier surveyors recognised only the *Nardus* and *Molinia* communities in "hill pasture", "grass heath" or "siliceous grassland", but it has become increasingly apparent that the most widespread semi-natural "grazing community" on these soils is dominated by bent and the fine-leaved fescues. Nardetum and Molinietum maintain themselves independently of grazing, and though dominated by a grass, Molinietum is closely allied, as we shall see, to the bog or moss formation.

Most if not all of this siliceous grassland is potential woodland, and was formerly covered with forest or scrub belonging to the Quercetum sessili-florae or related communities, which still exist here and there side by side with the grassland under identical conditions of exposure and subsoil. Felling combined with grazing, and very often probably grazing alone, are the agencies which have converted the forest area into grassland. Wood-land plants such as the bluebell (*Scilla non-scripta*), wood anemone (*Anemone nemorosa*), wood sorrel (*Oxalis acetosella*) and cow wheat (*Melampyrum pratense*) often occur in the grassland, especially in the shade of bracken, and doubtless represent relicts of former forest. "Grass heath" and bracken-covered slopes on the southern Pennines which are now quite treeless commonly bear Anglo-Saxon or Scandinavian place-names indi-cating the existence of former forest. And there is documentary evidence that the long Millstone Grit spurs and rock-terraces were covered with forest in Norman times (Woodhead, 1929, p. 26).

Invasion by heath. The heath formation often invades this grassland, the common ling (*Calluna vulgaris*) frequently colonising it when sheep grazing is diminished or excluded. This can be well seen along the line of the stone wall ("mountain wall") which commonly separates the grazing area from the moorland above. On the lower side is pure grassland, on the upper heather moor: if sheep are admitted to a limited area of the moor they quickly convert it into grassland; if they are excluded from the grassland this is rapidly colonised by *Calluna* (see Chapter VI, p. 130, and Fig. 35). Farrow (1916), working on the lowland sandy heaths of Breck-land in Norfolk and Suffolk, showed that ling heath is converted into *Agrostis-Festuca* grassland ("grass-heath") by intensive rabbit attack, and rapidly recolonises the ground when rabbits are excluded. The grassland is thus clearly a "biotic climax".

Soil. The soil is typically thin over the subjacent rock, well-drained and moderately acid, but shows little tendency to accumulate raw humus and form peat, except locally where drainage is impeded. The few pH values recorded vary somewhat widely: pH 5·9 (Leach, Longmynd, an unusually high value), 5·0 (Malvern Hills), 4·9 (Clee Hills), and 4·2 in two different Welsh localities, widely separated but both with very high rainfall where leaching must be extreme and *Nardus* readily invades. At Cahn Hill near Aberystwyth the pH value varies between 4·3 and 4·7. Thus pH 4–5 is probably the usual range.

The dominants. The turf is mainly composed of *Agrostis tenuis* (which may be replaced locally by *A. canina*) and of *Festuca ovina*, which occur in various proportions. Thus Stapledon (1914) states that on the sheep-walks of mid-Wales *Agrostis* may constitute from 6 to 45 per cent, *Festuca ovina* from 15 to 56 per cent of the herbage. The sheep's fescue is a more xerophilous grass than the bent and thus tends to become dominant in drier situations. On the lowland sandy heaths of Breckland, Farrow (1917) has shown experimentally that luxuriance of *Agrostis* and its dominance over *Festuca ovina* is directly induced by increased water supply. The bent is of moderate value as sheep feed, and the fescue, though it produces no great bulk of herbage, is nutritious and sheep thrive well on it, though they depend largely on accompanying plants.

Flora. A long list of species may be compiled from the bent-fescue upland pastures, since such factors of the habitat as altitude, exposure, angle of slope, depth and water content of the soil, physical and chemical nature of the underlying rock, as well as the intensity of grazing, vary very much; and no good purpose would be served by enumerating all the species that have been found in this kind of grassland.

Of grasses accompanying the dominants *Sieglingia decumbens* is often present in quantity and may form from 5 to 20 per cent of the herbage in some of the bent-fescue pastures of mid-Wales. Another grass often occurring in this community is the almost ubiquitous sweet vernal grass (*Anthoxanthum odoratum*). *Deschampsia flexuosa* occurs only locally where peaty humus accumulates. The field woodrush, *Luzula campestris*, is locally abundant. Tufts of *Nardus* frequently occur in the bent-fescue grassland, and the invasion of this species from adjacent Nardetum occurs when grazing is diminished, especially if there is a long series of wet years.

Of dicotyledonous herbs the heath bedstraw (*Galium saxatile*) is generally abundant and the tormentil (*Potentilla erecta*) almost as common: other species met with are *Campanula rotundifolia*, *Lathyrus montanus*, *Linum catharticum*, *Lotus corniculatus*, *Polygala vulgaris*, *Teucrium scorodonia*, *Viola riviniana*. *Euphrasia officinalis* (agg.), *Hieracium pilosella*, *Leontodon autumnalis* and *Veronica officinalis* which have very often been recorded from this community, are, according to Davies (*in litt.*) more characteristic in Wales of enclosed fields which have been arable and have reverted to grassland. The yellow mountain pansy (*Viola lutea*) is locally abundant in some regions, forming a beautiful florál feature at midsummer.

There is often a great deal of moss among the herbage in winter and early spring, the commonest species being:

Dicranum scoparium	Hypnum schreberi
Hypnum cupressiforme	Hylocomium squarrosum

On the drier and less peaty soils overlying the granite of Bodmin Moor in Cornwall *Festuca-Agrostis* grassland covers a very large area (Magor, unpublished). The flora is poor, the following species alone being recorded from two localities at 800 and 950 ft. (243 and 290 m.) altitude respectively.

Co-dominant

Festuca ovina and Agrostis tenuis

Abundant or very abundant

Agrostis setacea	Nardus stricta
Deschampsia flexuosa	Potentilla erecta

Less abundant

Carex binervis	r–f	Sedum anglicum	vr
Galium saxatile	o–f	Sieglingia decumbens	lva
Molinia caerulea	o		

Species of *Hylocomium*, *Hypnum* and *Polytrichum* are frequent. These areas are colonised by *Ulex gallii* (f–a) and by *Calluna vulgaris* (r–a), *Vaccinium myrtillus* (o–f) and *Erica tetralix* (r).

On the Staddon Grit and locally also on the slates of the same district *Festuca-Agrostis* grassland occurs on the drier soils and is always associated with locally dominant *Ulex europaeus*, *U. gallii* or *Pteridium* (Magor).

Pteridietum (Pl. 94). The bracken fern (*Pteridium aquilinum*) is a very aggressive species which has spread and is still spreading over much of the *Agrostis-Festuca* grassland and has seriously diminished its grazing value. Dense Pteridietum is said to occupy as much as 3·4 per cent of the total "agricultural area" (i.e. arable and pasture including "rough grazings") of Wales (Davies, 1936). In Scotland it is recognised as a serious menace to the sheep farmers. When the fronds are really luxuriant and closely set they shade the ground so deeply as to inhibit the vegetation of grasses and herbs. In extreme cases the soil is covered with a thick layer of bracken litter and becomes bare of other vegetation. But where the *Pteridium* is not so dense the grasses and herbs of the grassland survive below the overarching fronds of bracken. Since the *Pteridium* canopy is never fully developed till nearly the middle of June, and at higher altitudes and latitudes not till midsummer, the grassland plants are able to assimilate during the late spring and the beginning of summer. *Pteridium* requires a certain depth of soil for vigorous development, though its horizontally running rhizomes can grow at very various depths, from a few inches to 2 ft. or more. On the shallower soils it can scarcely compete successfully with other species, and it cannot penetrate waterlogged soils, stopping abruptly on the edge of marshy ground.[1] Its distribution is also limited by its demand for shelter. The bracken ascends to 1250 ft. on the Longmynd in Shropshire,[2] to 1500 or 1700 ft. on the southern Pennines, and to 2000 ft. in Scotland, in accordance with the greater height and mass of the mountains, but it is dominant only in valleys and on slopes sheltered from the more violent winds, often ceasing quite abruptly as the ridge is reached.

Pteridium is largely, if not primarily, a woodland plant, but is adaptable to a considerable range of light intensities, changing the habit of its fronds accordingly (see p. 280); and it is perhaps significant that the altitude it reaches corresponds more or less with the limit to which we may suppose forest at one time extended. It is also the fact that the woodland plants which often occur in siliceous grassland are very frequently (though not always) associated with the bracken. A notable instance of this is mentioned by Pethybridge and Praeger (1905), who describe, from the Wicklow

[1] Pl. 94, phot. 227.   [2] Pl. 94, phot. 228.

Mountains south of Dublin, glorious stretches of colour in April and May produced by sheets of bluebell, violet, germander speedwell, earthnut, lesser celandine and primrose, on areas which later in the season are covered by the fronds of the bracken. Just as these prevernal and vernal species vegetate and flower in the woods before the leafy canopy of the trees has unfolded, so they do here before the bracken fronds expand; and we may perhaps regard such vegetation as representing the lower layers of the forest community after the trees and shrubs have disappeared, the upper field layer, here composed of *Pteridium*, now dominating the community. But there is no doubt that in suitable soil the bracken spreads far and wide over siliceous grassland, where it attains a dominance much more over-whelming than in most woodlands. The normal method of spread is by the growth and branching of the rhizomes, but in certain years the germination of the spores and the growth of prothalli and young plants start fresh centres of invasion (Braid, 1937).

Not only does *Pteridium* thus invade *Agrostis-Festuca* grassland, but also Calluneta, Vaccinieta, Nardeta and even the drying peat of the edges of retrogressive Eriophoreta at relatively high altitudes, provided the situation is sheltered (Adamson, 1918, p. 104). This invasion is facilitated by rather deep sandy soil or drying peat, and is probably greatly aided by the burning of the moorland, since the bracken rhizomes, owing to their depth below the surface, are rarely injured by the fire, and they may easily occupy the ground before the moor plants have regenerated.[1]

Bracken is often cut for litter in the autumn, since it forms excellent bedding for cattle; but cutting after the fronds are dry and empty of reserves does nothing to check the growth of the rhizomes. Repeated cutting in successive years, if carried out after the fronds are mature but before the reserves have been transferred to the rhizomes, will of course diminish the luxuriance of the plant, but it takes a very long time to ex-terminate bracken in this way. Spraying with sulphuric acid, and more recently with sodium chlorate, has been employed to keep down bracken and is often recommended, but this procedure is both expensive and troublesome and is not free from danger. A better and quite effective method, though it takes longer to carry out, is to cut repeatedly during the season, as soon as new fronds appear. But this is seldom done owing to the cost of the labour required. Where bracken becomes overwhelmingly dominant and thus completely destroys the pasture, and where effective checks are too expensive, it is probably better to abandon the pasture and afforest the land (Stapledon).[2]

If we consider the bent-fescue community as replacing oak-birch wood of which *Pteridium* is an important constituent, then the Pteridietum is

[1] It is doubtful if bracken can invade vigorous undisturbed Callunetum (see p. 727)

[2] Quite recently considerable success in destroying bracken has been obtained by the use of various types of machine which crush the young fronds.

primarily a society which survives from the woodland association, protected from grazing because it is unpalatable, and finding the opportunity for fresh aggressiveness and dominance when it is exposed to the full light of the open.

Ulicetum. Gorse scrub composed of one or both of the two species, *Ulex europaeus* and *U. gallii,* is often a conspicuous feature in bent-fescue grassland, sometimes covering considerable areas. In the west the latter is the typical species. Locally the low gorse scrub is dense and forms a continuous canopy, but more usually the Ulicetum gallii consists of separate dwarf bushes more or less closely spaced, not more than 2 or 3 ft. high, and nibbled by sheep into a rounded form, set in a matrix of grassland (Pl. 95). According to Davies (1936) dense Ulicetum occupies about 0·2 per cent of the agricultural area of Wales.

Ulex europaeus, the common gorse, furze or whin, is mainly a plant of the lowlands, though isolated bushes may ascend to 2000 ft. in southern Scotland. Its commonest habitats are disturbed grassland on sandy or loamy soils belonging to the bent-fescue or neutral grassland types and mainly occurring on commons and waysides and in neglected pastures. It is also found on the edges of *Calluna* heath, often occupying in dense masses the disturbed ground between a village and a neighbouring heath, spreading along the lines of tracks and paths across the heath and sometimes invading the heath itself. The spread of the gorse along paths and tracks is probably due largely to transport of the seeds by ants, which bite and tear at the fleshy orange-coloured "caruncle", dragging the seeds along the ground. The common gorse cannot be regarded as a proper member of the heath association, and it is doubtful if it can invade *Calluna* heath unless the ground has been disturbed.

The common gorse was at one time much used as fodder. The soft young shoots are very palatable and nutritious, and formerly the hard mature shoots were ground up in special mills and used as cattle food in some parts of the country. In north and west Wales *Ulex europaeus* is said not to be native, but introduced (from Ireland) in the eighteenth century for cattle fodder, the plants being cut every alternate year (Alun Roberts). When this use was abandoned the gorse grew up, flowered and seeded, so that it has now spread widely along the roadsides and over the lower hill pastures, where it may be seen side by side with the native *U. gallii*. It is still occasionally planted to form hedges and rather widely in the midlands and south for fox coverts.

Ulex europaeus is a taller and freer growing shrub than *U. gallii* and commonly reaches a height of several feet, so that unless it is closely grazed or kept cut while still young its shoots are lifted out of the reach of grazing animals. Its main flowering season is April and May, though a few flowers open in mild weather from October to March, and quite a conspicuous show of blossom sometimes appears near the south coast in the

PLATE 94

Phot. 227. Pteridietum stopping abruptly at the foot of a well-drained slope. *Holcus lanatus* occupies the wet ground on the right. Wanister Hill, Co. Durham. *H. Jeffreys* (1916).

Phot. 228. A small valley on the Longmynd (Shropshire) with Pteridietum extending up the slope on the left but not on the right, which is more exposed to wind and occupied by Vaccinietum myrtilli. Callunetum covers the summit of the ridge above. *W. Leach* (1931).

PTERIDIETUM

PLATE 95

Phot. 229. Ulicetum gallii on the Wicklow Mountains south of Dublin. Cf. Fig. 89. *R. Welch.*

Phot. 230. Ulicetum gallii in *Agrostis-Festuca* grassland, Malvern Hills, Worcestershire.

SHEEP-GRAZED ULICETUM GALLII

middle of a mild winter. The pods ripen and explode so that the seeds are shot out during July.

When the two species grow together, as on the roadsides of Devon, Cornwall and Wales, they may be easily distinguished not only by the fact that they are never in flower at the same time, but also because the freer

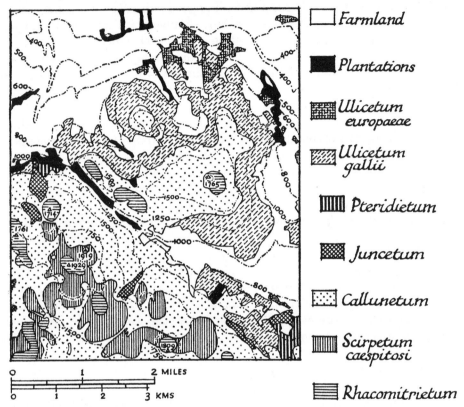

Farmland

Plantations

Ulicetum europaeae

Ulicetum gallii

Pteridietum

Juncetum

Callunetum

Scirpetum caespitosi

Rhacomitrietum

FIG. 89. ZONATION OF VEGETATION ON THE WICKLOW MOUNTAINS SOUTH OF DUBLIN

Ulicetum gallii forms a broad well-marked zone, mainly between 600 and 1300 ft. (Pl. 95, phot. 229). Below (to the north) are patches of Ulicetum europaeae between 400 and 700 ft. Pteridietum occurs locally in protected situations. Above is Callunetum (1250–1800 ft.— Chapter XXXVII), and the flat summit areas are occupied by Scirpetum caespitosi (above 1750 ft. —Chapter XXXV) often capped by Rhacomitrietum. Contours in feet. After Pethybridge and Praeger, 1905.

growing, taller bushes of *U. europaeus* are of a bluer green than those of the lower, more compact *U. gallii*, which are yellowish green.

On the bent-fescue grasslands of the northern and western hillsides *Ulex europaeus* usually occurs only in the lowest zone, especially along roadsides, tracks, and close to cultivated land, in the west often giving way above to *U. gallii* (Fig. 89 and Pl. 95). This species, as has been said, is of lower and more compact growth and on much-grazed land forms character istic rounded cushions from 1 to 2 ft. high. This growth form is mainly due to constant nibbling of the young shoots by sheep or rabbits, but in the most exposed places also to the action of wind in drying off shoots protruding from the surface of the cushion. In Wales, Cornwall, and to a less extent in other western English counties, *Ulex gallii* often forms a well marked zone on the grassy hillsides, alternating with areas dominated by bracken, and in some years covering the hills in late summer with sheets of golden bloom. In Ireland, where its distribution is similar, it sometimes ascends to 2000 ft. (610 m.). On the hills of the north of England it is much less general, and in the midland and eastern counties, while not wholly absent, it is quite rare. The shrub flowers from July to September, and the pods ripen in the following spring.

On sandy or loamy commons scrub of *Ulex europaeus* may protect the seedlings of colonising trees, much as juniper does on the chalk escarpments, and thus form the first stage of the seral development of woody vegetation; but more usually it forms, either alone or in company with other spiny shrubs, a dense and apparently permanent scrub community, casting so deep a shade that practically no vegetation can grow beneath it (cf. Chapter XXIV). How long such scrub communities actually survive has never been determined, but it is probable that they are relatively short-lived and thus more or less "migratory", the bushes dying in some places and seedlings starting in fresh areas of disturbed soil.

Gorse scrub is very often burned to clear the ground for the pasture grasses, but it has considerable power of springing again from the basal unburned parts of the stem, at or just below the soil surface, after destruction of the aerial shoots.

GRASS HEATH

"Grass heath", as it is developed on the sandy commons of the English lowlands, is a very characteristic though very variable community. It is called "grass heath" both because it occurs on the same soils as "true" (i.e. *Calluna*) heath, by which it is invaded and which it may replace, and also because it is inhabited by the "heath grasses" that also occur in Callunetum wherever there is room: among these *Agrostis* spp., *Festuca ovina*, *F. rubra*, *Deschampsia flexuosa*, and sometimes *Holcus mollis* are the commonest. The relation of grass heath to lowland *Calluna* heath is essentially the same as the relation of the bent-fescue hill pasture to upland heath or heather moor. Grazing will convert the heath into grassland, cessation of grazing leads to recolonisation by heath and the formation of a surface layer of raw humus or dry peat. Where the dominance of *Calluna* is incomplete the grasses of the grass heath and many of the associated species occur in the intervals. For this reason any extended floristic list of Callunetum will include many grass heath species. An essential difference between the two communities is that grass heath does not form the surface layer of raw humus (dry peat or mor) characteristic of Callunetum.

On the drier sandy soils in a relatively dry climate grass heath may contain a great number of "arenicolous" species, some of which are confined, in Britain at least, to this habitat. East Anglian heaths, particularly those of Breckland, are among the driest communities in England, and include many species which are not found in Callunetum, and several which do not occur outside East Anglia. Grass heath communities are also often developed on grazed or rabbit-infested fixed sand dunes.

The soil of grass heath may initially contain considerable amounts of calcium carbonate (e.g. on fixed sand dunes where the sand contains shell fragments and on the calcareous drift of Breckland), but its highly permeable nature leads to rapid leaching, so that the surface layers become markedly acid in reaction, unless buffered by abundant calcium carbonate in the subjacent soil as in many areas of Breckland. And the numerous sands which are more purely siliceous show an acid reaction from the beginning. In the driest and in disturbed places the community may be open, with bare patches of sand between the plants. This gives room for ephemerals and other annuals and also for the invasion of weeds from neighbouring sandy arable land.

Floristic list. For these reasons any comprehensive floristic list will be a long one. The following is a composite list made up from various sources, and it must not be supposed that the species included would all be found together on a particular heath. The species confined or nearly confined to Breckland are given in a separate list (p. 512). Most of them are not in fact grass heath species.

General dominants:

Agrostis tenuis

Festuca ovina (with F. rubra)

Dominant on the more acid heaths where thin peat is formed:

Deschampsia flexuosa

Cladonia spp.

Locally dominant:

Ulex europaeus (invading)

Pteridium aquilinum (invading)

Holcus mollis

Sarothamnus scoparius (invading)

Festuca capillata

Holcus lanatus

General list:

Achillea millefolium
Agrostis canina
A. setacea (south-west England)
A. stolonifera
Aira caryophyllea
A. praecox
Alchemilla arvensis
Allium vineale
Anthoxanthum odoratum
Arenaria serpyllifolia
Calluna vulgaris (invading)

Campanula rotundifolia
Carduus nutans
Carex arenaria (ld in Breckland)
C. binervis
C. caryophyllea
C. divulsa
C. pilulifera
Carlina vulgaris
Centaurium umbellatum
Cerastium arvense
C. vulgatum

C. semidecandrum
Cirsium arvense
C. lanceolatum
Conopodium majus
Corydalis claviculata
Daucus carota
Dianthus armeria
D. deltoides
Erica cinerea (occ. invading)
Erodium cicutarium
Erophila verna (ephemeral)
Euphrasia officinalis
Filago minima
Galium saxatile
G. verum
Genista anglica
G. pilosa
Geranium dissectum
G. molle
G. pusillum
Gnaphalium sylvaticum
Hieracium boreale
H. pilosella
H. umbellatum
Hypericum humifusum
H. perforatum
H. pulchrum
Hypochaeris glabra
H. radicata
Jasione montana
Koeleria cristata
Kohlrauschia prolifera (Dianthus
    prolifer)
Leontodon autumnalis
L. nudicaulis
L. hispidus
Linum catharticum
Lotus corniculatus
Luzula campestris
L. multiflora
Malva neglecta (rotundifolia)
Medicago arabica
M. hispida var. denticulata
M. lupulina

Minuartia (Arenaria) tenuifolia
Moenchia erecta
Myosotis arvensis
M. collina
M. versicolor
Ononis repens
Ornithopus perpusillus
Plantago coronopus
P. lanceolata
Polygala vulgaris
Potentilla argentea
P. erecta
Radiola linoides (damp ground)
Ranunculus bulbosus
Rumex acetosella
Sagina apetala
S. ciliata
S. procumbens
S. subulata
Saxifraga tridactylites
Scilla autumnalis
Scleranthus annuus
Sedum acre
S. anglicum
Senecio jacobaea
S. sylvaticus
Spergularia rubra
Taraxacum erythrospermum
Teesdalia nudicaulis
Teucrium scorodonia
Thymus serpyllum
Trifolium arvense
T. dubium
T. filiforme
T. procumbens
T. subterraneum
Veronica arvensis
V. officinalis
Vicia angustifolia
V. lathyroides
Viola canina
V. riviniana

In heavily rabbit-grazed areas (e.g. parts of Breckland) a number of tall rabbit-resistant plants standing up far above the level of the razed herbage are characteristic, for example:

Conium maculatum
Senecio jacobaea
Solanum nigrum

Urtica dioica
U. urens

together with smaller plants such as species of *Myosotis, Arenaria serpyllifolia*, etc.

Societies of *Pteridium* and *Ulex* frequently invade the heath, the bracken often covering wide areas with dense growth.

The following are among the common bryophytes on acidic grass heaths:

Campylopus flexuosus   Leucobryum glaucum
Ceratodon purpureus   Polytrichum piliferum
Dicranum scoparium   P. juniperinum
Hypnum schreberi   Tortula muralis

and of lichens the commonest are

Cetraria aculeata   Cladonia spp.

The grass heaths of Breckland

Climate and soil. Breckland is a well-defined physiognomic region in south-west Norfolk and north-west Suffolk (see Figs. 5, 31). It is of low elevation, varying from about 50 to about 200 ft. above sea-level (15–60 m.) with an average annual rainfall (seven stations) of 591 mm. or a little less than 24 in. The lowest record in the area is from Kilverstone Hall and is 551 mm. or about 22 in. The soil is derived from chalky glacial drift and interglacial sands, with local gravels and loess, and is almost everywhere sandy. There is local dune formation on a small scale, with blowouts, and considerable areas are covered with wind-blown sand. Despite its comparatively uniform texture the soil varies a great deal chemically. Where the chalky drift comes close to the surface the calcium carbonate and pH value are high, while in other areas there is much less carbonate but a relatively high base status, and others again are destitute of calcium carbonate, poor in bases generally and show very low pH values. These differences are reflected in the vegetation, which is floristically rich in the two former and very poor in the last-named type.

Vegetation. The vegetation is largely grassland with societies of *Pteridium* and *Carex arenaria* and considerable stretches of Callunetum. Farrow showed that the very heavy rabbit pressure to which some areas are subjected suffices to convert Callunetum into grassland (see pp. 136–8), but this is by no means the only factor involved. The calcareous soils are not invaded by *Calluna* and bear quite a different flora from the highly leached and podsolised sands. Watt (unpublished)[1] has in fact distinguished seven different types of grassland, one of which (A) is scarcely distinguishable from chalk grassland, another (B) has a wide range of pH values, a high base status and a rich flora (eighty species of flowering plants), while others again (D–G) have a low base status and a poor flora, the poorest (G) with only ten species of phanerogamic plants and lichens dominant.

Types D–G may fairly be classed as "grass heath" since it is only these which *Calluna* invades. Consideration of the status of types A–C must await the publication of Watt's work.

[1] Dr Watt has most kindly put his data, as yet unpublished, at my disposal.

Composite soil samples 0–6 *in.* (15 *cm.*)

	D	E	F	G
Loss on ignition	2·7	3·34	5·73	4·91
Loss on ignition or top 1½ in.* (3·75 cm.)	4·19	5·30	7·93	7·52
Exchangeable calcium†	0·76	0·13	0·07	0·0
pH value	4·3	3·8	3·6	3·5

* Soil black because of fine dispersion of humus.

† Figures almost negligible: of the better types of grassland A has 51·2, B 34·4, C 6·3 per cent of exchangeable calcium.

Species of selected areas. The numbers in brackets refer to the "constancy" of the species: 5 = occurrence in more than four-fifths, 4 = more than three-fifths, 3 = more than two-fifths, 2 = more than one-fifth, 1 = less than one-fifth of the areas listed.

It is to be noted that *Carex arenaria* and *Calluna vulgaris* can and do invade all these four types, but the areas chosen for examination were those in which *Agrostis* and *Festuca* were dominant.

	D		E		F		G	
			Spermaphytes					
Agrostis stolonifera ⎫ A. tenuis ⎬	f–a	(5)	o–a	(5)	o–a	(5)	o–la	(5)
Aira praecox	o–lf	(5)	o–la	(5)	lf–la	(5)	lf	(5)
Alchemilla arvensis	l	(5)	l	(3)	l	(2)		
Anthoxanthum odoratum	o	(1)						
Arenaria serpyllifolia	l	(3)	l	(1)				
Calluna vulgaris	r–o	(4)	r	(2)	vr	(1)		
Campanula rotundifolia	lf–la	(5)	r–lf	(2)	vr	(1)		
Carex caryophyllea	r–la	(3)						
Cerastium arvense	l	(1)						
C. semidecandrum	l	(4)	l	(3)	l	(2)		
C. vulgatum	l	(2)						
Cirsium arvense	o	(3)						
C. lanceolatum	r–o	(2)						
Erodium cicutarium	l	(1)						
Erophila verna	l	(2)						
Festuca ovina	a	(5)	f	(5)	o–f	(5)	o–f–la	(5)
Galium saxatile	o–a	(5)	o–la	(5)	a	(5)	o–lf–la	(5)
G. verum	o–f–la	(5)	r–o	(3)	vr	(1)	vr	(1)
Hieracium pilosella	r	(2)	vr	(1)				
Hypochaeris glabra	o	(5)	l	(2)				
Koeleria cristata	o	(3)	vr	(1)				
Luzula campestris	f–a	(5)	o–f–la	(5)	a	(5)	o–la	(5)
Myosotis collina	l	(5)	l	(3)	l	(2)		
Ornithopus perpusillus	r	(2)						
Poa pratensis	r–o	(1)						
Rumex acetosella	f–a	(5)	o–lf	(5)	f	(5)	o	(5)
Sagina ciliata	l	(4)	l	(2)				
S. procumbens	l	(2)						
Sedum acre	l	(3)						
Senecio jacobaea	o	(5)	r	(2)				
Stellaria media			l	(1)				
Taraxacum erythrospermum	o	(4)						
T. officinale	r	(4)						
Teesdalia nudicaulis	o–f–la	(5)	o–lf–la	(5)	l	(5)	l	(5)
Urtica dioica	l	(5)	l	(3)	l	(2)		
Veronica arvensis	l	(5)	l	(2)	l	(2)		
Vicia lathyroides	r	(1)						
Viola canina	o–f	(4)						
Total species of spermaphytes	38		23		16		9	

	D		E		F		G	
Bryophytes								
Brachythecium albicans	l	(5)	l	(2)	l	(2)		
Bryum sp.	l	(4)	l	(3)	l	(1)	l	(1)
B. capillare					l	(1)		
Cephaloziella sp.	o	(4)	l	(5)	l	(3)	l	(4)
Ceratodon purpureus	la	(5)	lf	(5)	la	(5)	lf	(5)
Climacium dendroides	l	(1)						
Dicranum scoparium	f–a	(5)	o–f	(5)	f	(5)	f–la	(5)
Hypnum cupressiforme	o	(5)	l	(3)	l	(3)	l	(2)
H. schreberi	lf	(3)						
Lophocolea bidentata	vr	(1)						
Lophozia barbata	o–f	(5)	l	(3)	l	(1)		
L. excisa	o	(5)	l	(2)	l	(1)		
Polytrichum juniperinum	o–lf	(5)	o–ld	(5)	la	(5)	lf	(4)
P. piliferum	lf	(4)	lf–la	(5)	lf	(2)	lf	(4)
Ptilidium ciliare	f–a	(5)	o–f	(5)	f	(5)	o–f	(5)
Tortula ruraliformis	l	(1)						
Species of bryophytes	15		11		12		8	
Lichens								
Biatora granulosa	vr	(1)					r	(2)
B. uliginosa			l	(5)	l	(5)	o	(5)
Cetraria aculeata	o	(5)	f	(5)	o–f	(5)	o–f	(5)
Cladonia alcicornis	r–o	(4)	o–l	(4)	r–o	(4)	r–o	(1)
Cl. bacillaris			l	(2)	l	(2)	o	(3)
Cl. coccifera	r	(2)	l	(3)	o, l	(2)	o, l	(3)
Cl. fimbriata	o	(5)	o, l	(4)	o	(3)	o	(5)
Cl. floerkiana			r	(1)				
Cl. furcata	f	(5)	f	(5)	f–a	(5)	f	(5)
Cl. pityrea	vr	(1)	r	(1)	r	(1)	vr	(1)
Cl. pyxidata	o	(5)	l	(5)	o	(4)	o	(5)
Cl. rangiformis	l	(4)	l	(2)	l	(3)		
Cl. sylvatica	f–a–d	(5)	d	(5)	d	(5)	d	(5)
Cl. uncialis	o	(4)	o		o–lf	(5)	o	(5)
Peltigera polydactyla	o	(5)						
Species of lichens	12		13		12		12	
Total of all species	65		47		40		29	

In these grass heaths, highly acid in reaction, very poor in bases and increasingly rich in humus, it will be seen that the phanerogamic and to a less degree the bryophytic flora decreases in passing from the less poor to the poorest type, while the number of species of lichens remains practically constant. Of the phanerogams most species are absent from G, but of those which remain nearly all are quite constant (5), though they may be only occasional and local in occurrence. These are of course the most unexacting and acid-tolerant species. The lichens on the other hand, mainly Cladoniae, are as numerous in species in G as in D, and *Cladonia sylvatica* is definitely dominant. All these types might indeed fitly be called Cladonietum sylvaticae with scattered flowering plants here and there in the *Cladonia*-mat, while the more calcareous grasslands A–C (not described here) are closed communities of grasses and herbs with a more or less interrupted under-storey of mosses and lichens.

Species peculiar to Breckland. Breckland is notable for the occurrence of a number of species which are entirely or nearly confined to this region within the British Isles. They are all species of definitely "continental" distribution, either with headquarters in the steppes of south-eastern Europe or with the centre of distribution not so definitely localised (Matthews, 1937).

The following is a list of British species either confined to Breckland or largely centred in and characteristic of that region:

‡*Artemisia campestris	‡Phleum phleoides
*Carex ericetorum[1]	*Scleranthus perennis
Herniaria glabra[2]	Silene conica[4]
*Holosteum umbellatum[3]	‡*S. otites
†*Medicago falcata	Tillaea muscosa[1]
†M. minima	*Veronica praecox
*M. varia (sylvestris)	‡*V. spicata
*Muscari racemosum	V. triphyllos[4]
Ornithogalum umbellatum	†*V. verna

* Confined or practically confined in Britain to Breckland.

† "More or less essentially steppe plants" (Matthews, 1937).

‡ "Definitely have their headquarters in the Sarmatian-Pontic steppes" (Matthews).

[1] "Continental northern" (Matthews).

[2] "Continental southern" (Matthews).

[3] "Continental" (Matthews), perhaps introduced but possibly native in Breckland.

[4] "Continental" (Matthews).

It is noteworthy that none of these species appear in Watt's list on p. 510, and that in fact all of them, except *Tillaea*, occur only in the better types of Breckland grass communities, whose soils contain calcium carbonate or have a high base status. They are not, therefore, grass-heath species and have been mentioned here only as a matter of interest. Consideration of the status of these grassland communities on the better soils must await the publication of Dr Watt's results.

Succession. Very recently also Watt (1938) has studied the development of the *Agrostis-Festuca* community in Breckland on very infertile sandy soil largely derived by leaching out of the lime from Chalky Drift containing much sand, and partly on sand which has been blown over the surface.

Soil. The development of this "Festuco-Agrostidetum" takes place on sand bared by "blow-outs" following degeneration of preceding vegetation. In the particular area studied this soil, which overlies Chalky Boulder Clay at a depth of 1·25 m., contains 93 per cent or more of "coarse" and "fine" sand, the silt and clay fractions being negligible. It is devoid of calcium carbonate and the surface layer has a pH value of 4·2.

The sere. In main outline the succession is simple (Fig. 90), but in detail very complex, each stage "slipping back" from the position it has won and progress starting again on a lower plane. The first effective colonist is *Polytrichum piliferum* (Fig. 90, "early stages"), which brings about the initial stabilisation of the loose blown sand. This is followed by *Cetraria* *aculeata* and later by *Cladonia sylvatica*. At the same time plants of *Festuca ovina* appear, scattered through the moss and lichen communities, and of *Agrostis* (*A. stolonifera* and *A. tenuis*) in rare patches. The moss accumulates sand, growing up through the freshly deposited layers, and the two lichens settle down on this substratum. Under the *Cetraria-Cladonia* mat the *Polytrichum* dies, and the mat, having lost its anchorage, disintegrates in whole or part, exposing the soil to partial erosion. Further

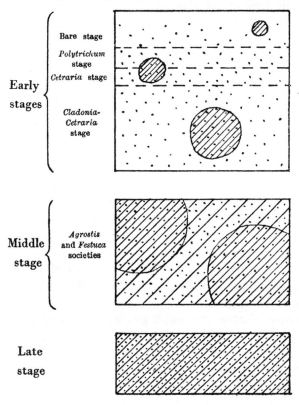

FIG. 90. DEVELOPMENT OF FESTUCO-AGROSTIDETUM IN BRECKLAND

Only the early progressive and "peak" stages are shown. The dots represent individuals of *Festuca ovina*, the diagonal lines *Agrostis* spp. (*Polytrichum* and the lichens are not represented by symbols.) From Watt, 1938.

development is conditioned by the presence of the higher plants, whose effect is to expedite the progress of the succession. In a narrow ring round each *Festuca* tuft, and throughout the patches of *Agrostis*, the dominance of *Cladonia* is established. The islands of *Festuca* and *Agrostis* are set in a background of *Cladonia-Cetraria*.

The next (Fig. 90, "middle stage") shows patches of abundant *Agrostis* with a little *Festuca* set in a background of much *Festuca* with a little *Agrostis*, the whole having an almost continuous carpet of *Cladonia*.

The third or adult stage (Fig. 90, "late stage") shows an intimate mixture of abundant *Agrostis* with numerous but small *Festuca* plants, set in a carpet of *Cladonia*. The humus content of the soil, very low in the early stages, now reaches 2·9 per cent.

The other lichens present during the succession are Cladonias (*C. furcata*, which rises to 13 per cent of cover in the *Cladonia-Cetraria* stage, *C. uncialis*, *crispata*, *coccifera*, *verticillata*, *pityrea*, *bacillaris*, *fimbriata*, *floerkeana*), *Alectoria chalybeiformis*, and *Diploschistes scruposus*, but they play no considerable part in the changes of vegetation; and of higher plants *Aira praecox*, *Rumex acetosella*, and *Teesdalia nudicaulis*, a few scattered individuals of which appear sporadically in one or other of the early stages. It appears that the poorest Breckland soils exclude all annuals except *Aira* and *Teesdalia*. *Carex arenaria* invades the middle stage in quantity but does not alter the general course of the succession, being subject to the same influence as *Festuca* and *Agrostis*; and *Luzula campestris* appears only in the adult stage, which also contains *Polytrichum juniperinum*, *Ptilidium ciliare*, *Lophozia excisa* and *Cladonia alcicornis*. These additions are perhaps due to the increased humus content.

None of the three "peak stages" described is stable. *Agrostis* may die in whole or in part, the lichen cover disrupts, leaving the soil open to erosion, and retrogressive stages, similar to the progressive, can be traced, followed by fresh progression towards the next peak. A series of dry years favours the retrogressions but is not their primary cause, since progressions and retrogressions take place side by side, and Watt thinks that the cyclic behaviour of each stage is primarily due to a still obscure cause related to cycles of lichen development.

The whole area is subject to rabbit grazing and the grassland climax is no doubt determined by this factor. Neither *Calluna* nor any other woody plant enters into the communities studied.

For extensive quantitative details and analyses of the data Watt's original paper (1938) must be consulted.

II. NARDUS GRASSLAND (NARDETUM STRICTAE)

(Pls. 96–7).

The very characteristic and widespread community dominated by the mat-grass or "white bent" (*Nardus stricta* L.) is intermediate in habitat and character between the bent-fescue grassland and the drier peat vegetation dealt with in Chapters xxxiv and xxxv. In *Types of British Vegetation* it was in fact described both as a typical form of "siliceous grassland" in the southern Pennines (Moss) and also, along with Molinietum, as a facies of "grass moor" in Scotland and elsewhere (W. G. Smith). While it is properly described here as a characteristic type of acid grassland, it has more intimate genetic relationships, as Adamson (1918) has shown, with acid peat vegetation than with bent-fescue grassland.

Distribution, habitats and growth form. *Nardus stricta*, which has a wide
geographical distribution, extending from the mountains of southern
Europe and the plains of western Asia to western Greenland, occurs in a
great variety of habitats, whose common character seems to be a soil of
acid and damp but not permanently wet raw humus or peat, with stagnant
or nearly stagnant soil water. In this country it ranges from near sea level
to quite high altitudes, attaining more than 3000 ft. in Scotland, though it
is much more commonly dominant on the hills than in the lowlands, and
decreases in abundance above 2000 ft. On many of our northern and western
mountains it forms a zone between the bent-fescue grassland and the peat
vegetation of the plateaux (Fig. 91). The whitish areas of Nardetum on the
shoulders and plateaux of the hills between 1000 and 2000 ft. are conspicuous
features seen from a distance. *Nardus* is very often associated with the
silver hairgrass (*Deschampsia flexuosa*).

"The vegetative organs form very compact tufts (or tussocks) made up
of a system of horizontal rhizomes almost on the surface of the soil. The
leafy and flowering shoots are firmly enclosed in thick tough basal sheaths
which persist long after the foliage is dead, and these form the double or
triple series of comb-like teeth so characteristic of the rhizomes. The
frequent branching of the rhizomes and the closeness of the shoots leads to
considerable congestion, and as new organs are formed they become piled
on the remains of the old ones; thus tussocks are formed which we have
seen knee-high. Plant remains and wind-borne detritus are added, and
hence a colony of *Nardus* generally forms a thick mat or sod which extends
peripherally and ousts all but a few rivals such as *Juncus squarrosus*,
Vaccinium myrtillus and in wet places *Molinia caerulea*. The humus mat
furnishes, however, a substratum for shallow-rooted, humus-frequenting
species, such as *Deschampsia flexuosa*, *Agrostis* spp., *Anthoxanthum*,
Luzula campestris (vars.), *Galium saxatile*, *Potentilla erecta*, etc. The roots of
Nardus are thick, and cord-like, and the finer lateral roots are mycorrhiza
with an endotrophic fungus" (W. G. Smith, 1918, p. 6).

Adamson (1918, p. 103) says that in the southern Pennines the *Nardus-
Deschampsia* grassland is perhaps the most extensive vegetation type,
occurring at altitudes of 400–1700 ft. (120–520 m.) or more, and that its
composition varies considerably with variation in the habitat. "On
relatively gentle slopes or level ground—where peat can accumulate and
where surface drainage is poor—*Nardus* becomes truly dominant almost to
the exclusion of other species." In such situations 6 in. (15 cm.) of peat
may be present. "On rather steeper slopes or where the peat is thin and
drier, dominance is shared by *Deschampsia flexuosa*,[1] which fills the inter-

[1] The silver hair-grass (*D. flexuosa*), according to Moss, is so abundant in the
Nardetum of the southern Pennines that in early summer, when its tall purple scapes
rise above the general level of the herbage, they give tone and colour to the whole
community.

stices between the tufts of *Nardus* and is not only frequently numerically
far more abundant but owing to its freely branching superficial rhizomes
has as much or more controlling effect on the other species of the association....On steep slopes where only a thin peat is present...liable to
periods of great drought, and in very exposed situations, *Nardus* is often
quite absent and *Deschampsia* is present alone....The distribution of the
two plants is well brought out in their topographic relations on many of the
lower grass-covered hills: the grassland dominated by *Nardus*, the escarpment face by *Deschampsia* alone, while below the 'edge', on the more
moderate slope, co-dominance of the two plants occurs" (Adamson, 1918).

W. G. Smith (*Types of British Vegetation*, 1911, pp. 282–4) writes that the
soil of the "grass moor in Scotland consists of a sod of about 6–9 in. thick,
made up of shoot bases and rhizomes matted with mosses and decaying
herbage, and resting on an impervious subsoil (glacial till or boulder clay)
from which it is easily separable", but which is penetrated by the cord-like
roots of the dominant. One of the chief habitats of Nardetum is the zone of
redistributed peat derived by erosion from and deposited round the edge of
a plateau peat moor (Pl. 96, phot. 232), and it may also occupy the lower
summits which have been denuded of peat (Pl. 97, phot. 234): *Nardus* rarely
colonised untouched peat.

Marginal Nardetum on the Moorfoot Hills. The best detailed account of
the community is given by W. G. Smith (1918) from this redistributed
marginal peat on the Moorfoot Hills in south-eastern Scotland (Fig. 91).
The most abundant and constant plants associated with the dominant in
this Nardetum were *Deschampsia flexuosa* and *Agrostis* spp. (*A. tenuis* the
commonest, but also *A. canina* and *A. stolonifera*). The other less constant
grasses were *Anthoxanthum odoratum*, *Festuca ovina* and *F. rubra*. *Galium
saxatile* and *Potentilla erecta* had a high constancy, as in the bent-fescue
grassland; and *Vaccinium myrtillus* an even higher one. *Calluna vulgaris*,
Juncus squarrosus and *Luzula campestris* had a moderate constancy, while
Molinia caerulea was much less constant. All of these varied from occasional
to abundant in different samples. Less abundant species were *Carex
goodenowii*, *C. binervis*, *Eriophorum vaginatum* and *Blechnum spicant*. Of
mosses *Polytrichum commune* was not at all constant but was abundant
where it occurred. *Aulacomnium palustre*, *Sphagnum*, *Dicranum* and
Campylopus spp. were less abundant. These data were taken from nineteen
samples in which *Nardus* was definitely dominant.

Welsh Nardeta. In Wales Nardetum is widely distributed on the hill
slopes, mainly between 1000 and 2000 ft. (300–600 m.). Davies (1936)
estimates that Nardetum together with Molinietum occupies 16·8 per cent
of the Welsh "agricultural area". It invades insufficiently grazed bent-
fescue grassland wherever the conditions favour the formation of acid
peaty humus without being too wet. It is of little use as general sheep
pasture, though the old wethers can graze it, and the disappearance in

PLATE 96

Phot. 231. Nardetum with *Juncus effusus* (Pennines).

Phot. 232. Upper margin of Nardetum invading the eroded edge of a peat-covered plateau. Moorfoot Hills, Southern Uplands, south of Edinburgh, 600 m. *Donald Macpherson.* From W. G. Smith (1918).

NARDETUM STRICTAE

PLATE 97

Phot. 233. Zonation of Nardetum with peat-covered plateau above and Callunetum on steeper slope below. Moorfoot Hills, 450–530 m. alt. Cf. Fig. 91. *Donald Macpherson.* From W. G. Smith (1918).

Phot. 234. Nardetum occupying the summit of a spur in the eastern Moorfoots. Flushed grassland with sparse *Pteridium* on the slope below, Callunetum on a rocky patch towards the left. In the foreground Nardetum on the near side of the valley. *Donald Macpherson.* From W. G. Smith (1918).

NARDETUM STRICTAE, MOORFOOT HILLS

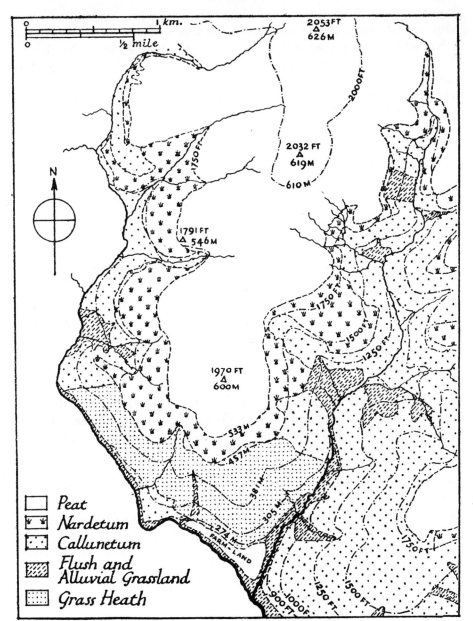

FIG. 91. ZONATION OF VEGETATION ON THE SLOPES OF THE MOORFOOT
HILLS (SCOTTISH SOUTHERN UPLANDS)

The peat-covered plateaux are eroded at their edges, and the redistributed peat below is
colonised by Nardetum (about 1300–1800 ft.), on the whole following the contours pretty
closely. Below this is Callunetum alternating with "grass heath" (*Festuca-Agrostis*), broken
by flush and alluvial grassland in the valleys (cf. Pl. 97). After W. G. Smith, 1918.

recent years of these wethers and also of the ponies from the hill pastures helps the invasion of the bent-fescue grassland by Nardetum.

Nardetum on Cader Idris. An example of high-level Nardetum (1500–1800 ft. = 457–548 m.) is described by Price Evans (1932) from the northern face of Cader Idris in Merionethshire. "The soil is almost everywhere of a peaty nature and in some parts there are deposits of thick peat." The following is a list of species:

Nardus stricta	d	Lycopodium alpinum	o
Galium saxatile	a	L. clavatum	o
Erica tetralix	la	Luzula campestris	o
Juncus squarrosus	la	Anthoxanthum odoratum	o
Empetrum nigrum	f	Sieglingia decumbens	o
Potentilla erecta	f	Agrostis tenuis	o
Vaccinium myrtillus	o–f	Cirsium lanceolatum	r
Festuca ovina	o–f (la)	Hypnum schreberi	f
Lycopodium selago	o–f	Rhacomitrium lanuginosum	f
Deschampsia flexuosa	o–f	Polytrichum commune	o–f
Carex binervis	o	Sphagnum sp.	o
C. panicea	o	Cladonia sylvatica	o–f
Polygala vulgaris	o	C. uncialis	o

It will be noted that, as in Smith's list from the Moorfoots, the heath bedstraw and the tormentil, two of the most constant species of the bent-fescue grassland, are abundant and frequent, while sheep's fescue and bent are themselves frequent or occasional. Here we have in addition three species of clubmoss, two of which are "Highland" species, corresponding with the high level of this Nardetum and indicating its relationship with adjacent arctic-alpine grassland (see Chapter XXXVIII). Thirteen species however are common to this Nardetum and that of the Moorfoots. *Cirsium lanceolatum* is of course a chance invader.

Nardus stricta is widely distributed in southern England, but now at least is very rarely dominant over any considerable lowland area. It is characteristic, as Horwood (1913) remarks, of a certain type of wet heath, and is occasionally found as a relict where such heaths may have formerly existed. It occurs for example (or did occur a few years ago) in damp hollows on the suburban "West Heath" at Hampstead, together with much more abundant *Molinia caerulea*, where a well-developed wet-heath community undoubtedly once existed.

III. *MOLINIA* GRASSLAND (MOLINIETUM CAERULEAE)

(Pls. 98–9)

Grassland dominated by the purple moor-grass or "flying bent" (*Molinia caerulea*) has several characters in common with the Nardetum, which it often adjoins. Like the *Nardus* community Molinietum is closely related to the main peat vegetation of the northern and western hills, to which it is often marginal, and both *Nardus* and *Molinia* themselves form

PLATE 98

Phot. 235. Tussock habit of *Molinia caerulea*.
T. A. Jefferies (1915).

Phot. 236. *Molinia* tussock with *Des-champsia flexuosa*. *T. A. Jefferies* (1915).

Phot. 238. *Molinia* seedlings colonising sun-cracks in bare peat. *T. A. Jefferies* (1915).

Phot. 237. Root system over 40 cm. long.
T. A. Jefferies (1915).

MOLINIA CAERULEA

PLATE 99

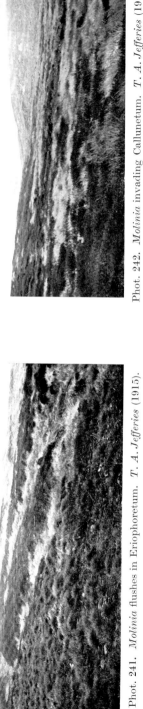

Phot. 240. *Molinia* flush in Callunetum.
T. A. Jefferies (1915).

Phot. 242. *Molinia* invading Callunetum. *T. A. Jefferies* (1915).

Phot. 239. *Molinia* sward in winter. *T. A. Jefferies* (1915).

Phot. 241. *Molinia* flushes in Eriophoretum. *T. A. Jefferies* (1915).

MOLINIETUM ON THE SOUTHERN PENNINES

peat to a certain extent. Molinietum is indeed often regarded (as by Fraser, 1933) as part of the "moorland". *Molinia* freely colonises the drier parts of the "blanket bog" of western Ireland and Scotland (Chapter XXXIV), and its associates are predominantly wet peat plants, as can be seen from the lists given below. It invades and eventually replaces degenerate oak-birch woods in the wet climate of the Pennines, and under some conditions may also invade Callunetum. Like *Nardus* again, *Molinia* occurs on lowland heaths and moors, and much more abundantly; but unlike the mat-grass it also occurs and may become dominant not only on acid fens but on those which are alkaline in reaction (Chapters XXXII, XXXIII).

Molinia bog and Molinia meadow. Stapledon (1914) and Jefferies (1915, p. 94) distinguish between "*Molinia* bog" and "*Molinia* meadow", the former dominated by large *Molinia* tussocks, the latter more lawn-like, smooth, and uniform, and of some value as pasture. A well-known variety of *Molinia caerulea* is "forma *depauperata*" with smaller leaves and one-flowered spikelets, and this may perhaps always be the dominant of "*Molinia* meadow".

Habitat factors. In his excellent paper on the Molinietum of the Pennines near Huddersfield (Pls. 98–9), T. A. Jefferies (1915) showed that while *Molinia* is adapted to a wetter soil than *Nardus* (as indeed was already well recognised) the decisive factor of its habitat is a supply of fresh water, relatively, though not of course extremely, rich in soluble salts, and this is entirely in accordance with the occurrence of the grass in other regions. Stagnancy and increasing acidity of the water enable the cotton grass (*Eriophorum vaginatum*), the main dominant of the wet moors or mosses of the Pennine plateaux, to supplant *Molinia*, while decreasing water supply brings in the Nardetum, and better drainage, with increase of mineral salts and of pH value, will favour the development of grassland of the bent-fescue type.

FIG. 92. BRANCH OF *MOLINIA CAERULEA* IN SPRING

The two leaf-bearing shoots are growing at the expense of the food in three leaf bases. The right-hand leaf base has lost the whole of its food store. From Jefferies, 1915.

Vegetative structure. *Molinia* has a condensed branching sympodial rhizome, and its aerial shoots are very characteristic (Fig. 92). Each bears two sets of leaves separated by a basal internode, swollen at the base and tapering upwards, and packed with reserve food material. The leaves themselves are long and thick, and taper to a fine point which usually withers during the summer—rather definitely a "mesomorphic" type. The leaves are deciduous, separating by a definite abciss layer. The root system[1] is very well developed, consisting of "cord roots" and finer branching roots covered with root hairs to their bases and reaching a depth of 2 ft. Thus the roots penetrate the soil below the surface peat, which, according to W. G. Smith, shows a strikingly constant depth of about 9 in. in the Moorfoot Molinieta, and according to Jefferies rarely exceeds 12 in. in the Pennines and is usually a good deal thinner.

Molinia peat. G. K. Fraser (1933) considers Molinietum specially characteristic of the southern Scottish moorlands and points out that the plant forms amorphous peat, in which, unlike the fibrous *Sphagnum* and *Eriophorum* peat and the "pseudo-fibrous" *Scirpus* peat, the decay of the plant remains has destroyed all trace of structure. This amorphous peat "varies from a black, mud-like peaty mass to a brown peat of spongy texture".

Southern Scottish Molinieta. W. G. Smith (1911, pp. 284–5) gives the following list for the Molinietum of Scottish "grass moor".

	Molinia caerulea	d	
Agrostis stolonifera	a	Hydrocotyle vulgaris	f
Carex canescens	f	Juncus articulatus	ld
C. dioica	f	J. effusus	ld
C. flava	f	Myrica gale	la
C. goodenowii	a	Narthecium ossifragum	f
C. panicea	a	Oxycoccus quadripetalus	f
C. stellulata	a	Parnassia palustris	f
Cirsium palustre	f	Pedicularis palustris	f
Comarum palustre	f	Ranunculus flammula	f
Deschampsia caespitosa	ld	Scirpus caespitosus	la
Erica tetralix	la	Triglochin palustre	f
Eriophorum angustifolium	la		
E. vaginatum	la	Aulacomnium palustre	f
Galium palustre var.		Sphagnum spp.	a
witheringii	f		

Pennine Molinieta. The Molinietum studied by Jefferies (1915) on the south Pennine moors, situated at about 1200 ft. (365 m.) above sea-level on the edge of the cotton-grass plateau, showed remarkably few species associated with the dominant, which occupies the ground very completely. Stretches occur in which no other species can be found except a few algae

[1] Pl. 98, phot. 237.

and liverworts.[1] The most constant associate is *Deschampsia flexuosa*. *Eriophorum vaginatum* (and more rarely *E. angustifolium*) occurs in more swampy, *Calluna* in drier, conditions, while *Vaccinium myrtillus* is frequent, especially on ridges, and occasionally colonises the *Molinia* tussocks themselves. Beyond these five species, according to Jefferies, none is more than occasional where *Molinia* is well developed. In addition to these Moss (1913) gives twenty-eight other species, nearly all wet heath or moorland forms and about half occurring in Smith's Moorfoot list. Among these are the sedges *Carex goodenowii*, *C. panicea*, *C. stellulata* and *C. flava*. And Stapledon (1914) mentions *Scirpus caespitosus* and *Erica tetralix* as typical constituents of the corresponding Molinietum in central Wales, where this community is particularly extensive, occupying, together with Nardetum, 16·8 per cent of the total "agricultural area" (Davies, 1936).

Cornish Molinieta. Magor (unpublished) describes Molinietum as "the typical wet facies of the grass moor" which is the main vegetation of the granite of Bodmin Moor in Cornwall. "It is developed extensively over the peat in all but the wettest parts of the bogs, and is rich in species, though actually made up of scattered societies and not very homogeneous." The following species were recorded from the "Pillar Marsh" near Roughton:

| Molinia caerulea | d |
| Sphagnum spp. | ld |

Very abundant

| Juncus conglomeratus | Holcus lanatus |
| Festuca ovina | |

Abundant and locally subdominant

| Erica tetralix | Eriophorum angustifolium |
| Juncus squarrosus | |

Abundant

Calluna vulgaris	Potentilla erecta
Narthecium ossifragum	Luzula multiflora var. congesta
Juncus bulbosus forma uliginosus	Aulacomnium palustre
Viola palustris	

Frequent

Agrostis tenuis	Lotus uliginosus
Drosera rotundifolia	Menyanthes trifoliata
Eriophorum vaginatum	Polygala serpyllifolia
Galium palustre	Ranunculus flammula
Hydrocotyle vulgaris	R. hederaceus
Hypericum elodes	Rhynchospora alba
Juncus bufonius	Succisa pratensis

[1] Watson (1932) suggests that the floristic poverty of this Molinietum, as of other communities of the region, is largely caused by the smoke arising from neighbouring industrial centres, but Davies remarks that the Molinieta of central and western Wales, where smoke pollution cannot be a factor, are equally poor in species. Recent experimental work at Manchester finds that smoke pollution has relatively little adverse effect on vigorous pasture.

Occasional

Carex binervis	Nardus stricta
C. goodenowii	Pedicularis palustris
C. stellulata	Sieglingia decumbens
Deschampsia flexuosa	Vaccinium myrtillus

Rare

Eleocharis multicaulis	Ranunculus lutarius
Orchis ericetorum	Prunella vulgaris
Pinguicula lusitanica	

In the same district, on the very shallow black gravelly soil overlying the Staddon Grit (Devonian) a Molinietum, floristically much poorer and obviously a good deal drier, is developed. This contains the following species:

Very abundant

Ulex gallii	Agrostis setacea
Erica tetralix	

Abundant

Agrostis canina	Festuca ovina
Deschampsia flexuosa	

Frequent

Erica cinerea	Carex diversicolor
Calluna vulgaris	

Occasional

Polygala serpyllifolia	Orchis ericetorum

Rare

Eriophorum angustifolium

Highland Molinieta. Crampton (1911) reported Molinieta from the channels of wide flushes arising from springs beneath the peat of the Caithness heather moors, the flow from which had caused the peat surface to collapse. *Molinia* is dominant chiefly where a sudden decrease of slope causes partial arrest of the water flow, leading to silting, *Nardus* on the other hand where the slope is steeper and the soil consequently drier. Molinieta are said to be more extensive in Sutherland and the west of Scotland generally than in Caithness. Fraser (1933) describes *Molinia* flushes in the "moorland" (blanket bog) vegetation and also in the grassland of Argyllshire. These are intermediate between the somewhat richer *Juncus* flushes and the poorer *Eriophorum* flushes.

Irish and other Molinieta. Molinietum occupies extensive areas in and about the blanket bogs of the west of Ireland, for example in Connemara. Here it freely colonises the drier parts of the bog, and contains more

Calluna and *Erica tetralix*, but still retains some of the same associates as the wetter Rhynchosporetum (see p. 713).

The Molinietum which forms a well-marked belt on the edges of the raised bogs in north Lancashire and in the New Forest, between the wet bog or "moss" dominated by *Sphagnum* or *Eriophorum* on one side, and a reedswamp or alder wood lining a stream or marginal "ditch" on the other, is clearly a transitional community occupying a habitat of intermediate edaphic character, less acid than the "moss" and drier than either moss or reedswamp (Rankin in *Types*, pp. 251, 262–3).

Pearsall (1918) has described Molinieta which form stages in the hydrosere leading to raised bog in the vicinity of Esthwaite Water in the Lake District. These, together with those of the East Anglian fens, will be dealt with in Chapters XXXII and XXXIII, where the successions from fresh water are described.

REFERENCES

ADAMSON, R. S. On the relationships of some associations of the Southern Pennines. *J. Ecol.* **6**, 97–109. 1918.

BRAID, K. W. Herbicides, with special reference to sodium chlorate and its effect on bracken. *Scot. Farmer and Farming World*, **41**, No. 2131, p. 1484. 1933.

BRAID, K. W. The Bracken eradication problem. *Agri. Prog.* **14**, 38. 1937.

CRAMPTON, C. B. *The vegetation of Caithness considered in relation to the geology.* Com. for Survey and Study of British Vegetation, 1911.

DAVIES, WILLIAM. "The Grasslands of Wales" in Stapledon, *A Survey of the Agricultural and Waste lands of Wales.* 1936.

EVANS, E. PRICE. Cader Idris: a study of certain plant communities in south-west Merionethshire. *J. Ecol.* **20**, 1–52. 1932.

FARROW, E. P. On the ecology of the vegetation of Breckland. II. *J. Ecol.* **4**, 57–64. 1916.

FARROW, E. P. On the ecology of the vegetation of Breckland. IV. *J. Ecol.* **5**. 1917.

FRASER, G. K. Studies of Scottish moorlands in relation to tree growth. *Bull. For. Comm., Lond.*, No. 15. 1933.

HORWOOD, A. R. Vestigial Floras. *J. Ecol.* **1**, 100–2. 1913.

JEFFERIES, T. A. Ecology of the purple heath-grass (*Molinia caerulea*). *J. Ecol.* **3**, 93–109. 1915.

JEFFREYS, H. On the vegetation of four Durham Coal-Measure fells, I. *J. Ecol.* **4**, 174–95. 1916.

LEACH, W. The vegetation of the Longmynd. *J. Ecol.* **19**, 34–45. 1931.

MAGOR, E. W. Geographical distribution of vegetation in Cornwall: Camelford and Wadebridge district (unpublished).

MOSS, C. E. *Vegetation of the Peak District.* Cambridge, 1913.

PEARSALL, W. H. The aquatic and marsh vegetation of Esthwaite Water. *J. Ecol.* **6**, 53–74. 1918.

PETHYBRIDGE, G. H. and PRAEGER, R. LL. The vegetation of the district lying south of Dublin. *Proc. Roy. Irish Acad.* **25**, 124–80. 1905.

SMITH, W. G. "Grass moor association." In *Types of British Vegetation*, pp. 282–6. 1911.

SMITH, W. G. The distribution of *Nardus stricta* in relation to peat. *J. Ecol.* **6**, 1–13. 1918.

SMITH, W. G. Notes on the effect of cutting bracken (*Pteris aquilina* L.). *Trans. Proc. Bot. Soc. Edinb.* **30**. 1928.

STAPLEDON, R. G. *The Sheep Walks of Mid-Wales* (privately printed). 1914.

WATSON, W. The bryophytes and lichens of moorland. *J. Ecol.* **20**, 284–313. 1932.

WATT, A. S. Studies in the ecology of Breckland. III. Development of the Festuco-Agrostidetum. *J. Ecol.* **26**, 1–37. 1938.

WEISS, F. E. The dispersal of the seeds of gorse and broom by ants. *New Phytol.* **8**, 81–9. 1909.

WOODHEAD, T. W. History of the Vegetation of the Southern Pennines. *J. Ecol.* **17**, 1–34. 1929.

Chapter XXVII

BASIC GRASSLANDS

At the opposite extreme from the grasslands considered in the preceding chapter are those developed on basic soil, i.e. soil in which the pH value of the solution exceeds 7. The maintenance of such a condition requires a constant supply of alkaline salts, usually calcium carbonate, and this is typically maintained by the progressive solution of the underlying lime-stone rock through the agency of percolating rainwater. It is because of this alkaline "buffer" that these grasslands are relatively "stable", apart from invasion by woody plants; but grazing is the main stabilising factor.

CHALK GRASSLAND

The most sharply defined and typical of the basic grasslands is the characteristic chalk grassland developed on shallow soil directly derived from the Chalk (Upper Cretaceous) of the south and east of England. The Chalk is a very pure limestone, containing from 90 to 99 per cent of calcium carbonate and thus leaving a very small residue on solution. This residue **Distribution** forms a thin soil (rendzina, see pp. 93–5), varying in thickness **and Soil** from a mere film to a depth of about 30 cm. above the chalk rock, on slopes and narrow ridges. Twenty to thirty-five cm. (8–14 in.) is a very common depth to the weathered surface of the chalk. Increase in thickness is checked, in spite of the progressive solution of the underlying rock, by the constant slow creep of the soil towards the bottom of the slope. It must always be borne in mind, however, that much of the extensive chalk out-crop, where the ground is flat or nearly so on the wide plateaux and gentle dipslopes, is covered with a variable thickness of loam owing to accumulation of insoluble residue. This may take the form of a thick layer of "clay-with-flints" (formed from the residue of many times the depth of chalk together with debris from vanished Tertiary beds and sometimes Quaternary deposits) which overlies a great portion of the outcrop; but much of the area not mapped as clay-with-flints is actually covered with non-calcareous loam. Both clay-with-flints and loam yield a soil of totally different character from the chalk rendzina. It is mainly on the relatively steep slopes of the escarpments and valley sides and on the summits of the narrower ridges that a typical chalk soil and therefore typical chalk grass-land is encountered.

Typical chalk grassland, which forms by far the most extensive "rough grazing" of south-eastern England, occupies much of the chalk escarpment surrounding three sides of the Weald, and also a good deal of the main chalk escarpment which runs, with many irregularities and interruptions, north-eastward from the coast of Dorset to that of the North Sea. The steeper sided valleys and minor escarpments which are numerous in the broad outcrop of the Chalk in central southern England also bear this type of grassland. Wherever, in fact, the angle is too steep and the soil too thin for arable cultivation, and where the chalk slopes are not wooded, we find this characteristic plant community maintained by grazing.

The eastern and central Sussex Downs ("South Downs"), bounded on the north by the escarpment falling steeply to the Wealden Plain are one of the best developed and most typical chalk grassland areas. Plate 100, for which I am indebted to *The Times*, shows a long stretch of the central South Downs in Sussex. The western South Downs, extending into Hampshire, together with the escarpment facing east on the western edge of the Weald, are largely covered with beechwood, and the same is true of the "North Downs" of Surrey and Kent, though the grassland areas are not inconsiderable. In Dorset and especially in Wiltshire (Salisbury Plain, Marlborough Downs, etc.), Hampshire and Berkshire, there are great areas of chalk pasture. In south Oxfordshire and Buckinghamshire (Chiltern Hills) beech forest is more widespread, but much of the escarpment is grassland (Pl. 101, phot. 244). North-east of the Chilterns, in Bedfordshire, Hertfordshire and south Cambridgeshire the chalk outcrop diminishes greatly in height, the slopes are gentler and there is much less open down-land than in the southern counties, though it is not entirely wanting (Dunstable Downs, Royston Heath). Practically absent in east Suffolk and Norfolk, chalk grassland reappears on the Lincolnshire and south-east Yorkshire Wolds, where the chalk outcrop is cut off by the sea.

The characteristic chalk rendzina soil is very constant in general character, whether it is under grass, scrub or trees, though it is of course affected by the kind of vegetation it bears. The typical "mature" chalk grassland soil (though the shallow chalk soil can never become *pedologically* mature since it is constantly wasted and rejuvenated) is richly humous in the surface layer and passes below into several inches of brown loam (often decidedly clayey just below the surface humus) containing particles of chalk, minute near the surface and becoming larger at lower levels till the weathered chalk surface is reached at a depth of 8 to 12 or 14 in. (20–35 cm.).

A "young" or "primitive" chalk grassland recently colonised by vegetation, subject to relatively rapid "creep" or checked in its development by dry conditions, as on steep southern exposures, is whitish grey and powdery when dry, almost to the surface, while the solid rock may occur at a depth of 3 to 4 in. or even less.

Another type of soil is formed below chalk grassland on steep cool northern slopes, and is well seen in several places along the north-facing escarpment of the South Downs. Here the soil is very dark, or even black, and very rich in humus, so that it shrinks strongly on drying to a hard dark cake. The vegetation, particularly when not heavily pastured, is

PLATE 100

Phot. 243. Panoramic view of the escarpment of the central South Downs (Sussex), entirely covered with chalk grassland (except for a little scrub), from the Devil's Dyke on the extreme left to Chanctonbury Ring (clump of planted trees on the summit of the Downs) in the distance on the right—between 8 and 9 miles (c. 13·5 km.). Below the Ring the escarpment is wooded. The western South Downs—20 miles (32 km.) from the observer—just appear on the extreme right in the far distance. A narrow belt of scrub has been left between the arable cultivation of the Weald and the Downs grassland. Photograph from *The Times*.

ESCARPMENT OF THE SOUTH DOWNS

PLATE 101

Phot. 245. Unpastured grassland on the north-facing escarpment at Beddingham Hill south-east of Lewes. Tall *Bromus erectus* and *Avena pratensis* dominant. *Pimpinella saxifraga* in flower. Scattered *Crataegus*.

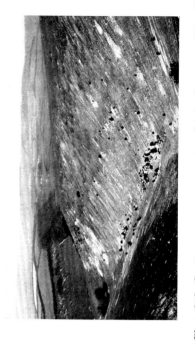

Phot. 246. Continuous grass cover not maintained on steep (34°–39°) southern slope of Ramsdean Down, Butser Hill, Hants.

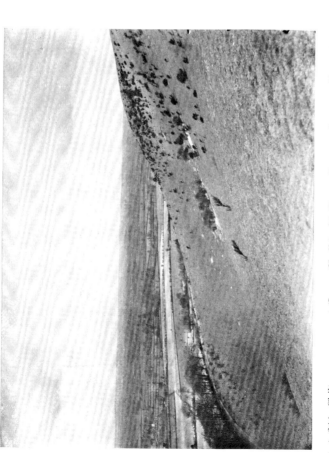

Phot. 244. Chiltern escarpment (Lower Chalk) with Upper Greensand terrace (light coloured arable fields) and the Plain of Aylesbury beyond. Scattered juniper scrub, wind-cut nearest the observer, and a group of alders associated with a rabbit burrow. Grassland in right-hand bottom corner not excessively rabbit grazed though not far from the nearest burrow. Photograph from *The Times*.

dominated by the taller grasses such as *Avena pratensis*, *A. pubescens*, *Bromus erectus*, *Arrhenatherum elatius*, and occasionally even *Deschampsia caespitosa* and *Festuca elatior*.

The average pH value of the top 2 or 3 in. of soil is about 7·5 in all three types. The CaO content is high, with an average of about 20 per cent; it decreases from the "young" grey to the dark humus soils, though it is still high in many of the last, and the average "loss on ignition" (humus content) correspondingly increases. Potash and magnesia, too, are in good supply throughout. Under these conditions, then, because of the alkaline buffer, leaching is ineffective in reducing the content in basic salts and substantially increasing the acidity, even of the top two inches, of most chalk soils, though in a few the pH value falls below the neutral point.

Table XIII. *Chalk grassland soils from Sussex Downs*

	Grey (6)	Brown (6)	Dark (7)
Mean CaO	29·1	18·1	15·0
Mean loss on ignition	24·6	25·9	31·3
Mean pH value	7·4	7·4	7·5

Table XIV. *Partial analyses of six typical chalk grassland soils*

	Grey and brown soils				Dark soils	
	A	B	C	D	E	F
No. of area (Tansley and Adamson, 1926)	3	22	12	5	21	19
Depth of sample	0–2″	0–3″	0–3″	0–5″	0–1″	0–3″
Loss on ignition	25·9	23·0	17·3	32·0	32·6	42·6
CaO	27·4	32·6	33·7	11·1	15·5	3·3
MgO	0·32	0·51	0·31	0·35	0·62	0·77
K_2O	0·58	0·28	0·26	0·29	0·33	0·35
pH	7·4	7·9	7·8	6·9	7·8	7·2

Water content. The water content of a shallow chalk soil, half way down a slope where the angle of inclination was about 23°, bearing typical chalk grassland vegetation, and its variations throughout the year, have been thoroughly investigated by Anderson (1927). "The surface soil was a brown friable loam; small lumps of chalk up to 12 mm. diameter occurred at a depth of 2·5 cm. and increased in size and frequency until the loam gave way to broken chalk at a depth of 20 cm. Solid chalk occurred at 25 cm." "The pH values ranged from 7·4 at the surface to 7·6 for the underlying chalk.... The upper 7 cm. of soil were abundantly occupied by roots, which were less frequent in the zone below, while the lowest zone considered was tapped only by roots of *Galium* (*verum*) and *Poterium* (*sanguisorba*)." This description represents quite a typical structure of southern English chalk grassland soil.

Determinations of the water content were taken at four levels, (1) 0–7·5 cm., (2) 15–22·5 cm., (3) 30–37·5 cm., (4) 68·5–76 cm., every few days from January 1924 to August 1925, covering periods of spring and summer drought as well as wet weather.

The average value of the water content (loss at 100° C.) of the surface 3 in. was 36·8 per cent, a value comparable with the averages of other well vegetated English soils of various textures. A maximum value of 61·29 per cent was recorded on 12 May 1924, after a period of heavy rain, and a minimum of 8·58 per cent on 6 July 1925, after a period of drought during which evaporation was considerable. Rising to 17·8 per cent on 7 July after a little rain it fell again to 8·8 and 8·9 per cent on the 13th and 19th as a result of further absence of rain. With good rainfall at the end of the month the percentage water content of the surface soil rose to something like normal (29·5 per cent) on 8 August. These extreme summer minimum values (comparing with 18·5 per cent in an oakwood) are probably an important factor in determining the composition of chalk grassland. The water content at greater depths shows much less extreme variation:

Table XV. *Percentage water contents of chalk grassland soil*

January 1924—August 1925

Depth in cm.	0–7·5	15–22·5	30–37·5	68·5–76
Mean	36·78	26·63	25·84	27·20
Maximum	61·29	35·59	34·05	34·93
Minimum	8·58	11·50	10·76	15·34
Range	52·71	24·09	23·29	19·59

Thus the water content at greater depths is comparatively eustatic, and there is some evidence that it is replenished from below by capillary rise of water. It is therefore clear that while the surface-rooting vegetation (Fig. 93, top line), which includes the dominant grasses and several species of **Rooting** characteristic chalk grassland herbs (Fig. 94), must be distinctly **depth** limited by summer droughts, the species whose roots penetrate deeply (Fig. 93, bottom line), like *Galium verum* and *Poterium sanguisorba* (Figs. 95, 96), can rely on a steady supply.

Drought is in fact the great danger to the surface-rooting vegetation on these very well-drained shallow soils overlying the extremely permeable chalk rock, especially on narrow ridges and southern and south-western exposures which are open to the full heat of the summer sun and also to prevailing winds. In a dry summer the leaves of the shallow-rooting grasses become parched and withered, though the deeper-rooting grasses and dicotyledons may remain fresh and green, for their roots penetrate to the clayey layer referred to and some to the fissures of the chalk below.

Floristic composition. The chalk grassland flora is only moderately rich in species, the average number (including mosses) occurring on 20 areas taken at random being 46. The smallest list (26) was taken from a southern slope with very shallow soil, the largest (65) from a northern slope.

The occurrence of about 20 of the commonest species of chalk grassland, which form the vast bulk of the herbage of grazed areas, is on the whole

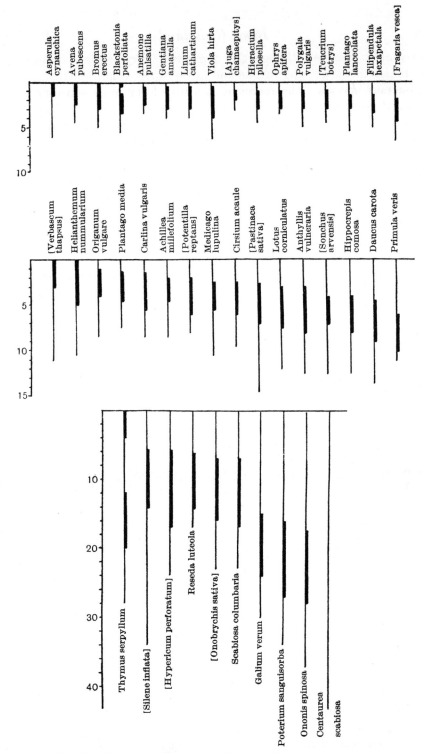

FIG. 93. ROOTING DEPTHS OF CHALK GRASSLAND PLANTS
(SHALLOW, MEDIUM AND DEEP)

The thin line indicates the maximum depth of penetration observed, the thick line the range
of maximum development of feeding roots. The names of plants not belonging to the chalk
grassland community proper are enclosed in square brackets. From Anderson, 1927.

remarkably constant, though the dominants vary, mainly according to differences in the grazing regime and in the water conditions.

Constant species. The following is a list of the more constant species compiled from 62 typical chalk grassland areas in Sussex, Hampshire, Wiltshire, Berkshire and Oxfordshire (of which however two-thirds were in Sussex and the majority of the remaining third in Wiltshire).

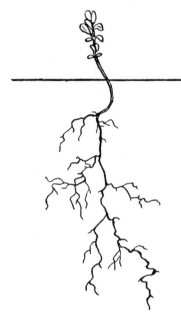

FIG. 94. *LINUM CATHARTICUM*

A typical shallow rooting chalk grassland species (young plant, natural size). From Anderson, 1927.

The first 15 species occurred in more than 80 per cent of the areas, representing the highest degree of constancy (5):

Grasses

Avena pratensis	Festuca rubra
Briza media	Koeleria cristata
Festuca ovina	

Other species

Carex diversicolor (flacca)	Pimpinella saxifraga
Cirsium acaule	Plantago lanceolata
Leontodon hispidus	Poterium sanguisorba
Linum catharticum	Scabiosa columbaria
Lotus corniculatus	Thymus serpyllum

The next seven species occurred in more than 60 and not more than 80 per cent of the areas (constancy 4).

Grass

Avena pubescens

Other species

Achillea millefolium	Ranunculus bulbosus
Asperula cynanchica	Trifolium pratense
Galium verum	Brachythecium purum

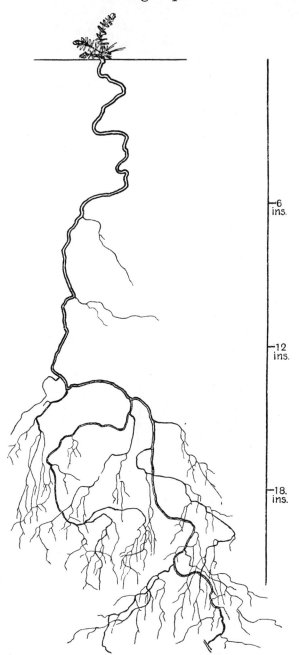

FIG. 95. *GALIUM VERUM*

A typical deep rooting chalk grassland plant (¼ natural size). From Anderson, 1927.

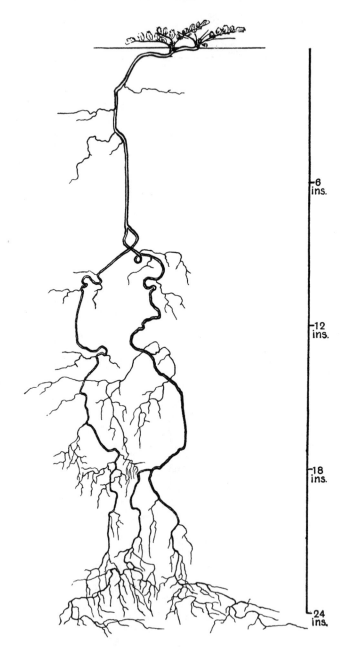

FIG. 96.   *POTERIUM SANGUISORBA*

A typical deep rooting chalk grassland plant ($\frac{1}{4}$ natural size).   From Anderson, 1927.

A larger number (twenty-three species) occurred in more than 40 and not more than 60 per cent of the areas (constancy 3).

Grasses

Anthoxanthum odoratum
Bromus erectus

Dactylis glomerata
Trisetum flavescens

Other species

Bellis perennis
Campanula rotundifolia
Carex caryophyllea
Carlina vulgaris
Centaurea nemoralis
Cerastium vulgatum
Chrysanthemum leucanthemum
Euphrasia nemorosa
Filipendula hexapetala
Hieracium pilosella

Medicago lupulina
Phyteuma orbiculare
Plantago media
Polygala vulgaris
Primula veris
Prunella vulgaris

Camptothecium lutescens
Hylocomium squarrosum
H. triquetrum

Exclusive species. Of these forty-five species which fall into the three highest constancy classes, some are also widely distributed on other soils (mostly other types of dry or medium grassland) while others are either confined or nearly confined to chalk grassland. This last category comprises the so-called *exclusive species*, and 5 degrees of exclusiveness may be recognised. The highest degree (5) includes species never (or very rarely) found in communities other than chalk grassland, the next (4) those which are far more common in chalk grassland, but also occasionally occur in other communities, the next (3) those which are common in chalk grassland but often occur in other communities also. The two lowest degrees—(2) and (1)—include the species which are in no way specially characteristic of chalk grassland, and those which are accidental aliens in that community. Species of the higher degrees of exclusiveness may occur in chalk grassland with any degree of constancy, for a species confined or nearly confined to this community may be found in a high percentage of areas (e.g. *Poterium sanguisorba*, or *Scabiosa columbaria*), or on the other hand it may be a comparatively rare species (e.g. *Senecio integrifolius*, *Thesium humifusum* or *Aceras anthropophorum*).

The following is a list of species belonging to the two highest degrees of exclusiveness which occurred in, but were not found in any large percentage of, the areas used as data (constancy less than 3):

Anacamptis pyramidalis
Aceras anthropophorum
Anthyllis vulneraria
Brachypodium pinnatum
Campanula glomerata
Helianthemum nummularium
    (vulgare)

Hippocrepis comosa
Ophrys apifera
Senecio integrifolius
Thesium humifusum

In addition the following species, which were not found in any of the areas investigated, are known to be confined or nearly confined to chalk grassland in other areas. Most of them are rare or local:

Anemone pulsatilla	l	Orchis ustulata	r
Herminium monorchis	l	Picris hieracioides	o
Hypochaeris maculata	r	Polygala austriaca	vr
Linum anglicum (perenne)	vl	P. calcarea	la
Ophrys fuciflora		Seseli libanotis	r
(arachnites)	r	Viola calcarea	l
O. sphegodes (aranifera)	r		

The grasses:

Characteristic species:

(1) Of greatest constancy:

Festuca ovina, very abundant, very often dominant.

F. rubra, very abundant, very often co-dominant, but rarely completely dominant.

Avena pratensis, abundant, often dominant.

Briza media, moderately abundant.

Koeleria cristata, moderately abundant.

(2) Of less constancy:

Avena pubescens, abundant, occasionally dominant.

Bromus erectus, abundant, often dominant.

Trisetum flavescens, moderately abundant.

(3) Of low constancy:

Arrhenatherum elatius, sometimes dominant, but perhaps only in disturbed soil, though it may persist for many years after disturbance has ceased.

Brachypodium pinnatum, locally dominant.

Non-characteristic species:

(1) Of medium constancy:

Agrostis stolonifera, Anthoxanthum odoratum, Dactylis glomerata, all moderately abundant.

(2) Of low constancy:

Cynosurus cristatus, occasional to frequent.

Deschampsia caespitosa, on damp soils only.

(3) Of very low constancy:

Agrostis tenuis, only on slightly acid soil.

Brachypodium sylvaticum, near wood edges, very local.

Festuca pratensis, locally abundant on damp slopes.

Lolium perenne, occasional.

Sieglingia decumbens, only on slightly acid soil.

The fine-leaved fescues. The most widespread, abundant and important member of the chalk grassland vegetation is almost certainly *Festuca ovina*, but *F. rubra* follows it as a close second. The latter species is only known in typical semi-natural chalk grassland as the subspecies *genuina* Hack. It has been stated more than once that *F. rubra* subspecies *fallax* plays an important part in chalk grassland, but this is a mistake which presumably must have originated on account of the very close association of *F. ovina* and *F. rubra* subspecies *genuina* in the majority of chalk grasslands, creating an intimate mixture of the two which is distinctly tufted and at the same time contains unmistakable *F. rubra* wherever one chooses to look. A hasty examination of such a turf might give the impression that it was composed of a caespitose form of *F. rubra*, i.e. of the subspecies *fallax*. But careful and widespread search of the chalk grasslands of Surrey, Sussex, Hampshire, Berkshire and Oxfordshire has so far not revealed any specimens of *F. rubra* that are not more or less stoloniferous.

The ecological relationship of the two fescues to one another on the downs has not yet been fully studied, but it seems evident that the areas where *F. ovina* preponderates most markedly over *F. rubra* occupy the driest habitats on south-facing slopes, and also in a good many cases surfaces where the parent rock has been exposed and soil formation and the grassland succession are not complete. The more active part played by *F. ovina* as a colonist may well be correlated with its apparently greater powers of drought resistance.[1]

While in the driest situations *F. rubra* may be very inconspicuous or even absent, in the moistest *F. ovina* is present in fair quantity. The two species seem to withstand grazing equally well. Differences in their relative abundance are not correlated with the grazing factor except where grazing has recently become slight or has ceased. Here *F. rubra* (which is the larger plant when not grazed) on the whole has the advantage in the long grass that develops especially on a north or east aspect, becoming codominant or dominant; but *F. ovina* remains abundant. This predominance of *F. rubra* is not so strikingly seen where grazing has been for long in abeyance, for here the place of both the fescues tends to be occupied by taller growing species, of which the two most important are *Bromus erectus*

Bromus erectus Hudson and *Avena pratensis* L.[2] (sometimes replaced by *A. pubescens* Hudson). *Bromus erectus* is intolerant of grazing but it can grow in very dry situations and may colonise bare chalk surfaces alongside of the sheep's fescue. In southern England it is mainly a plant of chalk and oolite grassland but of rather local occurrence, and this may be due to the grazing factor. Where it does occur *Bromus erectus* is very often dominant over considerable areas.[3] It is noteworthy that this

[1] Most of the information on this page from Hope Simpson.

[2] Pl. 101, phot. 238.

[3] Tüxen (1927) considers that the English chalk grassland belongs to the "Mesobrometum", a community in which it appears that *Bromus* is not necessarily present. Whatever we may think of such a nomenclature there is no doubt that the very constant floristic composition of English chalk grassland corresponds pretty closely with that described by Tüxen for the "Mesobrometum" of the Muschelkalk (Miocene) of Hanover.

grass is quite local on the western Sussex Downs where the rainfall is relatively high and the effect of rabbits very severe, but on the drier eastern downs where the grazing (sheep rather than rabbits) is less drastic, it is often dominant over considerable areas, as also in many localities in Berkshire and Wiltshire.

Avena pratensis, which has a much wider distribution in **The Avenae** Britain than *Bromus erectus*, since it extends to the north of Scotland and is not confined to calcareous soils, is nevertheless a very characteristic chalk grassland species. It is nearly as widespread on the chalk soils as *Festuca ovina*, but its average abundance is lower, though still considerable. Here and there it is perhaps co-dominant, and the same is true of *A. pubescens* though this species has a lower constancy than its congener. The factors which determine the occurrence and abundance of these two species of *Avena* are still obscure.

Another grass often closely associated with *Bromus erectus*, and locally dominant in chalk grassland, is the tor grass, *Brachypodium pinnatum*. On **Brachypodium pinnatum** many downs it appears exclusively in nearly pure, actively spreading, circular patches. Here it has all the appearance of a recent and aggressive invader of the community. It is believed to have increased substantially in abundance and distribution during the last half century, but is still confined to the south and east of England. Its rather harsh foliage is avoided by stock and by rabbits, so that the luxuriant patches of *Brachypodium*, bright yellowish-green in colour, often stand out conspicuously from the closely grazed surrounding grassland. The evidence suggests that this grass tends to become dominant where rabbit pressure is not excessive and sheep grazing is at a low ebb. Its recent spread is very likely connected with the reduction of the sheep flocks on the downs, for example in east Sussex. *B. sylvaticum*, a woodland and wood edge grass, sometimes occurs in chalk grassland, spreading on rather open soil from the borders of woods, or occupying sites where woodland has been felled, but it is not a normal constituent of the chalk grassland community.

Arrhenatherum elatius, the false oat-grass, is another species **Arrhenatherum elatius** which sometimes plays a conspicuous part in chalk grassland, though it is far from having a high constancy in any available set of chalk grassland lists, and is nearly always recorded as "local". It is said to be quite intolerant of grazing and trampling and is indeed apparently absent from all heavily pastured slopes. Nevertheless it attains dominance here and there on chalk grassland in Oxfordshire and Wiltshire, but probably only after disturbance of the soil. A more characteristic habitat of this grass is the highly calcareous roadside "verges" in chalk districts, where it is frequently dominant for long stretches, often in company with knapweed (*Centaurea nemoralis* = "*C. nigra*" pro parte) and wild parsnip (*Pastinaca sativa*). This vegetation is generally cut over in the late summer or early autumn, but is not grazed except quite casually.

Briza media, Koeleria cristata and *Trisetum flavescens*   The quaking grass (*Briza media*) is almost as constant in chalk grassland as *Festuca ovina*, but its average abundance is much less. It is by no means confined to chalk soils, occurring in many grasslands with poor soil, both dry and wet.

Koeleria cristata is another chalk grassland species of high constancy and relatively high average abundance. Though characteristic of chalk grassland it occurs on other soils of good base status, and ranges to the north of Scotland.

Trisetum flavescens, the golden oat-grass, is less constant but has a higher average abundance than *Koeleria*. In the south it is characteristic of chalk grassland, though occurring in many other good pastures, and it does not range so far north as the two preceding species.

The ten species of grass described are all that are really characteristic of chalk grassland as contrasted with neighbouring grassland communities— *Festuca ovina* and *F. rubra* because of their very high constancy and very great average abundance. Though extremely widespread and even co-dominant in the drier siliceous pastures and on sandy soils (Chapter XXVI), *F. ovina* is nowhere so markedly predominant as on limestone and chalk. Of the others *Bromus erectus* is nearly confined to chalk and oolite, while the Avenae, *Koeleria*, *Trisetum* and *Brachypodium pinnatum* are far commoner on chalk than elsewhere.

Other grasses. The other grasses occurring in chalk grassland are not in any sense "characteristic", because they are equally or more abundant in neutral grassland communities. Of these species the cocksfoot, *Dactylis glomerata*, has a relatively high constancy in chalk grassland (occurring in 72 per cent of the lists from the South Downs) and also a fairly high abundance. This is a very widespread and abundant plant of neutral grassland but tolerates relatively high alkalinity, and though a fairly tall grass it also stands grazing well.

The sweet vernal-grass, *Anthoxanthum odoratum*, a ubiquitous species of very wide edaphic tolerance, has about the same constancy and abundance in chalk grassland as *Dactylis*.

The crested dogstail, *Cynosurus cristatus*, is much less constant (32 per cent) and has a moderate average abundance. Timothy (*Phleum pratense*) has about the same constancy as *Cynosurus*, but is never more than an occasional constituent. *Poa pratensis* has about the same constancy as the last two species: on one ungrazed north-facing escarpment slope in Sussex it was very abundant in company with *Avena pubescens* and *Dactylis*: for the rest it is only local and occasional.

Lolium perenne (perennial rye-grass), a dominant of the best neutral grasslands, is decidedly rare on chalk soils. On the South Downs it is found occasionally on the deeper brown loams derived from the chalk.

Holcus lanatus, the "Yorkshire Fog", is an aggressive grass which spreads from centres of insemination in pastured grasslands by means of prostrate or semi-prostrate rosette-shoots. It occurs in about half the lists from the South Downs, varying from occasional to locally abundant. It becomes specially prominent when rabbit pressure is withdrawn from a rabbit grazed area (Pls. 103, phots. 252, 254). This species again is in no way specially attached to chalk grassland, having a very wide edaphic tolerance.

Deschampsia caespitosa is a species forming large rosettes or tussocks with long harsh leaves and tall spreading panicles. It is essentially a grass of wet places and especially of puddled surfaces. Together with *Festuca elatior*, which has also a high water requirement, its occurrence on chalk is limited to cool moist unpastured slopes of northern aspect, but in such situations it tends towards co-dominance with other grasses, notably *Festuca rubra*, in a thick herbage which eliminates most of the lower growing chalk grassland species.

Two grasses which are characteristic of acid soils, *Agrostis tenuis* and *Sieglingia decumbens*, are occasionally met with in chalk grassland, especially where surface leaching has taken place: in such cases the calcium content is much lower than usual. *Agrostis stolonifera*, on the other hand, is much commoner on the Sussex Downs, occurring with considerable frequency in about half the areas studied. In the more western chalk grassland areas however (Wiltshire, Berkshire, etc.) it has not been noticed. This is a species which occurs both in wet and in highly calcareous soils, as is true of several others, e.g. *Cirsium palustre*,[1] *Carex diversicolor*[2] (*flacca*), etc. Intermediates (presumably hybrids) between *Agrostis tenuis* and *A. stolonifera* are common and the autecology of the *A. tenuis—A. stolonifera* complex is by no means understood.

Herbs. Chalk grassland is well known for the relatively large number of highly or moderately exclusive species, and the attractiveness of many of their flowers.

Characteristic species, i.e. of high constancy or of high exclusiveness:

    (i) Of greatest constancy (5), roughly in order of abundance:

        Carex diversicolor (*flacca*), very abundant, sometimes locally dominant.

        Poterium sanguisorba, moderately exclusive, very abundant, locally dominant.

        Plantago lanceolata and *Cirsium acaule*,[3] abundant.

        Thymus serpyllum, Linum catharticum, Lotus corniculatus, Leontodon hispidus, Scabiosa columbaria (exclusive), *Pimpinella saxifraga*, all moderately abundant.

On account of their high constancy and considerable or great abundance

[1] Pl. 102, phots. 248, 250.
[2] Pl. 102, phots. 247–9.   [3] Pl. 102, phots. 247, 249; Pl. 103, phot. 251.

all the above species may be regarded as characteristic of chalk grassland, though several are abundant in quite different communities.

(ii) Of somewhat less but high constancy (4):

Asperula cynanchica, almost exclusive, moderately abundant.
Hieracium pilosella, abundant on relatively open soil.
Phyteuma orbiculare, exclusive, locally abundant on the southern chalk.

(iii) Of medium constancy (3):

Carlina vulgaris, Medicago lupulina, Plantago media (the last abundant on relatively open soil), *Polygala vulgaris*. Two species which might be called "characteristic" are *Gentiana amarella*, which just falls below constancy 3 (reaching it on the South Downs), and is moderately though not highly exclusive, and *Viola hirta*, almost exclusive to calcareous grassland and calcareous woodland, and also only just failing to reach constancy 3 in the 62 areas on which the figures are based.

(iv) Of low constancy but very high or relatively high exclusiveness:

Aceras anthropophorum, exclusive, local.
Anacamptis pyramidalis, almost completely exclusive, locally sometimes abundant.
Anthyllis vulneraria, almost completely exclusive, locally abundant.
Blackstonia perfoliata, moderately exclusive.
Campanula glomerata, almost completely exclusive.
Centaurea scabiosa, moderately exclusive.
Daucus carota, moderately exclusive, locally abundant, especially near the sea.
Filipendula hexapetala, almost completely exclusive, locally abundant.
Gymnadenia conopsea, moderately exclusive.
Helianthemum nummularium (vulgare), almost completely exclusive.
Hippocrepis comosa, almost completely exclusive, locally abundant.
Ophrys apifera, almost completely exclusive, sometimes locally abundant.
Orchis maculata Sm. et auct. (=*O. fuchsii* Druce), locally abundant and fairly characteristic.
Origanum vulgare, moderately exclusive, but mainly near wood margins, in seral stages, or where rabbits are numerous.
Senecio integrifolius, exclusive, locally frequent.
Thesium humifusum, exclusive, local.

Non-characteristic species:

(i) Of high, but not the highest constancy (4) and low exclusiveness:
Achillea millefolium, Galium verum, Ranunculus bulbosus, Trifolium pratense.

(ii) Of medium constancy (3) and low exclusiveness:

> *Bellis perennis, Campanula rotundifolia, Centaurea nemoralis, Cerastium vulgatum, Euphrasia nemorosa, Galium erectum, Primula veris, Prunella vulgaris, Rumex acetosa, Succisa pratensis, Trifolium repens, Veronica chamaedrys.*

The non-characteristic herbs of low constancy are very numerous and scarcely worth citing:[1] some are occasional grassland species, some alien invaders from woodland and arable.

Bryophytes.[2] The following are the most important bryophytes that have been found in chalk grassland. None is generally "characteristic" of the community except *Brachythecium purum*, because of its high constancy, though it is in no way exclusive, and *Camptothecium lutescens*, because it is a calcicole and moderately constant.

The most constant species, in approximate order of importance are:

Brachythecium purum	Dicranum scoparium
Hylocomium triquetrum	Camptothecium lutescens
H. squarrosum	Hypnum cuspidatum
H. splendens	Fissidens taxifolius

After these come a considerable number of species, falling roughly into two classes:

(a) Species generally absent, but playing an important part in one or two areas:

Neckera crispa	Frullania tamarisci
Rhacomitrium lanuginosum	

and a set of acrocarpous mosses which play an important part in some of the driest places where the turf is short or open, viz. species of *Weisia*, *Trichostomum* and *Barbula*.

(b) Species which are probably present in many areas, but which never play a conspicuous part in the community and are rarely recorded:

Brachythecium rutabulum	Mnium affine
Eurhynchium praelongum	M. undulatum
Fissidens adiantoides	Thuidium abietinum
Hypnum chrysophyllum	T. tamariscinum
H. molluscum	Scapania spp.

Hypnum cupressiforme and its variety *elatum* are intermediate in status between (a) and (b).

Among the mosses in the foregoing lists, the only one of really high constancy is *Brachythecium purum*, a very common grassland species which is found as an understorey in the great majority of the chalk grasslands. Next come the Hylocomia, woodland or grassland mosses, which appear

---

[1] Complete lists from the South Downs will be found in Tansley and Adamson (1925 and 1926).

[2] Most of this information from Hope Simpson.

especially in relatively damp chalk grassland, particularly on northern exposures protected against the direct rays of the sun, and are often very luxuriant: *Dicranum scoparium*, a very common moss, more generally found on acid soils: its rather frequent occurrence in chalk grassland is presumably to be attributed to its loose attachment to the herbage; and *Camptothecium lutescens*, one of the few "calcicole" mosses really common in this community.

Of the rarer mosses, *Hypnum molluscum*, *H. chrysophyllum* and *Neckera crispa* are calcicole, the last named however usually inhabiting bare limestone rock. *Rhacomitrium lanuginosum* is on the whole a mountain and moorland moss, and has been recorded from two localities only, the relatively lofty War Down (802 ft. = 243 m.) and Butser Hill (889 ft. = 270 m.) in Hampshire, where the rainfall is high. The liverwort *Frullania tamarisci* occasionally plays a conspicuous part on northern exposures which are unusually damp and have a relatively numerous moss flora. Mosses are at a minimum or absent altogether on dry southern slopes.

Other bryophytes besides these have been recorded in chalk grassland but they are not worth mention here. Some of them, and of those mentioned above, are much commoner in seral stages leading to chalk grassland (e.g. *Thuidium abietinum*).

Lichens. Lichens are infrequent, locally conspicuous only where the turf is short, and not always there. *Cladonia rangiformis* (about the commonest) and *C. sylvatica* are the most important species. Other species of *Cladonia*, as well as *Peltigera* and *Collema*, are found more rarely. *Collema pulposum* is fairly common in places disturbed by rabbits.

Structure of the vegetation. The structure of the grassland naturally depends upon the grazing. If it is left ungrazed and is not situated on a dry southern or western exposure it is colonised by the taller growing grasses such as the Avenae, which are the most widespread on soils of various degrees of moisture, sometimes by *Arrhenatherum*, very occasionally by the meadow grasses *Festuca pratensis*, *F. elatior* and *Lolium perenne*, locally over wide areas by *Bromus erectus* (which is not dependent on shelter or moisture), frequently by *Dactylis glomerata*, and fairly often, on damp northern exposures only, by the wet-loving species *Deschampsia caespitosa*. In such situations the tall herbage tends to suppress the dwarfer species of grazed grassland and even the mosses, which become weak and etiolated. *Centaurea nemoralis* and *Rumex acetosa* hold their own well among the tall grasses, but *Poterium sanguisorba* is the only species, other than grasses, which may become locally dominant in such situations. A lower storey is formed by the partially suppressed plants of the lower growing species and by mosses.

On well grazed grassland *Festuca ovina* (often intimately mixed with *F. rubra*) is most frequently dominant, or there is a mixture of numerous species, including the markedly "calcicolous" grasses and herbs of re-

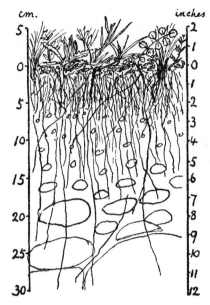

FIG. 97. PROFILE OF HERBAGE AND ROOT SYSTEMS ON PASTURED AND MODERATELY RABBIT GRAZED AREA

Herbage about 5 cm. deep. 0–7 cm. soil firmly held by roots of *Festuca ovina, *Avena pratensis and pubescens, Hieracium pilosella, Linum catharticum, Leontodon hispidus, Ranunculus bulbosus, *Plantago lanceolata, and most of the root systems of Carex diversicolor. 7–15 cm., soil lightly held by roots of *Asperula cynanchica, Thymus serpyllum, Trifolium pratense and some Carex diversicolor. 15–25 cm. loose soil with roots of *Poterium sanguisorba, *Cirsium acaule and some Lotus corniculatus. Occasional roots penetrate below 30 cm. * Recognisable in the herbage. From Tansley and Adamson, 1925.

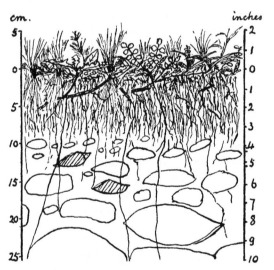

FIG. 98. PROFILE OF HERBAGE AND ROOT SYSTEMS ON SOMEWHAT LEACHED SOIL, MODERATELY RABBIT GRAZED

Herbage about 5 cm. deep. 0–9 cm. soil closely held by roots of Festuca ovina, Agrostis tenuis, Carex diversicolor, Euphrasia nemorosa, etc. 9–25 cm., loose soil mixed with chalk lumps and a few flints. Roots of Poterium sanguisorba and Cirsium acaule penetrate beyond 25 cm. From Tansley and Adamson, 1925.

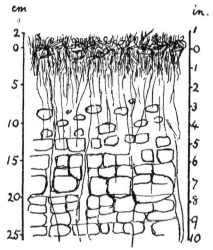

FIG. 99. PROFILE OF SEVERELY GRAZED HERBAGE EATEN DOWN TO 2 CM.

Festuca ovina, Lotus corniculatus, Thymus serpyllum, Plantago lanceolata, Prunella vulgaris, etc. Note the great reduction in the size of leaves and that the dense root layer extends here only to 4 cm. From Tansley and Adamson, 1925.

latively dwarf habit, but an absence of the non-calcicolous meadow grasses. If the herbage is 4 in. or so deep (moderate sheep grazing) there is very commonly an understorey of mosses in which *Brachythecium purum* is usually dominant or very abundant. If the herbage is reduced to an inch or less by intensive rabbit nibbling the species are not much **Effect of** reduced in numbers but the mosses are usually absent, presum-**rabbits** ably owing to increased dryness, except on steep northern slopes or in the shadow of woodland or tall scrub. In these last situations mosses are, on the contrary, luxuriant, especially the Hylocomia, and in rabbit-

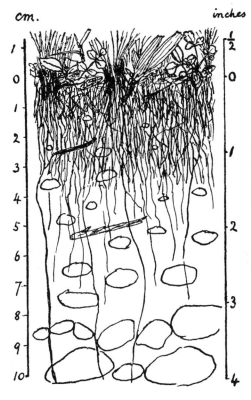

FIG. 100. A SIMILAR SEVERELY RABBIT EATEN AREA

Drawn to a larger scale. Herbage less than 2 cm. high: same species as in Fig. 99. Dense root layer barely extending to 4 cm. Soil shallower (brown loam). From Tansley and Adamson, 1925.

infested areas actually dominant, since they are not touched by the rabbits, and may form a continuous sward, through which here and there protrude the shoots of a few more or less rabbit-resistant flowering plants (Pl. 102, phot. 248).

Figs 97 and 98 give an idea of the structure of chalk grassland under conditions of moderate rabbit grazing, Figs. 99 and 100 under extremely heavy rabbit attack. Figs. 101–103 show the effect of the exclusion of rabbits in reducing the number of individual plants, after 6 and after 12 years. These three charts were made in the same enclosure but are not of the

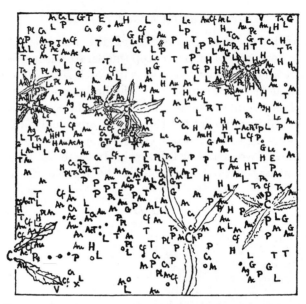

FIG. 101. CHART QUADRAT OF GRASSLAND UNDER MODERATELY HEAVY RABBIT PRESSURE (25 cm. square)

Herbage 4–5 cm. high. 441 shoots of 21 species are shown. The leaf rosettes of *Cirsium acaule* (C) and *C. palustre* (Cp) are sketched in. All the interspaces between the plants shown are filled with *Festuca ovina* (and ?*F. rubra*) which is definitely dominant. From Tansley and Adamson, 1925.

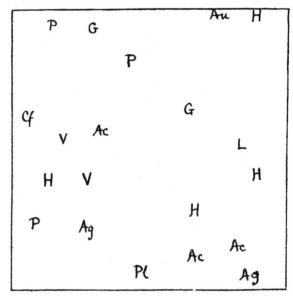

FIG. 102. CHART QUADRAT OF THE SAME SIZE AFTER EXCLUSION OF PASTURING AND RABBITS FOR 6 YEARS

Herbage 18–20 cm. high. 20 shoots of 10 species are shown. *Festuca ovina* (and ?*F. rubra*) still dominant filling all the space between the shoots of the other plants on the surface of the vegetation, though there are concealed bare spaces of soil. Note the enormous effect of competition in diminishing the number of shoots (and species) when all grazing is excluded. From Tansley and Adamson, 1925.

PLATE 102

Phot. 248. Grassland on north slope shaded by wood and exposed to rabbits. *Hylocomium squarrosum* and *H. splendens* dominant (untouched by rabbits). The sparse flowering plants include *Carex diversicolor* (abundant) and *Cirsium palustre* (both rabbit-resistant).

Phot. 250. Another quadrat 50 cm. square. Same species. with *Cirsium palustre* and *Viola hirta*.

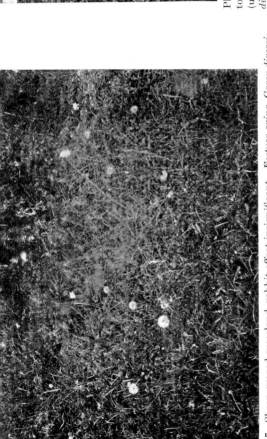

Phot. 247. Sheep-grazed grassland, rabbit effect insignificant. *Festuca ovina, Carex diversicolor, Galium verum* (centre), *Leontodon hispidus* (fruit), *Cirsium acaule, Poterium sanguisorba, Scabiosa columbaria* (fl.), *Spiranthes spiralis.* Base of juniper bush at the back.

Phot. 249. Quadrat 50 cm. square fully exposed to rabbits. *Festuca ovina, Poterium sanguisorba, Cirsium acaule, Carex diversicolor.*

CHALK GRASSLAND AND RABBIT EFFECTS

PLATE 103

Phot. 251. Uneven ground with rabbit hole and short turf of *Festuca ovina*. *Helianthemum nummularium* (rabbit-resistant) dominant and flowering; also *Cirsium acaule*, *Poterium sanguisorba*, *Galium verum*, etc. Herbage barely 3 in. deep.

Phot. 252. Rabbits excluded for 6 years. *Holcus lanatus* (left and centre), *Galium verum* (right) dominant. *Poterium sanguisorba*, *Agrostis tenuis*, etc. Herbage at least 8 in. deep. *Festuca ovina* subordinate.

Phot. 253. Rabbits excluded for 6 years. *Avena pratensis* (front) mainly dominant, with *Galium verum*, *Calluna vulgaris* (fl., centre) and *Potentilla erecta* (fl., right centre) have appeared. Herbage about 6 in. deep.

Phot. 254. Metre quadrat protected from rabbits for 12 years. Herbage 8–10 in. deep. *Poterium sanguisorba*, *Holcus lanatus*, *Avena pratensis*, *Helianthemum*, etc. Species and individuals much more numerous than in Phot. 253. Cf. Figs. 102–3.

same quadrat. Pl. 103, phots. 251–4, illustrate the appearance of the herbage under these different conditions.

On southern and south-western exposures open to the full strength of the summer sun the vegetation is apt to be stunted whether it is grazed or not, the number of species is reduced, and mosses are absent or negligible.

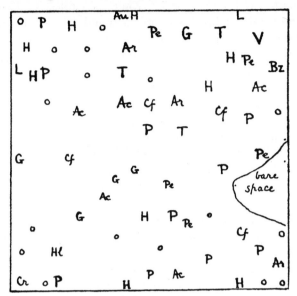

FIG. 103. CHART QUADRAT OF THE SAME SIZE AFTER EXCLUSION OF PASTURING AND RABBITS FOR 12 YEARS

Herbage 20–25 cm. high, 51 shoots of 14 species are shown. The herbage is less dense and there are many bare spaces on the soil though none on the surface of the vegetation. *Festuca ovina* is still very abundant but no longer dominant. Competition has proceeded further and has led to the proportional increase of *Galium verum*, *Poterium sanguisorba* and *Holcus lanatus*, which tend to local dominance (cf. Pl. 103, phot. 252). The increase in the number of species and individuals during the second six years shows that equilibrium is not yet reached. From Tansley and Adamson, 1925.

SUCCESSION

Succession from bare chalk to chalk grassland may be studied on the surfaces of chalk rock on the sides of road cuttings and chalk pits, and on the heaps of "spoil" consisting of larger or smaller fragments which are often accumulated when chalk is quarried.

On bare chalk rock The actual surface of the rock may be colonised in quantity by the strictly calcicolous moss *Seligeria calcarea*, whose embedded protonemata give a grey colour to the exposed rock surface. Accompanying the *Seligeria* are sometimes filaments of an alga, *Chroolepus* sp. But it is only where fine particles of chalk accumulate in the joints and bedding planes that higher plants can get a hold for their roots, and in such situations a number of species of spermaphytes occur. On these easily disintegrated chalk surfaces *Festuca ovina, Echium vulgare, Senecio jacobaea* and *Tussilago farfara* have been found as pioneer colonists (with *Bromus erectus* in Wiltshire) as well as a number of other species, mostly belonging to chalk grassland. No less than sixty species of flowering plants, besides eight species of moss, were seen on the horizontal floor of a quarry on the Hampshire-Sussex border, where a stretch of bare chalk had been partially dissolved, and then had dried and cracked. About 10 per cent of this surface was occupied by mosses, of which four species were markedly calcicolous, and *Barbula cylindrica* was the most abundant. The sixty species of angiosperms were all present as isolated individuals, no species except *Poa annua* being more than frequent, and they probably did not occupy between them more than about 1 per cent of the surface. This was a very miscellaneous collection (including three woody species) derived from woodland, arable and chalk grassland, all of which existed in the neighbourhood. If left undisturbed there can be little doubt that scrub and woodland would ultimately develop, but the course of such a succession has never been observed.

In quarries

Stages of succession to chalk grassland were observed in 1920 and again in 1936 on the slopes of three fine spoil banks in an abandoned quarry at War Down in Hampshire. The first observations were made in 1920 by Tansley and Adamson (1925), the later by Hope Simpson (unpublished) who ascertained important facts about the history of the quarry. This was abandoned about 1895 and the two youngest spoil banks probably date from that time. An older bank was estimated by a quarryman to have been made about 1880. If these dates are correct the newer banks were about 25 and 40 years old, and the older bank 40 and 55 years old when the observations were made. No rabbit grazing was noticed in 1920, but during the later years at least, i.e. some time after 1920, the banks were grazed by rabbits, which had made a warren in the quarry. In 1932 War Down was acquired and planted by the Forestry Commission and rabbits were excluded. Thus the plant communities developed for at least part of their history under the influence of rabbit grazing, from which they were freed during the last 4 years.

On spoil banks

The coarse quarry talus, consisting of larger blocks and fragments mixed with small to minute chalk particles, was found in 1920 to be colonised largely by *Echium vulgare* and *Senecio jacobaea*, while on the fine spoil banks *Festuca ovina* and *Tussilago farfara* were the leading pioneers. Some of the colonists of the latter were "weeds" of open soil and there were one or two chalk scrub plants, but the great majority of the species belonged to chalk grassland, with which the quarries were surrounded. Among the most abundant of these were *Festuca rubra, Lotus corniculatus, Agrostis stolonifera, Carex diversicolor, Thymus serpyllum*.

Hope Simpson found that the number of species present on the two younger slopes had approximately doubled between 1920 and 1936, and very nearly doubled on the older slope.[1] The change was for the most part a "filling up" of the community by new species and the process was evidently slow. The percentage areas that were bare in 1920 had considerably decreased in 1936, but even on the oldest bank the vegetation cover was not quite continuous after a presumed lapse of 55 years. The flora of the three banks was by no means uniform and each clearly showed fluctuations in its plant covering not related to the general successional process. Thus *Echium vulgare, Sonchus oleraceus* and *Tussilago farfara*, all plants of open soil, still persisted after the lapse of 16 years, and two of them had even appeared where they had not been present in 1920; while *Arctium* sp., *Cirsium arvense, Crepis capillaris, Epilobium angustifolium, Heracleum sphondylium*, and a few other species which are not chalk grassland plants, had freshly arrived. And a few species, some of which often occur in this community, such as *Anthoxanthum odoratum, Cirsium palustre, Ranunculus bulbosus*, had disappeared from the banks.

But apart from these fluctuations, a definite progress towards a more complete development of the community may undoubtedly be recognised on the three slopes taken together, just as the vegetation of the oldest bank represented in 1920 a more complete development of the community than that of the younger banks. Taking all three slopes together, *Agrostis siolonifera, Avena pratensis, Festuca ovina* and *F. rubra, Hieracium pilosella*, had not fluctuated much, while the following had increased on balance:

Arrhenatherum elatius	Linum catharticum
Asperula cynanchica	Lotus corniculatus
Carex diversicolor	Plantago lanceolata
Gentiana amarella	Thymus serpyllum
Leontodon hispidus	Brachythecium purum

nearly all species of high or considerable constancy in chalk grassland.

The following, of lower average constancy but including some highly constant and characteristic species, had freshly appeared:

Avena pubescens	Plantago media
Briza media	Polygala vulgaris
Galium erectum	Poterium sanguisorba
Holcus lanatus	Prunella vulgaris
Koeleria cristata	Trifolium pratense
Leontodon autumnalis	T. repens
Medicago lupulina	Trisetum flavescens
Orchis maculata	
Pimpinella saxifraga	Hylocomium triquetrum

Highly characteristic species which are still absent are *Scabiosa columbaria* and *Hylocomium squarrosum*.

Woody plants were rare on the banks in 1920, *Clematis vitalba* appearing on one and *Rosa canina* on another. Both of these were still present in

---

[1] It is possible that the proportional increase of species was not quite so great as this because Hope Simpson's search was conducted in a more thorough and leisurely manner than the earlier one.

1936, though on different banks, while *Crataegus monogyna, Rubus fruticosus, Solanum dulcamara* and *Sorbus aria* had freshly appeared, though all were still rare. Their appearance represents a progress towards the development of chalk scrub which is checked or inhibited by rabbit grazing. Where sufficient seed parents of shrubs and trees are present in the neighbourhood there is no doubt that scrub and eventually forest would develop in the absence of grazing, before the grassland became mature. In other words, chalk grassland, as we know it, is a subclimax or plagioclimax determined by grazing.

The data that have been obtained from the War Down spoil heaps do not enable us to trace the details of development of this plagioclimax, but they do indicate the pioneers and the gradual "filling up" of the community in the presence of rabbits. The exclusion of rabbits during the four years from 1932 to 1936 may be responsible or partly responsible for the appearance of the new woody plants: also of *Arrhenatherum*, and perhaps of *Deschampsia, Heracleum* and *Orchis*.

"Primitive" chalk grassland. Table XVI on p. 549 gives the floristic composition and soil analyses of three closely adjacent small areas of chalk grassland, each of about 2 sq. m., on very shallow soil occupying irregular ground near the quarries at Buriton Limeworks in Hampshire (Tansley and Adamson, 1925). The chalk was probably bared at some stage in the working of the quarries and has been completely colonised by grassland species forming a "primitive" chalk grassland community which shows a simpler structure and composition than mature grassland. It appeared at the date of the visits to be practically unaffected by grazing, though it is of course possible, and indeed likely, that grazing had taken place at some time previously:

"A comparison of the three soils shows a steady increase from (*a*) to (*c*) in water content and in humus and a decrease in total carbonates and in calcium. The nitrate determinations, made some four months after the samples were collected, are probably rather an index of the abundance of nitrifying or nitrogen fixing organisms in (*c*) than a trustworthy measure of the amount of nitrates in the soil as it exists in the field.

"The characteristic feature of the vegetation of (*a*) is its extreme dwarfing, due no doubt primarily to drought. The shallow rooting *Festuca ovina* is markedly dominant, forming more than half of the whole herbage. The other grasses are few (four species) and not abundant. The rest of the vegetation is made up of a small selection of the common herbs of dry grassland, of which *Hieracium pilosella, Leontodon hispidus, Lotus corniculatus, Thymus serpyllum* and *Asperula cynanchica* were most conspicuous.... The only two species found on this area and not on (*b*) or (*c*) are *Taraxacum erythrospermum* and *Phleum pratense* var. *nodosum*, both plants of markedly dry soils. Of the seven mosses four are distinctly xerophilous forms, and three are marked by Watson as calcicolous.

Table XVI. *Vegetation and soils of three small areas of "primitive" chalk grassland in close proximity*

(a) Turf scarcely continuous. Herbage ½–1 in. (1·25–2·5 cm.) in height. Slope to S.S.W. 18–20°. Soil 2½–3 in. (6·25–7·5 cm.) in depth to chalk *in situ*, grey, very dry and powdery, included in a mass of fine rootlets.

(b) Turf more continuous. Herbage 4 in. (10 cm.) in height. Very slight slope to south. Soil 4 in. (10 cm.) in depth to chalk *in situ*, grey-brown.

(c) Close herbage 5 in. (12·5 cm.) in height. Bottom between two hummocks. Soil 4 in. (10 cm.) to chalk *in situ*, brown.

Soil samples	Water loss of fresh soil on air drying	Water loss of dry soil at 100° C.	Loss on ignition	Total carbonates	Nitrates as NaNO₃	P₂O₅	K₂O	MgO	CaO	Insoluble residue	"Lime requirement"
(a)	17·0	3·0	6·2	71·1	0·0048	0·160	0·327	0·201	54·6	14·6	Nil
(b)	27·0	4·0	19·7	65·9	0·0037	0·164	0·640	0·183	32·1	5·6	Nil
(c)	33·8	4·6	28·6	62·8	0·0500	0·170	0·301	0·062	36·25	22·0	Nil

	(a)	(b)	(c)
Agrostis stolonifera	f	f	a
Asperula cynanchica	r	a	f
Avena pratensis	—	—	o
Bellis perennis	o	o	—
Briza media	o	f	o
Campanula rotundifolia	—	f	o
Centaurea nemoralis	r	f	o
Cynosurus cristatus	—	—	a
Dactylis glomerata	—	—	f-la
Euphrasia nemorosa	d	va	o
Festuca ovina	—	f	a
Galium verum	f.la	a	o
Hieracium pilosella	—	—	f
Holcus lanatus	f	f	f
Leontodon hispidus	o	—	f
Linum catharticum	—	—	o
Lolium perenne	—	o	—
Lotus corniculatus	f	a	f
Medicago lupulina	f	a	f
Origanum vulgare	—	—	r
Phleum pratense var. nodosum	—	—	—

	(a)	(b)	(c)
Plantago lanceolata	r	o	f
Primula veris	o	o	r
Ranunculus bulbosus	o	o	o
Taraxacum erythrospermum	f	a	f
Thymus serpyllum	e	f	f-a
Trifolium pratense	—	o	a
T. repens	o	a	a
Trisetum flavescens	—	—	—
Barbula cylindrica*	a	f	—
B. unguiculata	o	—	—
Brachythecium purum	—	—	f
B. rutabulum	—	o	o
Bryum capillare	o	—	o
Camptothecium lutescens*	f	o	o
Fissidens taxifolius	o	—	—
Hypnum cuspidatum	—	f	o
H. molluscum*	o	—	o
Mnium undulatum	—	—	—
Thuidium abietinum	a	—	o
T. tamariscinum	—	—	—
	25	25	31

* Mosses marked by Watson (1918) as calcicolous.

"Comparing (*b*) and (*c*) there is an increase in the number of species of grass from 5 to 6 and 8 respectively, and in (*c*) a great increase in the bulk of the grass herbage. This is to be correlated with the great increase in water content and the much greater nitrifying power of the soil of (*c*). *Agrostis stolonifera*, *Cynosurus cristatus*, *Dactylis*, *Lolium* and *Holcus lanatus* appear, *Trisetum* increases in abundance, while *Festuca ovina* progressively decreases, though it still remains abundant. Of the herbs *Trifolium repens* appears, *T. pratense* and *Plantago lanceolata* increase in abundance, while some species, such as *Asperula cynanchica*, *Campanula rotundifolia*, *Hieracium pilosella*, *Lotus corniculatus*, and *Thymus*, show their maximum frequency in (*b*), perhaps owing to the competition of the taller grasses in (*c*). *Euphrasia nemorosa*, *Origanum vulgare* and *Primula veris* appear for the first time in (*c*).

"Of the mosses, *Barbula cylindrica*, *B. unguiculata* and *Bryum capillare* decrease and disappear with complete closure and increasing depth of the turf; *Brachythecium purum*, one of the most ubiquitous of chalk grassland mosses though not a "calcicole" species, appears in (*b*) and increases in (*c*); while *B. rutabulum*, *Mnium undulatum* and *Thuidium tamariscinum* first appear in the damper conditions of (*c*)." (Tansley and Adamson, 1925, pp. 184–5.)

Stabilisation by grazing. Chalk grassland, like most of the other grasslands of this country, is maintained as grassland by pasturage or mowing for hay, the latter being very rare on the chalk. The downs have been used as sheep walks for many centuries and probably from Neolithic times, and horned stock are now frequently pastured upon them. The feeding of sheep in folds and of cattle on cake or in "tended" pastures between their visits to the downs may have some effect on the natural grassland community by the transference of foodstuffs through the dung to the down grassland, but such possible effects have not been investigated.

The extreme abundance of rabbits in some chalk grassland areas has an overwhelming effect on the vegetation (see pp. 542–5 and Chapter VI, p. 134), but this, though very widespread, is always local, the razing of the herbage to within half an inch of the soil surface never extending very far from a collection of burrows.

The grazing factor, then, is the primary factor which maintains and stabilises grassland, but the successional data given on previous pages seem to show that a vegetation dominated on the whole by grasses, though largely consisting of dicotyledonous herbs, comes into existence on bare chalk independently of grazing. Under natural conditions, i.e. in the absence of more or less intensive grazing, the grassland would pass more or less rapidly (according to the degree of proximity of parent shrubs and trees) into scrub and woodland, for woody plants may begin to come in as soon as the herbs and grasses, but no examples of this complete sequence have been observed. The succession to woodland on stabilised chalk grassland where grazing is insufficient to exclude the colonisation of woody plants has been described in Chapter XVIII.

Formation of loam. On the flat or gently sloping ground of the chalk plateaux and dipslopes non-calcareous soil, even where it is not formed of definite Quaternary deposits, may reach a considerable depth as a result of the gradual accumulation of residual material from the dissolution of the chalk. Loams of this origin occur in all stages of development and degrees of thickness. They show the characteristic "brown earth" rather than the "rendzina" profile, are distinctly leached in the upper layers and bear vegetation different from that of the shallow chalk soil. When this loam reaches a considerable thickness it is known as "clay-with-flints", from the numerous flints originally embedded in the chalk which form part of the chalk residue. A good deal of it however probably contains the remnants of Tertiary and Quaternary deposits. These loams cover very great areas of the chalk outcrop and bear beechwood or oakwood where they are not cultivated.

Chalk heath. The earlier stages of loam formation are constantly met with on flat or slightly inclined parts of the chalk plateau, close to an escarpment or chalk valley. The leaching of the upper layers gives rise to a slightly acid soil, which is inhabited by a mixture of typical chalk grassland plants, with other species indicative of acid conditions. Of the latter the two grasses *Agrostis tenuis* and *Sieglingia decumbens* are those which first appear in such situations, and they are followed by species like *Potentilla erecta* and *Calluna vulgaris*. Of the chalk grassland species some, like *Festuca ovina*, which are surface rooting, can tolerate the acidification of the surface layers, while others, such as *Poterium sanguisorba* have deep roots which descend to the calcareous soil below. Thus is developed the beginnings of a vegetation which has been called *chalk heath*, because it is marked by a mixture of calcicolous plants rooting in calcareous soil with indifferent and calcifuge heath plants rooting in the acid surface soil. Large areas of plateau and dipslope are dominated by *Calluna* or *Erica cinerea*, accompanied by such characteristic southern heath plants as *Ulex minor*.

An analysis of the vegetation of five chalk heaths shows that the main list of species is the same as that of the normal chalk grassland, but that the occurrence and abundance of such characteristic species as the Avenae, *Koeleria*, *Scabiosa columbaria*, *Pimpinella saxifraga* and *Phyteuma orbiculare* are considerably diminished. At the same time *Aira praecox*, *Calluna*, occasionally *Erica cinerea* and *Galium saxatile*, make their appearance, while the occurrence and abundance of *Agrostis tenuis* greatly increase, and to a less degree, though still considerably, the frequency of *Carex caryophyllea*, *Trifolium repens*, *Potentilla erecta*, *Veronica officinalis*, *Viola riviniana* and *Dicranum scoparium*. The CaO content of the upper layers of soil of these five chalk heaths varied from 0·53 to 2·0 per cent, and the pH values from 6·0 to 6·9, increasing with depth.

It seems possible that "chalk heaths" of the kind described develop into typical heaths by further leaching and the accumulation of acid humus at the surface, but nothing of the kind was obvious in the district investigated (South Downs). More typical heaths do occur (rarely) in the district, but they are situated on shallow sandy Quaternary deposits overlying the chalk. Such heaths occur much more extensively on the plateaux of other chalk districts.

Under natural conditions, i.e. in the absence of regular grazing, the incipient heaths, like the chalk grassland itself, would undoubtedly be invaded by shrubs and trees and progress to woodland.

INFERIOR OOLITE GRASSLAND

The Inferior Oolite of the Cotswold Hills in Gloucestershire is a less pure and harder limestone than the chalk, but bears much the same vegetation. Thus the Cotswold beechwood community does not differ widely from that of the chalk escarpment, and the natural grassland of the Cotswolds is almost identical with that of the chalk downs. Though they have not been intensively studied there can be little doubt that both beechwood and grassland communities are very close to those of the chalk. The general lists of grassland species and their relative frequencies are very much the same as on the chalk, though on the flat tops of the hills, even near the escarpment, there is a greater depth of soil, probably correlated with the lesser purity of the oolitic limestone giving a greater proportion of residual material; and the surface layers of soil are poorer in free calcium carbonate. With this goes a greater frequency of species like *Agrostis tenuis* and *Sieglingia decumbens*. Certain other species, unknown or rare on the south-eastern chalk, also figure more or less conspicuously on the oolite, for example *Solidago virga-aurea* and *Spiranthes spiralis*. *Bromus erectus* is very widespread and very frequently dominant in the grassland of the Cotswolds, and it is often closely associated with *Brachypodium pinnatum*; and this is also true of some of the Wiltshire and Berkshire chalk downs. The natural grassland of the oolite has not, however, been closely studied, and it is not possible to give a fuller description.

GRASSLAND OF THE OLDER LIMESTONES AND THE BASIC IGNEOUS AND METAMORPHIC ROCKS

The grassland of the older (Palaeozoic) limestones has much in common with that of the chalk, but also marked differences. The thin soil has the same fundamental relation to the rock below, whose continual solution acts as a basic buffer to the soil water: the turf is typically dominated by *Festuca ovina* (probably associated with *F. rubra*), with several of the same accompanying species, both grasses and dicotyledonous herbs, as in chalk grassland; and like it the grassland of the older limestones is traditional sheep pasture.

PLATE 104

Phot. 255. Limestone grassland on the steep western slope of Cross Fell (northern Pennines), with *Festuca ovina, Sesleria caerulea* and many calcicolous species, but also *Agrostis tenuis*. To the right, above, is a limestone "scar" and small scree: to the left, below, is Callunetum on a gently inclined grit slope. *Mrs Cowles.*

Phot. 256. Nardetum on slightly inclined flagstone platform above the limestone scar shown in Phot. 255. *Mrs Cowles, I.P.E.* 1911.

LIMESTONE GRASSLAND AND NARDETUM

Features of the older limestones. On the other hand the much greater hardness of the older limestones leads to the formation of crags—"scars" as they are called in the north—which often support a flora of their own: a large number of chalk species are absent from the older limestones, while one or two "calcicoles" (e.g. *Sesleria caerulea*) occur that are not found on the chalk; and finally the cooler and damper climate in which the older limestones are exposed leads to widespread formation of acid humus, so that a number of "calcifuge" species are usually present, rooted in this humus, and often mixed or alternating with the calcicoles.

The grasslands of the older limestones were briefly described by several of the pioneer students of British vegetation in the early years of the present century. Thus Robert Smith (1900*b*) gives a short list of the chief species occurring on a limestone exposure at Balnabodach, Loch Tummel, in Perthshire: W. G. Smith and C. E. Moss (1903) describe the "limestone hill pasture" (Mountain or Carboniferous Limestone) in the Harrogate and Skipton district of Yorkshire as "consisting chiefly of *Festuca ovina*", and give a list of species: F. J. Lewis (1904) gives a similar list of the "bright green closely cropped pasture" where the limestone is "free from peat" in the northern Pennines:[1] C. E. Moss (1907) describes though he does not give a list of the limestone pastures (also on Mountain Limestone) of the Mendip Hills in Somerset; and later (1911, 1913) deals with the grassland of the same geological formation in the southern Pennines of Derbyshire.

Very similar grassland occurs on the igneous and metamorphic rocks "rich in mineral salts, especially calcium and magnesium" (basalt, andesitic lavas and schists) described by Robert Smith (1900*a*) from the Edinburgh district, and by W. G. Smith (1904–5) from Fife.

More than one of the earlier surveyors point out that the "calcicole" plants growing on these geological formations typically occur where the soil is thin above the rock, and are thus associated with the "scars" of the limestone or with exposed knolls of basalt. In hollows between the knolls, where the surface is kept damp, humus accumulates and calcifuges such as *Calluna* and *Vaccinium myrtillus* are present, belonging to quite a different vegetation. And in the regions of higher rainfall the surface soil of the general limestone grassland itself becomes leached, accumulates acid humus, and contains a number of species which are not characteristic of limestone soils. Thus the turf may contain abundance of *Agrostis tenuis* (which, as we have seen, only occurs very locally in chalk grassland) so that the limestone grassland may even come to approximate to the bent-fescue pastures of the siliceous rocks. As we shall see in a later chapter this leaching and accumulation of acid humus may go so far on limestone plateaux as to lead to the development of typical *Calluna* heath immediately above the limestone rock, while "limestone heaths", with mixtures of calcicolous and calcifuge species, quite comparable with the chalk heaths

[1] Pl. 104, phot. 255.

previously described, are common in many limestone areas. Nevertheless where the drainage is free, as on the slopes of the rounded Mountain Limestone hills, a uniform pasture, not containing calcifuge species and approximating more or less closely to chalk pasture, though always with fewer species, is developed over wide areas.

The following composite list (Table XVII) is constructed from those given for the older limestones (mainly Mountain Limestone) and other basic rocks by the earlier workers mentioned above. The absence of species from the shorter lists does not necessarily mean that they do not occur in the corresponding areas, since the thoroughness of the listing varies a great deal. But at least a fair idea can be obtained of the dominant, characteristic, and most abundant species:

Table XVII

(1) Moss (1913), Derbyshire, Mountain Limestone.
(2) Smith and Rankin (1903), West Yorkshire, Mountain Limestone.
(3) Lewis (1904), Northern Pennines, Mountain Limestone.
(4) R. Smith (1900a), Edinburgh district (basaltic rocks).
(5) W. G. Smith (1904–5), Sidlaws, volcanic rocks.
(6) W. G. Smith (1904–5), Central Fife, basalt.
(7) R. Smith (1900b), N. Perthshire, limestone at Loch Tummel.

	1	2	3	4	5	6	7
Grasses							
Agrostis spp.	a	.	+	+	+	+	.
Anthoxanthum odoratum	o, la	+	.	+	+	+	.
Arrhenatherum elatius	la	.	.	.	.	.	.
Avena pratensis	o	+	.	.	.	.	.
A. pubescens	l	+	.	.	.	.	.
Brachypodium pinnatum	vr	.	.	.	.	.	.
B. sylvaticum	a	.	.	.	.	.	.
Briza media	o–a	+	.	.	.	.	.
Bromus erectus	?	.	.	.	.	.	.
Cynosurus cristatus	la	+	.	.	+	.	+
Deschampsia flexuosa	.	.	.	.	.	+	.
Festuca elatior	r	.	.	.	.	.	.
F. ovina	a–d	d	+	+	+	+	+
F. rubra	r	.	.	.	.	.	.
Holcus lanatus	.	.	.	.	.	.	+
Koeleria cristata	o	+	+	.	.	.	+
Poa pratensis	l	.	+	+	+	.	+
P. trivialis	.	.	.	+	.	.	.
Sesleria caerulea	.	+	+	.	.	.	.
Sieglingia decumbens	.	.	.	+	.	.	.
Trisetum flavescens	l	.	.	.	.	.	.
Herbs							
Achillea millefolium	a	.	.	.	+	.	.
Alchemilla vulgaris	o	.	+	.	.	.	+
Allium vineale	vr	.	.	+	.	.	.
Anacamptis (Orchis) pyramidalis	vr	.	.	.	.	.	.
Antennaria dioica	la	+	.	.	.	.	.
Anthyllis vulneraria	o	+	.	+	.	+	.
Asperula cynanchica	la	.	.	.	.	.	.
Astragalus danicus	.	.	.	+	.	.	.
Bellis perennis	o	.	.	.	+	.	.

Campanula glomerata	o	.	.	.	.	.	.
C. rotundifolia	o	.	+	+	+	+	.
Carduus nutans	o	+	.	.	.	.	.
Carex caryophyllea	la	.	.	.	.	.	.
C. diversicolor (flacca)	o	.	.	.	.	.	.
C. ornithopoda	r	.	.	.	.	.	.
C. pilulifera	l	.	.	.	.	.	.
Carlina vulgaris	o	+	+	.	.	.	.
Centaurea nemoralis	a	.	+	.	.	.	.
C. scabiosa	o	.	.	.	.	.	.
Centaurium umbellatum	r	.	.	.	.	.	.
Chrysanthemum leucanthemum	a	.	.	.	.	.	+
Cirsium eriophorum	l	+	.	.	.	.	.
C. palustre	l	.	.	.	.	.	.
Clinopodium vulgare	r	.	.	.	+	.	.
Cochlearia alpina	l	.	.	.	.	.	.
Coeloglossum viride	r	+	.	.	.	.	.
Conopodium majus	o	.	.	.	+	.	.
Crepis capillaris	a	.	.	.	.	.	.
Erophila boerhaavii	la	.	.	.	.	.	.
E. verna	la	.	.	.	.	.	.
Euphrasia officinalis	a	.	+	.	.	.	+
Filipendula hexapetala	r	+	.	.	.	.	.
Galium cruciata	la	.	.	.	.	.	.
G. pumilum (sylvestre)	la	+	?	.	.	.	.
G. saxatile	r	.	+	+	+	+	+
G. verum	a	+	.	+	+	+	+
Gentiana amarella (agg.)	o–a	+	.	.	.	.	+
Geranium lucidum	la	.	+	.	.	.	.
G. sanguineum	r	.	.	+	.	+	.
G. sylvaticum	.	+	.	.	.	.	.
Gymnadenia conopsea	r	+	.	.	.	.	.
Helianthemum nummularium	a	+	+	+	+	+	+
Hieracium murorum	la	.	+	.	.	.	.
H. pilosella	a	+	+	.	+	.	+
Hippocrepis comosa	r	+	.	.	.	.	.
Hutchinsia petraea	l	.	.	.	.	.	.
Hypericum hirsutum	lf	.	.	.	.	.	.
Lathyrus montanus	r	.	.	+	+	.	.
L. pratensis	o	.	.	.	.	.	.
Leontodon autumnalis	la	.	.	.	.	.	.
L. hispidus	a	.	.	.	.	.	.
L. nudicaulis	a	.	.	.	.	.	.
Leucorchis (Habenaria) albida	.	+	.	.	.	.	.
Linum catharticum	a	+	+	.	+	.	.
Lotus corniculatus	a	+	+	.	+	.	+
Luzula campestris	a	+	.	+	.	.	.
Minuartia (Arenaria) verna	l	+	.	.	.	.	.
Ononis repens	.	.	.	+	.	.	.
Ophrys apifera	r	.	.	.	.	.	.
Orchis maculata	l	.	.	.	.	.	.
O. mascula	la	.	+	.	.	.	.
O. ustulata	r	.	.	.	.	.	.
Origanum vulgare	o–a	.	.	+	.	.	.
Oxalis acetosella	.	.	.	.	.	+	.
Picris hieracioides	o	.	.	.	.	.	.
Pimpinella saxifraga	o	.	.	+	+	+	.
Plantago lanceolata	a	.	.	.	.	.	+
P. media	la	+	.	.	.	.	.
Platanthera bifolia	vr	+	.	.	.	.	.
P. chlorantha	.	+	.	.	.	.	.
Polygala vulgaris	a	+	+	.	+	.	+
Potentilla erecta	.	.	.	+	+	+	.
P. verna	l	.	l	+	.	.	.

Poterium officinale	l	+	.	.	.	.	.
P. sanguisorba	a	+	.	.	.	.	.
Primula veris	a	.	.	.	.	.	.
Prunella vulgaris	o	.	.	.	.	.	.
Ranunculus bulbosus	a	+	.	.	.	.	.
Rumex acetosa	o	.	+	.	.	+	.
R. acetosella	r	+	+	.	+	.	.
Saxifraga granulata	la	+	.	.	.	.	.
S. hypnoides	l	+	+	.	.	.	.
Scabiosa columbaria	o	+	.	.	.	.	.
Sedum acre	a	.	+	.	.	.	.
Spiranthes spiralis	vr	.	.	.	.	.	.
Stachys (Betonica) officinalis	o	+	.	.	.	.	.
Succisa pratensis	r	.	.	.	.	+	.
Teucrium scorodonia	l	.	+	.	+	+	.
Thalictrum minus	l	.	.	.	.	.	.
Thymus serpyllum	a	a	+	.	+	+	+
Trifolium dubium	r	.	.	.	.	.	+
T. pratense	o	.	.	.	.	+	.
T. repens	r–o	.	.	+	.	.	+
Veronica chamaedrys	o	.	.	.	+	+	.
V. officinalis	.	.	+	.	.	+	.
Vicia angustifolia	vr	.	.	.	.	.	.
V. hirsuta	.	.	.	+	.	.	.
V. lathyroides	.	.	+	.	.	.	.
Viola hirta	o	.	.	+	.	.	.
V. lutea	la	la	+	+	+	+	.
V. riviniana	o	.	+	+	+	+	.
Botrychium lunaria	r–o	+	.	.	.	.	.
Ophioglossum vulgatum	la	+	.	.	.	.	.
Pteridium aquilinum	r	+	.	+	.	.	+
Totals	110	45	31	27	26	22	20

If these lists are compared with those of chalk grassland it is obvious at once that many of the southern species are absent, while others become increasingly sporadic or disappear altogether as we travel farther north. The much greater length of the Derbyshire list (110 species) is due partly no doubt to the more thorough investigation of the grassland there than in the more northern areas, but mainly to the failure of many species which occur, though often rarely, in the southern Pennines, to reach the northern Pennines or Scotland. Several plants, however, which are absent from the southern Scottish basalts come in again on the limestone in Perthshire.

Certain species, very characteristic of the southern chalk, are regularly found both on the limestones and on the basic igneous rocks in many of the northern areas, and of these *Helianthemum nummularium (vulgare)* is the most constant. Others, such as *Thymus* and *Pimpinella saxifraga*, occur in several, though not in all. *Astragalus danicus*, local on the chalk of the eastern counties of England, appears again on the basaltic rocks of the Edinburgh district.

Among positive characters of the northern limestone and other basic grasslands we may note the appearance of certain species which do not occur in the south: for example *Viola lutea* is abundant in several of the northern localities, and is found also in the bent-fescue grassland on siliceous rocks, but not in the south. *Sesleria caerulea* does not occur in the south, nor even in Derbyshire, and is very local in Scotland, but is a characteristic grass of the northern Pennine limestones. Besides these we have the frequent occurrence on limestone, already referred to, of species characteristic of acid soils, e.g. *Agrostis tenuis*, sometimes abundant, *Potentilla erecta*, and *Galium saxatile*, the last quite absent from chalk grassland.

Limestone heaths. These are formed in just the same way as the chalk heaths described on pp. 551–2. Smith and Rankin (1903) record the following species as occurring together in a limestone heath in west Yorkshire:

Calcicolous or tolerant	Calcifuge
Sesleria caerulea	Calluna vulgaris
Festuca ovina	Vaccinium myrtillus
Viola lutea	Nardus stricta
Thymus serpyllum	Polytrichum sp.

and Moss (1907, 1913) records many more from similar mixtures of heath and calcicolous vegetation on the Mountain Limestone of the Mendips in Somerset and of the Pennines in Derbyshire. (Cf. also p. 473.)

Limestone swamps. Unlike the Chalk, the Mountain Limestone often supports swamps, owing to high rainfall and locally obstructed drainage, and these contain a characteristic vegetation. Moss (1913) gives the following list for the limestone swamps of the southern Pennines (Derbyshire):

Caltha palustris	a	Filipendula ulmaria	a
Carex acuta	a	Geum intermedium	la
C. disticha	l	G. rivale	la
C. diversicolor (flacca)	o	Juncus compressus	vr
C. pendula	l	J. inflexus (glaucus)	la
C. strigosa	r	Mentha spp.	a
C. sylvatica	l	Orchis maculata	o–a
Chrysosplenium alterni-		Parnassia palustris	la
folium	la	Pedicularis palustris	r
C. oppositifolium	la	Petasites hybridus	la
Cirsium heterophyllum	a	Polemonium caeruleum	la
C. palustre	a	Scirpus compressus	r
Epilobium hirsutum	a	Thalictrum flavum	r
Epipactis palustris	vr	Trollius europaeus	la
Eupatorium cannabinum	la	Valeriana dioica	la
Festuca arundinacea	vr	V. officinalis	la
F. elatior	r		

REFERENCES

ANDERSON, VIOLET L. Studies of the vegetation of the English chalk. V. The water economy of the chalk flora. *J. Ecol.* **15**, 72–129. 1927.

HOPE SIMPSON, J. F. The War Down spoil heaps (unpublished).

LEWIS, F. J. Geographical distribution of the vegetation of the basins of the Rivers Eden, Tees, Wear and Tyne. Part I. *Geogr. J.* 1904.

MOSS, C. E. Geographical distribution of vegetation in Somerset: Bath and Bridgewater District. *Roy. Geogr. Soc.* 1907.

MOSS, C. E. "Limestone grassland association" in *Types of British Vegetation*, pp. 154–60. 1911.

Moss, C. E. *The Vegetation of the Peak District*. Cambridge, 1913.

Smith, Robert. Botanical survey of Scotland: I. Edinburgh District. *Scot. Geogr. Mag.* 1900 a. II. North Perthshire District. *Scot. Geogr. Mag.* 1900 b.

Smith, W. G. Botanical survey of Scotland: III. Forfar and Fife. *Scot. Geogr. Mag.* 1904–5.

Smith, W. G. and Moss, C. E. Geographical distribution of vegetation in Yorkshire. Part I. Leeds and Halifax district. *Geogr. J.* 22. 1903.

Smith, W. G. and Rankin, W. M. Geographical distribution of vegetation in Yorkshire. Part II. Harrogate and Skipton district. *Geogr. J.* 1903.

Tansley, A. G. and Adamson, R. S. Studies of the vegetation of the English Chalk. III. The chalk grasslands of the Hampshire-Sussex border. *J. Ecol.* 13, 177–223. 1925. IV. A preliminary survey of the chalk grasslands of the Sussex Downs. *J. Ecol.* 14, 1–32. 1926.

Tüxen, R. Bericht über die pflanzensoziologische Excursion der floristisch-soziologischen Arbeitsgemeinschaft nach dem Plesswalde bei Göttingen. *Mitt. d. floristisch-soziologischen Arbeitsgemeins. in Niedersachsen*, pp. 25–51. 1927.

Watson, W. The bryophytes and lichens of calcareous soils. *J. Ecol.* 6, 189–98. 1918.

Chapter XXVIII

NEUTRAL GRASSLAND

The term "neutral grassland" was first used in *Types of British Vegetation* (p. 84) to include semi-natural grasslands whose soil is not markedly alkaline nor very acid, mostly developed on the clays and loams which occupy so much flat lowland country in the midlands and south of England as well as on many of the tracts of alluvium in the valleys of the north and west—broadly speaking the soils which also bear the lowland oakwood dominated by *Quercus robur* (Chapter XIII). Most of these soils are actually on the acid side of neutrality—perhaps the majority between pH 6 and 7—owing to slight leaching and humus accumulation, though many of the mother clays from which they are derived are distinctly alkaline. Their general characters, however, and the vegetation they bear, differ completely from those of the acidic sands and of the upland siliceous rocks which bear the bent-fescue community (Chapter XXVI) and Quercetum sessiliflorae (Chapter XV). Typically they are much richer in available nitrogen and the mineral elements necessary for nutrition, though bad treatment may result in serious impoverishment in any or all of these. And they are inhabited by many species of grass and a wide variety of other herbaceous plants not found on the extremely acid soils, nor on the thin basic soils dealt with in the last chapter.

On the other hand some alluvial meadow soils which bear "neutral grassland" vegetation, for example the Thames-side meadows described on pp. 568–70, are actually as alkaline as typical chalk or limestone soils. The pH value is by no means the only determining factor, and the term neutral was originally intended to apply (in fact is better applied) to characterisation of this grassland neither by markedly "calcifuge" nor markedly "calcicole" species, rather than to neutrality of the soil solution.

"Neutral grassland" is thus a very comprehensive term, including a great variety of vegetation, on one side passing into that of the more acid soils dominated by bent and sheep's fescue, on the other into the grassland of markedly calcareous soils with the characteristic calcicolous species. The alluvial soils often have a high water table and may be periodically flooded ("meadow soils", p. 88), passing towards marsh when they tend to be waterlogged, but some are well drained and fairly dry throughout the year. The soils also show a wide range of texture and of natural fertility.

The chief difficulty in the way of a satisfactory treatment of neutral grassland is the rarity of really "natural" examples, a rarity due to the fact that by far the greater part of this type of grassland is enclosed and

subject to more or less intensive grazing (or to mowing for hay), often also
to surface cultivation and manuring. Most of the "permanent
grass" of our agricultural returns falls into the category of
neutral grassland, and "permanent grass" represented, in 1919,
53 per cent of the farmlands of England and Wales, or nearly 40 per cent
(in 1936 about 42 per cent) of the total area of the country. A great deal of
this land was at one time arable, as may be seen from the "lands" (ridge
and furrow) in many grass fields. This land has been sown down or allowed
to "tumble down" to grass. There is, however, a certain amount of neutral
grassland, as on the edges of old woods and on commons and village greens,
which has never been sown or ploughed and which is doubtless derived from
original forest land by felling followed by grazing, or by degeneration and
ultimate disappearance of the woods through grazing alone, in the same
way that bent-fescue grassland has been derived from the original forest on
the hillsides of the uplands.

Treatment the important factor. It is important to recognise that the
existing state of any particular piece of enclosed and tended neutral grassland is very little affected by its origin, whether it is derived from forest by
felling and grazing or whether it has been ploughed and sown down, or has
"tumbled down" to grass; nor when it has been sown does the composition of the original seed mixture greatly affect its present condition. Even
the *original* nature of the soil, i.e. of the underlying rock, may make little
difference. According to the climate, situation and soil and (most important of all) the treatment of the field, i.e. the kind, amount and duration of
grazing and/or manuring, some of the sown species will die out, others will
increase, and invaders from outside may enter and establish themselves
till a condition of approximate equilibrium is attained between the plant
population and all the incident factors. Of these factors the regime to
which the field is subject is of decisive importance. The classical "park
grassland" experiments at Rothamsted demonstrated the profound effect
of differential manuring on the grassland flora, and more recently Stapledon
and his colleagues have frequently shown that different kinds of grazing
alone, in the absence of any manuring, will completely alter the character
of the vegetation and the species present. The kinds and degrees of cultivation (or neglect) are almost infinitely various. Thus a permanent grass field
may be intensively grazed, or it may be undergrazed, by sheep, cattle or
horses, or by mixed stock, and at different times of year. It may be
manured with dung, or with various artificial manures in any proportion,
it may be harrowed or rolled. And a pasture field may be "put up" for hay
at various intervals. Every combination of these factors will produce a
somewhat different, often a widely or almost totally different result, and
all these results will naturally also be conditioned by soil, climate and
situation.

Theoretically it would be possible to ascertain the assemblage of grasses and other herbaceous plants that would naturally colonise a given soil type in the absence of any agricultural treatment except moderate grazing, and then the various modifications which would follow different kinds of grazing and manuring. In fact the occasional fragments of untreated and apparently unmodified grassland on clays and loams are too limited to enable us to draw any satisfactory general conclusions even if they had been closely studied ecologically, which they have not. We have to content ourselves at present with such information as can be gained from general observation, from recorded lists of species, and from data derived from studies whose primary object was agricultural.

Commons, greens and verges. Probably the least modified of the neutral grasslands are those on the edges of woods and on commons and village greens on medium or heavy soil. These last are, it is true, practically always grazed, but they are not subject to intensive exploitation; and in the remoter parts of the larger commons the grassland, which often passes into open scrub or woodland, is probably as near as we can get to "natural" neutral grassland.

The grass "verges" at the sides of the country roads are also instructive in this connexion. They are generally cut over in the early autumn when the hedges are trimmed, sometimes earlier in the summer, and locally they may be grazed by donkeys, horses or goats; but they are not manured, though their vegetation is frequently modified by the piling of road metal, or the throwing up of road scrapings or the material dug out from ditches, and, before the introduction of the modern motor road surface, by the road dust. The grassland of such verges would normally be quickly succeeded by scrub and woodland if parent trees are available in the neighbourhood, and here and there colonisation by shrubs and trees and the process of succession may actually be observed (see pp. 295–6). In general the succession is held up by the annual cutting of the verge and the grazing of the commons. Such habitats would undoubtedly repay much closer study than they have received. The effect of treading on verges and footpaths has been considered in Chapter XXV, pp. 496–8.

The grasses of neutral grassland. Of the grasses which dominate neutral grassland all the species are indigenous, though they are mostly better known as more or less "cultivated" plants in tended pasture or meadow, or sown in temporary "leys", than in "natural" situations. The perennial rye grass (*Lolium perenne*) is undoubtedly the most valuable agriculturally, both in pasture and for hay. It grows best on the richest soils, where it is typically co-dominant with "wild" white clover (*Trifolium repens*). The vigour of these two species on really good soils tends to exclude other species and thus to reduce the total flora. Perennial rye-grass is very much commoner in the best-tended pastures, and sown with white clover in temporary leys, than in "wild" situations. It is of tufted habit and when luxuriant its flowering stem may be as much as 2 ft. (60 cm.) in height.

Cocksfoot (*Dactylis glomerata*) is another valuable grass both for hay and in pasture, and is very widely distributed in a great variety of soils and situations, as much in "natural" situations, such as wood edges, commons and roadside verges, as in tended grassland. It is a strongly tufted grass of rather coarse habit, and its flowering stalks may reach a height of 4 ft.

Meadow foxtail (*Alopecurus pratensis*) grows in large, rather loose tufts and is of coarse tall growth, the vegetative shoots, which develop early in the season, as much as 2 ft. high, and the flowering stems with their terminal cylindrical spikes often exceed 3 ft. It flourishes best and is often dominant in meadow land regularly cut for hay, and is less prominent in pasture.

Timothy (*Phleum pratense*) is also a tall tufted grass, reaching a height of 4 ft., and flourishing best in hayfields on good, rather heavy soils, usually disappearing under pasture. It develops much later in the season than meadow foxtail.

Crested dogstail (*Cynosurus cristatus*) is a shorter grass and unlike the two preceding species flourishes better in pastures than in permanent hayfields. It occurs on a considerable range of soils, both dry and damp. Crested dogstail is a valuable grass on second-class pastures, where it is often dominant or co-dominant, taking the place of the perennial rye-grass of the best pastures. The hard glumes of the characteristic mature inflorescences are refused by stock, so that unless it is heavily grazed early in the season the dry inflorescences remain in the pasture.

The two meadow-grasses in the narrow sense (*Poa pratensis* and *P. trivialis*) are rarely dominant, but often grow freely mixed with other grasses, the former on rather drier, the latter on moister or even wet soils, but the two may be found together in alluvial meadows. *Poa pratensis* is the only British meadow grass (if we exclude *Holcus mollis*) with an underground creeping rhizome, while *P. trivialis* has creeping stolons. The former, though widely distributed, is not very productive and but seldom really abundant in pastures in this country, though it is highly esteemed in America as "Kentucky blue grass", and has invaded most of the prairies, where it is not native. *P. trivialis* is more valuable on suitable soils.

Several species of fescue occur in neutral grassland. *Festuca pratensis* (meadow fescue) is a widely distributed and valuable grass in the lowlands. *F. elatior*, as its name implies, is a much taller grass, coarser and more robust, though well taken by stock. It is said to be more hardy and to do better on poorer and drier soil than *F. pratensis*, and also does well on the heaviest clays.

The fine-leaved fescues, often dominant both in acidic and basic grasslands are not characteristic of the neutral type. Thus the red fescue (*Festuca rubra*) has very numerous varieties, some of which occur in salt marsh (and spray-washed maritime grassland), and one on sand dunes, and this species is also abundantly mixed with sheep's fescue on chalk downs. But some varieties occur in neutral grassland, both meadow and pasture, and especially the former. *F. rubra* is indeed occasionally locally dominant in pasture, forming a dense well-grazed turf of shortly creeping shoots: other forms grow in isolated tufts. The red fescues have much narrower, finer leaves than any of the species previously described, the leaves of some being bristle shaped like those of sheep's fescue.

The sheep's fescue (*F. ovina*), with bristle-shaped leaves, also has numerous varieties, some of which are now separated as species. The commonest form (*F. ovina* sensu stricto) is one of our most widely distributed and abundant grasses, being frequently dominant, as we have seen in the last two chapters, both in acid grasslands of the "bent-fescue" type, and in chalk and other limestone grasslands. It is essentially a plant of dry well-drained soils, and though it can flourish under very high rainfall it cannot stand water-logged soil nor the competition of more luxuriant grasses, and under such conditions becomes subordinate or suppressed altogether. Hence it is not at all characteristic of, and is usually absent from, neutral grassland, but occurs in some of the poorest and driest pastures where it is comparatively free from this kind of competition. It actually forms more than 10 per cent of the herbage in the *unmanured* park grassland at Rothamsted (p. 573). *Festuca ovina* differs from *F. rubra* in being strictly tufted, with no creeping shoots escaping from the base of the tuft ("extra-vaginal" shoots) and typically bristle-shaped leaves.

The common bent (*Agrostis tenuis*) is another very abundant and widely spread grass, probably the most numerous species of all in semi-natural grassland vegetation: on the poorest pastures, where it is very generally dominant over great areas, it sometimes forms 90 per cent of the herbage on a great range of soils, from sandy loams to heavy clays. It also occurs freely in medium pastures which are somewhat acid, mixed with the better grasses. In neutral pasture it is of limited grazing value, though, as we have seen, it is one of the staples, together with sheep's fescue, of upland sheep pasture, especially in regions of higher rainfall. It has rather short, somewhat scabrid leaves and matures late in the season: in the poor neglected pastures of which the common bent is so characteristic, its shoots become withered and quite useless for pasture in late summer.

The white bent, *Agrostis stolonifera* (*alba*), is commoner on moister and better soils, the typical form being of distinct agricultural value in wet meadows and on heavy soils.

The sweet vernal grass (*Anthoxanthum odoratum*) is another very widely distributed grass on a great variety of soils, though it is not a dominant species. It develops very early in the season, flowering in late April or early May before any of the other grasses. It is tufted, but of very variable size and habit. Sweet vernal grass is well grazed and quite useful in the upland bent-fescue sheep-walks.

The Avenae are not characteristic of neutral grassland, though both *Avena pubescens* and *A. pratensis* sometimes occur. The closely allied golden oat grass (*Trisetum flavescens*) is also much more characteristic of chalk soils, but occurs, sometimes abundantly, in good grassland on the lighter loams, where it does better under haying conditions than in pasture. It is a tufted, slightly creeping species with flowering shoots 1–2 ft. high. The false oat-grass (*Arrhenatherum elatius*) is a much taller, strong-growing, rather coarse grass, up to 4 ft. in height, which grows on a variety of soils, and is often dominant on the grass verges of roads, especially on calcareous soils. It persists well under cutting, and produces abundant hay though of poor quality, but cannot endure grazing.

Two very common grasses which, on the better grasslands, are weeds from the standpoint of the pastoralist, are the Yorkshire fog (*Holcus lanatus*) and the coarse, tufted, rosette-forming *Deschampsia caespitosa*. The former is a ubiquitous, very aggressive grass on good soils, and often forms a large proportion of the herbage of very good pastures, increasing under manuring. It is covered with soft hairs and is generally refused by stock, except the young shoots. *Deschampsia caespitosa* forms large tufted tussocks or stools with long narrow harsh leaves finely serrated along the margins. It affects waterlogged soil, especially if puddled so that water is held up in the surface layers.

Briza media, the common quaking grass, with its very loose panicle of ovate spikelets, is much more characteristic of chalk than of neutral grassland, but it occurs on various poor soils in both dry and wet situations. *Briza* is worthless agriculturally and rapidly disappears under manuring.

Non-gramineous species. The non-gramineous species of the herbage are generally classified by agriculturists into "leguminous" and "miscellaneous", because the former are extremely valuable pasture plants—some as valuable as the best grasses—while many though not all of the latter are worthless as food for stock. The Leguminosae contain on the average more protein, much more calcium, and less carbohydrate than the grasses, and are of quite comparable nutritive value, but in addition they are of great importance agriculturally because they fix the free nitrogen of the air and leave it in the soil in an available form. They are practically always represented, though not specially numerous in species and individuals, on most neutral grassland, but they are much encouraged by phosphatic and potash manures such as "basic slag" and "kainit", as the Rothamsted park grassland manuring experiments show.

Taking neutral grassland as a whole the "miscellaneous species" of herbs are very numerous. The actual number of species on any given area is more or less inversely proportional to the "fertility" of the grassland—the better the soil and the better it is grazed the fewer the species. The primary cause of this relation is that the best and the best managed soils encourage the growth of certain species—notably *Lolium perenne* and *Trifolium repens*—to such an extent that the miscellaneous species are crowded out and are unable to re-enter the community. But on the very poorest and most unfavourable soils the number of species is again reduced, because only a few are able to endure the conditions.

Leguminous plants. Of leguminous species the clovers (*Trifolium*) are some of the most widely distributed as well as the most important agriculturally. The three species most characteristic of neutral grassland are *T. pratense*, *T. repens* and *T. dubium*.

The red clover (*T. pratense*) is a free-growing tufted biennial or perennial. Though it is a slow coloniser it occurs naturally on many different types of soil, and is more abundant in old meadowland than in old pasture, though it is of use in both and a valuable accumulator of nitrogen. The cultivated varieties are often sown as pure crops.

The white clover (*T. repens*) is also found on a wide variety of the better soils, and propagates very freely by means of runners, especially in pastures. It is one of the most valuable agricultural plants, and is characteristically associated with *Lolium perenne* in the best pastures. The native strains ("wild white clover") are much more desirable than the so-called "Dutch clovers".

T. dubium (yellow suckling clover) is a somewhat tufted annual, with semi-erect shoots and small heads of minute yellow flowers, widely distributed in various kinds of neutral grassland. It is of some use in pasture though not nearly so valuable as the red and white clovers, being a smaller plant.

Medicago lupulina (the black medick or trefoil) is also an annual, with weak trailing shoots and yellow flowers, and has a strong superficial resemblance to *Trifolium dubium*. It is also very widely distributed, but is absent from the more acid soils in regions of high rainfall. Being a small plant it is not of great importance in mixed herbage, but may have considerable value when sown.

One of the most widely distributed of leguminous species is bird's foot trefoil (*Lotus corniculatus*), which occurs on a much greater variety of soils than the clovers and medick, not only in neutral grassland (though it is absent from the best soils) but in chalk and limestone grassland, where it is constant, in grass heath, and in many of the bent-fescue upland sheep pastures, where it is often the only legume present. It has a long tap-root and is thus able to obtain water from deeper soil layers when the surface becomes dry. Doubtless it has considerable value in pastures where it is naturally abundant, but *Lotus* is not sown agriculturally. Like white clover it responds freely to phosphatic manures.

The meadow vetchling (*Lathyrus pratensis*) is another leguminous plant common in neutral grassland, particularly hayfields, and on roadsides, and the same may be said of several species of vetch (*Vicia*).

"Miscellaneous" species. With the horde of "miscellaneous" species inhabiting neutral grassland it is not necessary to deal individually. Three or four are however worth mention.

The milfoil or yarrow (*Achillea millefolium*) is an abundant plant in many kinds of grassland, particularly on the lighter and drier soils. It is deep-rooting and extremely drought-resistant, the leaves remaining fresh and green at the end of hot dry summers when most of the herbage is dried up. When growing gregariously it forms a dense "sole", excluding other plants. It is frequently well cropped by stock, and is thus of considerable value.

The ribwort plantain (*Plantago lanceolata*) is another very abundant plant on almost all kinds of soil except acid peat, and like the milfoil, is deep-rooting and drought-resistant. It certainly has a value on many pastures, though it may increase so much at the expense of better pastoral plants as to become a troublesome weed. It is useless for hay because its leaves are not easily caught by the machine, and even when harvested are difficult to dry.

The "burnet saxifrage" (*Poterium sanguisorba*) is characteristic of calcareous soils, but is also found in some kinds of neutral grassland, both meadow and pasture. Its shoots are not much eaten by stock, except sometimes when quite young. Its congener *P. officinale*, is more characteristic of damp alluvial meadows.

Carex diversicolor (*flacca*) is the most widely distributed of the grassland sedges. It is a chalk grassland constant and also occurs, often abundantly, on low-lying and waterlogged soils.

Many of the "miscellaneous" species of neutral grassland are both palatable and rich in protein and minerals, especially calcium, so that they may be heavily grazed in winter. On the other hand they contain on the whole less dry matter than the grasses and do not shoot again so readily when grazed down.

Meadow and pasture. The broad classification of grasslands according to their use, i.e. into *meadows* which are regularly mown for hay, and *pastures* which are grazed, is ecologically of great importance, because the alternative uses lead to very considerable differences in the vegetation. Meadows are specially characteristic of alluvial soil in which the water table is seldom far below the surface, and which may be periodically flooded by the rise in the waters of neighbouring streams. Sometimes they are *irrigated*, i.e. artificially flooded by means of irrigation channels. Regularly inundated meadows are known as *water meadows*. Natural or artificial irrigation by river water fertilises the soil by addition of fresh mineral salts, and the alternate motion of the water from the streams to the meadows in time of flood and back to the river when the water level sinks keeps the soil well aerated. If there is any considerable fall in the course of the river bed the ground water below the meadow soil is also constantly moving. Added to the high initial fertility which is a common character of alluvial soils this combination of factors favours the dominance of moisture-loving grasses of tall luxuriant habit and the consequent relative suppression or absence of lower growing species. The flora of water meadows is remarkably constant because of the constancy of conditions dominated by the inundation factor and the regular mowing.

Grassland cut for hay is not of course by any means confined to alluvial meadows, any more than pasture is confined to the uplands. And much grassland is frequently used for the two purposes in different years, or in the same year by grazing the aftermath of the hay crop.

The three following lists contain the species which, according to Stapledon (1925), are on the whole characteristic of (1) old meadows, (2) old pastures, and (3) are equally common in both.

(1) *Plants specially characteristic of old meadows*

Grasses

Alopecurus pratensis
Bromus hordeaceus (mollis)
Dactylis glomerata
Festuca rubra
Holcus lanatus
Trisetum flavescens

Herbs

Allium vineale
Anthriscus spp.
Cardamine pratensis
Centaurea nemoralis (nigra)
Chrysanthemum leucanthemum
Colchicum autumnale
Geranium molle
Heracleum sphondylium
Lathyrus pratensis
Leontodon hispidus
Lychnis flos-cuculi
Malva sylvestris
Orchis maculata
O. morio
Platanthera bifolia
Polygonum bistorta
Poterium officinalis
Rhinanthus spp.
Taraxacum officinale
Trifolium dubium
T. pratense
Vicia cracca
V. hirsuta

(2) *Plants specially characteristic of old pastures*

Agrostis tenuis
Cynosurus cristatus
Festuca ovina

Carex spp.
Cirsium arvense
C. lanceolatum
C. palustre
Conopodium majus
Crepis capillaris
Filipendula ulmaria
Genista tinctoria
Juncus spp.
Linum catharticum
Ononis spp.
Poterium sanguisorba
Senecio jacobaea

(3) *Plants equally common in old meadows and old pastures*

Agrostis stolonifera
Lolium perenne

Achillea millefolium
Ajuga reptans
Alchemilla vulgaris
Bartsia odontites
Bellis perennis
Cerastium vulgatum
Convolvulus arvensis
Daucus carota
Galium verum
Geranium dissectum
Hypochaeris radicata
Leontodon autumnalis
Luzula campestris (agg.)
Plantago lanceolata
P. major
Primula veris
Prunella vulgaris

Pulicaria dysenterica   Rumex acetosella
Ranunculus acris   Veronica chamaedrys
R. repens   V. serpyllifolia
Rumex acetosa

With these lists it is interesting to compare the data obtained by H. Baker (1937) from some alluvial grassland bordering the Thames just above Oxford. These meadows are of great interest because it is known that one of them (Port Meadow, about 400 acres) has been continuously pastured by horses and cattle (not sheep) at least since the time of the Domesday survey (1085), while the others (Yarnton, Oxey and Pixey Meads, together about 270 acres) have been regularly mown for hay at least for centuries. The two grasslands may, therefore, be considered to have come into equilibrium with these contrasted biotic factors.

Thames-side meadows and pastures

The soil of both is river alluvium of the same general type. It is subject to annual flooding, and before the flow of the river was regulated by locks and weirs it was probably submerged for six months or so in winter. Since the control the floods have been smaller but more frequent. The pH values of the Port Meadow soil vary from 6·6 and 6·8 at 3 in. depth in the middle area (B), which lies about 2 ft. above the summer water level in the adjacent river and is subject to the greatest leaching, to 7·9 and 8·0 at 6 in. in the lowest part (A) which slopes gradually from about 1 ft. above the summer water level to the river. All the "Meads" (cut for hay) show very constant values, between 7·3 in the surface soil in summer and 8·0 to 8·3 at 6 in. in the autumn and winter. The pH value of Thames water is about 8·0.

The list of ninety-five species occurring in these grasslands includes 26 found only in Port Meadow (grazed) and thirty-nine only in the Meads (mown), while thirty are common to both—striking evidence of the differential effect of the two biotic factors. In Port Meadow there is a further differentiation between the lowest lying region (A) which slopes from river level to about 1 ft. above, containing marsh species, and the highest area (C) whose surface is about 3 ft. above water-level. Nineteen of the twenty-six species occurring in Port Meadow only are confined to one of these two areas (in A 10, in C 9):

Port Meadow (*grazing only*)

A (wettest)	C (driest)
Alopecurus geniculatus	Koeleria cristata
Apium nodiflorum	Achillea millefolium
Hippuris vulgaris	Carex caryophyllea
Nasturtium officinale	Cirsium acaule
Ranunculus drouetii	C. lanceolatum
R. flammula	Leontodon nudicaulis
Sium erectum	Plantago media
Veronica anagallis-aquatica	Ranunculus bulbosus
V. beccabunga	Trifolium dubium
V. scutellata	

Nearly all the species limited to (A) are marsh plants which exist in the short turf in miniature forms, prostrate or about 1 in. high, closely cropped but flowering.

The remaining seven species occur in B alone (the intermediate area) or extend through two or all three of the areas, AB, AC or ABC:

Poa annua (A, B, C)

Bellis perennis (A, B, C)
Crepis capillaris (A, B)

Luzula campestris (B)
Potentilla anserina (A, C)
P. reptans (B)
Prunella vulgaris (A, C)

Hay Meads (*mowing*)

Drier

Alopecurus pratensis
Arrhenatherum elatius
Bromus hordeaceus
Briza media

Carex hirta
Chrysanthemum leucanthemum
Linum catharticum
Ophioglossum vulgatum
Rumex acetosa
Succisa pratensis

Wetter

Agropyron repens
Festuca elatior
F. loliacea
Hordeum pratense

Carex acutiformis
Lychnis flos-cuculi
Lysimachia nummularia
Oenanthe silaifolia
Pedicularis palustris
Rumex crispus
Senecio aquaticus
Tragopogon pratensis
Triglochin palustre
Valeriana dioica

Throughout

Bromus commutatus
Caltha palustris
Carex disticha
C. goodenowii
C. panicea
Equisetum palustre
Filipendula ulmaria
Lathyrus pratensis

Leontodon hispidus
Poterium officinale
Rhinanthus crista-galli
Silaum silaus
Stellaria graminea
Thalictrum flavum
Vicia cracca

Most of the grasses confined to these hayfields are among the tallest and most abundant species of the mead vegetation. The great majority of the dicotyledons are tall scapose hemicryptophytes intolerant of heavy grazing.

The thirty following species occur both in Port Meadow and the hay meads:

*Agrostis stolonifera
Anthoxanthum odoratum
Cynosurus cristatus
Dactylis glomerata
*Deschampsia caespitosa
Festuca pratensis
F. rubra

Glyceria fluitans
*Holcus lanatus
Lolium perenne
Phleum pratense
Poa pratensis
P. trivialis

* The species marked with an asterisk occur throughout the entire alluvial area examined.

Cardamine pratensis
Carex diversicolor (flacca)
Centaurea nemoralis
Cerastium vulgatum
Galium palustre
Hypochaeris radicata
Juncus articulatus
Leontodon autumnalis

Lotus corniculatus
Myosotis scorpioides (palustris)
Oenanthe fistulosa
Plantago lanceolata
Ranunculus acris
R. repens
Trifolium pratense
T. repens

Magdalen Meadow. As an example of a low-lying alluvial meadow with a medium loamy soil and mixed treatment the flora of Magdalen Meadow, Oxford, may be given. This is not strictly a water meadow since it is not regularly irrigated, but the ditch which surrounds it connects with the River Cherwell, and the meadow is flooded during the winter or early spring of some years, rarely as late as May. The summer water table in the bordering ditches may fall to 3 ft. or more below the surface of the meadow. Thus there is considerable movement of water in the soil due to the rise and fall of the water table. The meadow is usually pastured in summer and autumn (deer or cattle), but is sometimes put up for hay. As a pasture it would probably be reckoned second or third class.

Flora of Magdalen Meadow (not exhaustive)

Grasses

Agrostis stolonifera	o	Holcus lanatus	la
Alopecurus pratensis	a, ld	Hordeum nodosum	
Anthoxanthum odoratum	la	(pratense)	la
Dactylis glomerata	ld	Lolium perenne	l
Festuca pratensis	lf	Poa pratensis	lf
F. rubra	la	P. trivialis	lf

Herbs

Ajuga reptans	o	Ranunculus acris	a
Bellis perennis	l	R. bulbosus (only on highest	
Cardamine pratensis	f	ground)	
Cerastium vulgatum	l	R. repens	l
Ficaria verna	o	Rumex acetosa	f
Fritillaria meleagris	la	Taraxacum officinale	o
Galium palustre	o	Trifolium pratense	l
Lychnis flos-cuculi	o–f	T. repens	o–f
Lysimachia nummularia	f–a	Veronica serpyllifolia	o
Nepeta hederacea	o		
Potentilla reptans	f	Ophioglossum vulgatum	o
Prunella vulgaris	o		

Alopecurus and *Dactylis* are co-dominant in some areas, forming herbage 2 ft. high in June. In other places the species of *Festuca* and *Poa* with *Hordeum nodosum* dominate herbage which is shorter and less coarse, but luxuriant. The fritillaries (Pl. 105, phot. 257) for which the meadow is famous are confined to these types of herbage and are absent from slightly higher lying, more closely grazed, areas of mixed grasses in which *Bellis* and *Cerastium* occur. Grazing between March, when their shoots appear above

PLATE 105

Phot. 257. *Fritillaria meleagris* ("snake's head") in neutral alluvial grassland. Magdalen Meadown, Oxford. *N. F. G. Cruttwell.*

Phot. 258. *Narcissus pseudo-narcissus* in Lias clay pasture, borders of Dorset and Devon. Photograph from *The Times.*

NEUTRAL GRASSLAND

the ground, and mid-June, when the seeds are distributed, suppresses the fritillaries.

General flora of neutral grassland. The following list (Table XVIII) contains 126 species in all. Of these eighty-two are recorded by Fream (1888) from the water meadows at North Charford, bordering the Christchurch Avon in south Hampshire (first column). These meadows are developed on a clayey alluvial loam above the Upper Chalk. Almost the same number (85) are recorded by Lawes, Gilbert and Masters (1882) from the very old park grassland at Rothamsted in Hertfordshire on a heavy loam (clay-with-flints) resting on the Chalk (second and third columns). An area of this park grassland was divided into a number of plots in 1856, and different plots were consistently treated with different manures for a long series of years, two of them being left without any treatment. Every year the hay is cut and sorted into the constituent species for each individual plot separately.

The first column (W) gives the Hampshire water-meadow species recorded by Fream, the second (P) the species from all the plots, manured and unmanured, of the park grassland at Rothamsted, the third (P^1) from the unmanured plots only. Of the letters in the third column a indicates that the species formed more than 10 per cent, b between 1 and 10 per cent and c less than 1 per cent of the whole bulk of hay from the two unmanured plots. It will be noticed that twenty-four species have appeared in various manured plots which are not present in the unmanured, though the latter contain more species than any of the *individual* manured plots. This is in accordance with the general principles that the poorer the land (within limits) the more numerous is the flora, and that one-sided manuring favours particular species.

Of the exclusively water meadow species a few (marked *) are definitely reedswamp or waterside species and scarcely belong to the meadow community as such; while in the park land some of the recorded species (marked §) are woodland or wood edge plants which have established themselves in the grassland.

Forty-one of the whole list of species occur on the water meadows only (thirty-three if we exclude waterside plants not belonging to the meadow community), forty-four on the Rothamsted park land only (thirty-nine if we exclude woodland and wood-edge plants), while forty-one are common to both lists.

These combined lists contain a representative selection of species, including a wet and a relatively dry habitat, and the 41 grasses and herbs which are common to the two lists include the most constant species of neutral grassland, which have a wide range of habitat in respect of soil moisture.

The kernel of neutral grassland. The following sixteen species—eight grasses and eight herbs—which occur in the Charford alluvial meadows, in

Table XVIII

Species	W	P	P[1]		Species	W	P	P[1]
Agrostis stolonifera	+	.	.		Lathyrus pratensis	+	+	b
A. tenuis	+	+	a		Leontodon autumnalis	+	+	c
Alopecurus geniculatus	+	.	.		L. hispidus	+	+	c
A. pratensis	+	+	b		Linum catharticum	+	.	.
Anthoxanthum odoratum	+	+	b		Lotus corniculatus	+	+	b
Arrhenatherum elatius	+	+	c		L. uliginosus	+	+	.
Avena pubescens	.	+	b		Luzula campestris	.	+	b
Briza media	+	+	b		Lychnis flos-cuculi	+	.	.
Bromus hordeaceus (mollis)	+	+	c		Lysimachia nummularia	+	.	.
B. racemosus	+	.	.		*Lythrum salicaria	+	.	.
Cynosurus cristatus	+	+	c		Medicago lupulina	+	.	.
Dactylis glomerata	.	+	b		Myosotis scorpioides (palustris)	+	.	.
Deschampsia caespitosa	+	+	c		Ononis repens	.	+	.
Festuca elatior	+	.	.		Ophioglossum vulgatum	.	+	c
F. loliacea	+	+	c		Orchis morio	.	+	c
F. ovina	.	+	a		Ornithogalum umbellatum	.	+	.
F. pratensis	+	+	b		Pimpinella saxifraga	.	+	b
F. rubra	+	.	.		Plantago lanceolata	+	+	b
*Glyceria maxima (aquatica)	+	.	.		P. major	+	.	.
G. fluitans	+	.	.		P. media	.	+	.
Holcus lanatus	+	+	b		Polygonum persicaria	+	.	.
Lolium perenne	+	+	b		Potentilla anserina	+	.	.
*Phalaris arundinacea	+	.	.		P. reptans	+	+	c
Phleum pratense	.	+	c		§P. sterilis	.	+	c
Poa annua	+	.	.		Poterium sanguisorba	.	+	c
P. pratensis	+	+	c		Primula veris	.	+	c
P. trivialis	+	+	b		Prunella vulgaris	+	+	c
Trisetum flavescens	+	+	b		Ranunculus acris	+	+	c
					§R. auricomus	.	+	.
Achillea millefolium	+	+	b		R. bulbosus	+	+	b
Agrimonia eupatoria	.	+	c		R. repens	.	+	b
Ajuga reptans	+	+	c		Rumex acetosa	+	+	b
Alchemilla vulgaris (agg.)	.	+	c		R. crispus	+	+	.
Anthriscus sylvestris	.	+	.		R. longifolius (aquaticus)	+	.	.
Bellis perennis	+	+	c		R. obtusifolius	.	+	.
Caltha palustris	+	.	.		§Scilla non-scripta	.	+	c
Cardamine pratensis	+	+	.		Scrophularia aquatica	+	.	.
*Carex paludosa	+	.	.		Scutellaria galericulata	+	.	.
C. caryophyllea	.	+	c		Senecio aquaticus	+	.	.
Centaurea nemoralis (nigra)	.	+	b		S. erucifolius	.	+	.
Cerastium vulgatum	+	+	b–c		Sonchus oleraceus	.	+	.
Chrysanthemum leucanthemum	+	+	c		Stellaria graminea	.	+	c
Cirsium arvense	.	+	.		§S. holostea	.	+	.
C. palustre	+	.	.		Symphytum officinale	+	.	.
Conopodium majus	.	+	b		Taraxacum officinale	+	+	c
Daucus carota	.	+	.		Thalictrum flavum	+	.	.
*Eleocharis palustris	+	.	.		Thymus serpyllum	+	+	c
*Epilobium hirsutum	+	.	.		Tragopogon pratensis	.	+	c
E. parviflorum	+	.	.		Trifolium dubium	.	+	.
E. tetragonum	+	.	.		T. pratense	+	+	b
Eupatorium cannabinum	+	.	-		T. procumbens	.	+	.
Ficaria verna	.	+	.		T. repens	+	+	c
Filipendula ulmaria	+	+	.		Valeriana dioica	+	.	.
Fritillaria meleagris	.	+	.		V. officinalis	+	.	.
Galium aparine	.	+	c		*Veronica anagallis-aquatica	+	.	.
G. palustre	+	.	.		*V. beccabunga	+	.	.
G. verum	.	+	c		V. chamaedrys	+	+	c
Geum rivale	+	.	.		V. officinalis	.	+	.
Heracleum sphondylium	+	+	c		V. serpyllifolia	.	+	c
Hieracium pilosella	.	+	c		Vicia cracca	+	+	.
Hypericum perforatum	.	+	.		§V. sepium	.	+	.
Hypochaeris radicata	.	+	.		Veronica scutellata	+	.	.
Juncus acutiflorus	+	.	.					
J. inflexus (glaucus)	+	.	.			82	85	61
Knautia arvensis	.	+	c					

* Reedswamp or waterside plants not forming part of the meadow community.

§ Woodland or wood-edge plants which have established themselves in the park grassland.

The total number of species is 126, or 113 if we exclude species not properly belonging to the grassland communities.

the Rothamsted park grassland, and in both the grazed and the mown alluvial meadows at Oxford may be taken as the kernel of the neutral grassland community.

Grasses

Anthoxanthum odoratum
Cynosurus cristatus
Deschampsia caespitosa
Festuca pratensis

Holcus lanatus
Lolium perenne
Poa pratensis
P. trivialis

Herbs

Cardamine pratensis
Cerastium vulgatum
Leontodon autumnalis
Lotus corniculatus

Plantago lanceolata
Ranunculus acris
Trifolium pratense
T. repens

It is a striking fact that none of these species, but only *Agrostis tenuis* and *Festuca ovina*, the two dominants of the acid hillside pastures and of the grass heaths, form more than 10 per cent of the *unmanured* plots at Rothamsted, emphasising their tolerance of poor soil and their inability to compete successfully with the more luxuriant grasses and herbs of manured or grazed neutral grassland.

Besides the plants in the preceding lists the following may also be met with fairly often in neutral grassland:

	Achillea ptarmica		Polygala vulgaris
	Bartsia odontites		Potentilla erecta
damp	Carex diversicolor (flacca)	damp	Poterium officinale
	Centaurium umbellatum	damp	Pulicaria dysenterica
	Cerastium glomeratum	wet	Ranunculus flammula
	Crepis capillaris		Rhinanthus crista-galli (agg.)
	Euphrasia officinalis (agg.)		Saxifraga granulata
	Galium cruciata		Senecio jacobaea
	G. erectum		Serratula tinctoria
	G. mollugo		Silaum silaus
	Geranium pratense	dry	Spiranthes spiralis
	Hypericum acutum (tetrapterum)		Succisa pratensis
	Lathyrus nissolia		Trifolium medium
	Malva moschata		Vicia angustifolia
	Narcissus pseudo-narcissus (Pl. 105, phot. 258)		V. hirsuta
	Nepeta hederacea		V. lathyroides
damp	Orchis maculata		V. tetrasperma

Juncetum effusi. On flat heavy grassland used as pasture, and in which the soil is often waterlogged, species of *Juncus* (*J. effusus*, the commonest dominant, *J. conglomeratus*, often with *J. inflexus* (*glaucus*), *J. acutiflorus* or the annual *J. bufonius*) almost always occur and are locally dominant, and where the surface is wettest any of the following species may be found in the turf between the rushes:

Agrostis stolonifera
Alopecurus geniculatus
Deschampsia caespitosa

Glyceria fluitans
Poa trivialis

Bidens cernua
B. tripartita
Caltha palustris
Cardamine pratensis
Carex flava
C. goodenowii
C. leporina
C. panicea
C. remota
Callitriche stagnalis
Eleocharis palustris
Epilobium palustre
E. parviflorum
Galium uliginosum
Gnaphalium uliginosum
Hydrocotyle vulgaris

Lotus uliginosus
Mentha aquatica (agg.)
Pedicularis sylvatica
Polygonum hydropiper
P. persicaria
Ranunculus flammula
R. hederaceus
R. sceleratus
Stellaria uliginosa
Triglochin palustre
Veronica beccabunga
V. scutellata

Equisetum fluviatile (limosum)
E. palustre

The water meadows and this Juncetum have several species in common and form transitions from neutral grassland to marsh, but while the former are well aerated the water of the latter is usually stagnant, leading to decided differences in the floristic list. Acid soil conditions sometimes develop in the Juncetum, and occasionally bog plants, such as bog moss and cotton grass, may occur in the wetter places, forming a transition to bog (Chapter xxxiv).

Grades of pasture. Very much the greater area of enclosed English grassland is used as pasture, and the conditions of pasture are very different from those of meadow, resulting in decided differences in the flora, as we have already seen. The dung of the stock is an important manurial factor, especially if cake is fed to the animals or they have access to other sources of food extraneous to the pasture itself. In the best practice the dung is collected and spread evenly over the turf. On the other hand the pasture loses the increase in live weight of the animals fed upon it, and extra manure of some kind is required according to the soil and the kind and amount of grazing contemplated. The proper regulation of the grazing is of the utmost importance. The great desideratum is to secure a constant supply of fresh young shoots of the more nutritious grasses, and to avoid the production of harsh coarse growth and of flowering stems. A continuous turf or "sole" with a minimum of bare spaces (which are generally colonised by worthless "weeds") and an avoidance of the formation of "mat" (dead plant debris) should be aimed at.

From the agricultural point of view Stapledon (1925) classifies the "tended pastures" (i.e. excluding "rough grazings") in a descending series of grades of fertility: (1) fatting pastures, (2) dairy pastures, (3) general purposes and sheep pastures, and (4) "outrun" pastures. The fourth class are practically "untended" pastures, mostly on old arable land that has tumbled down to grass and cannot support stock unless the animals have repeated access to cultivated land or tended grassland.

(1) *Fatting pastures*, so called because they are capable of fattening beasts during the summer months without the addition of feeding stuffs. The best known are the famous pastures of Leicestershire and Northamptonshire (boulder clay and lias), but there are others on Romney Marsh (marine silt), in Blackmoor Vale, Dorset (alluvium), and elsewhere. The soil is always a loam or silt.

The great floristic feature of all the best pastures is the prominence of perennial rye-grass (*Lolium perenne*) and white clover (*Trifolium repens*), and the paucity of "weeds"—in other words, the small number of miscellaneous species. The total flora usually comprises less than twenty-five species of flowering plants (of which the majority are grasses), and sometimes no more than twenty. Of useful grasses (other than *Lolium*), *Cynosurus cristatus*, *Dactylis glomerata*, *Hordeum nodosum* (*pratense*) and *Trisetum flavescens* are usually well represented, while of the less valuable species *Agrostis* spp. and *Holcus lanatus* are sometimes present in considerable quantity. In the Leicestershire pastures rye grass and white clover constitute from half to nearly three quarters of the herbage, in the Blackmoor Vale from a quarter to a half, and on Romney Marsh from a third to nearly nine-tenths. Of the non-gramineous plants milfoil or yarrow (*Achillea millefolium*) is generally present.

(2) *Dairy pastures* occur on a greater variety of soil, and where this is calcareous the number of species present may be more than 30. The chief grasses are the same as in (1), but the bents may exceed 50 per cent of the herbage and *Cynosurus* may reach 20 per cent or more. In addition to white clover, which is often very abundant, red clover (*Trifolium pratense*), *Lotus corniculatus* and *Lathyrus pratensis* are more prominent.

(3) *General purposes and sheep pastures* cover a greater area than (1) and (2) together, occurring on all soil types and managed in every conceivable manner. They form the most important class for the country generally, and would especially repay more skilled management.

Lolium is very seldom dominant, at most co-dominant with *Cynosurus* and *Agrostis*, which are usually much more prominent. *Holcus*, *Poa trivialis*, *Festuca rubra*, *F. ovina* and *Dactylis* make up a good deal of the herbage. White clover is not so consistently dominant as in (1) and (2), though in some fields it is very abundant. Other Leguminosae are more plentiful and varied, and "weeds" contribute as much as 15–30 per cent of the herbage. The total number of species often exceeds 40, and may reach 50.

(4) The untended pastures, generally completely neglected, are very frequently dominated, especially on dry poor soil, by *Agrostis tenuis*, a shallow-rooting grass whose shoots tend to dry up in the summer and give very little feed. Its dominance may be so complete as to reduce the total number of species to less than 20. Thus the most restricted flora is found in the best and the worst pastures.

REFERENCES

BAKER, H. Alluvial meadows: a comparative study of grazed and mown meadows. *J. Ecol.* **25**, 408–20. 1937.

FREAM, W. On the flora of water-meadows, with notes on the species. *J. Linn. Soc. (Bot.)* **24**, 454–64. 1888.

JENKIN, T. J. "Pasture Plants". In *Farm Crops*, **3**, 1–42. 1925.

LAWES, J. B., GILBERT, J. H. and MASTERS, M. T. Results of Experiments on the mixed herbage of permanent meadow. Part II. Botanical results. *Philos. Trans. Roy. Soc.* 1882.

STAPLEDON, R. G. "Permanent grass." In *Farm Crops*, **3**, 74–136. 1925.

TANSLEY, A. G. "Neutral grassland." In *Types of British Vegetation*, pp. 84-7. 1911.

Part VI

THE HYDROSERES

FRESHWATER, MARSH, FEN AND BOG VEGETATION

Chapter XXIX

THE HYDROSERES.

VEGETATION OF PONDS AND LAKES

Hitherto we have been dealing with land vegetation: first the climax forests, or rather such remains of them as still exist, nearly all in a more or less modified condition, representing the natural climatic formations of the lowlands and lower hill slopes of the greater part of these islands; secondly the semi-natural grassland formation into which most of the forest land has been converted by the direct or indirect activities of man. Of the primary xeroseres leading to forest we have been able to recognise fragments here and there, though it is impossible to find complete examples owing to constant human interference with the natural sequences of vegetation. Various subseres initiated by felling, grazing or burning have also been described.

When we turn to the hydroseres, the other great primary type of successional sequence, we are able to present a more complete picture. The aquatic and subaquatic habitats are less easily accessible and utilisable than the plains and hill slopes of the dry land; and though man has not been backward in draining lakes and marshes and in canalising rivers he has left enough relatively undisturbed aquatic habitats in which the vegetation is essentially natural to make possible a fairly full reconstruction of the processes involved in primary succession.

Conditions of life in water. Life in water is subject to conditions radically different from those of terrestrial subaerial life. Aquatic organisms, unlike land plants, are in no danger of suffering from lack of one of the prime necessities of existence. At the same time green plants are exposed to the risk of deficiency of the necessary gases, oxygen and carbon dioxide, and also of mineral nutrients. The amounts of the two gases actually available to submerged plants growing in natural waters varies enormously according to a number of different conditions, and no general statement

Oxygen can be made except that oxygen is much more likely to be deficient than carbon dioxide. Rooted aquatic plants have been shown to be more dependent on nutrient ions absorbed from the soil than on absorption of these from the water surrounding their shoots. Rooted aquatic vegetation therefore varies very much in amount and composition according to the nature of the substratum below the water, and this usually depends on the presence or absence, and the composition, of *silt*. Floating

Silt plants are of course entirely dependent upon the water itself for all their supplies, and since different waters contain very different amounts of dissolved substances, floating vegetation may be scanty or altogether absent from these causes.

Decreasing light intensity limits the depth to which green plants can descend on a lake bottom, and some can exist with less light than others; but apart from light the two chief factors affecting aquatic vegetation are the presence or absence of sufficient dissolved oxygen and the deficiency or adequacy of the mineral nutrients derived from silt. Where abundant green vegetation has been able to establish itself photosynthesis amply covers the supplies of oxygen necessary for the respiration of both plants and animals, but in poorly vegetated lakes and rivers the oxygen supply depends largely on solution from the air by current and wave action. Silt is also one of the most important factors, except in ponds situated in rich mineral soil where abundant nutrients are always present and the floor is covered by a rich organic mud composed of rainwash mixed with humus derived from decayed plant remains. Unless the water is definitely fouled from any cause a luxuriant aquatic vegetation is maintained. A pool of similar size in a district of hard rock or sterile sand may support a very scanty vegetation of specialised type, or be almost barren of life. In rivers the rate of flow and the nature of the rocks through which the course of the stream lies determine the existence and nature of the silt, and the rate of flow is also one factor in the supply of dissolved oxygen.

Rivers. Rivers too, just because of the current, are subject to less variation than still waters in such factors as temperature and dissolved gases which are critical for the existence and luxuriance of vegetation; but when floods bring down a lot of silt, sweeping away existing vegetation, and increasing turbidity, the conditions may nevertheless change completely within a short time. The vegetation of rivers with a very slow current flowing through alluvial soils resembles that of canals and of ponds and small lakes situated in similar soils.

Habitat classification. The provisional classification of fresh-water plant communities adopted in *Types of British Vegetation* was mainly based on the above-mentioned factors, and ran as follows:

Communities of
(a) Foul waters.
(b) Waters rich in mineral salts:
   (i) Nearly stagnant waters.
   (ii) Slowly flowing waters.
(c) Waters poor in mineral salts.
(d) Quickly flowing streams:
   (i) Non-calcareous streams.
   (ii) Calcareous streams.

This classification according to habitat is on the whole natural so far as it goes, and has been followed by some subsequent writers, but it is not quite logically coherent, and is especially defective because it neglects to take proper account of the important "silting factor".

The silting factor. Pearsall (1918) pointed out that rooted water plants had been shown to depend mainly on solutes derived from the substratum rather than on salts dissolved in the water surrounding them, and that water poor in mineral salts may bring down vast quantities of sediment, especially when the streams are in flood. Deposited in a lake in the form of silt this is important as furnishing a favourable habitat for many water plants which do not grow on very coarse or on highly organic substrata. On the other hand most of our waters rich in mineral salts occur in geologically stable areas largely composed of relatively soft Secondary, Tertiary or Quaternary beds overlaid by more or less finely divided soil. Such waters, when in flood, and often normally, will contain a large proportion of material in suspension and in consequence lakes, pools, "broads" and the river beds themselves, if not constantly dredged, tend to become silted up. Thus they do not show the "primitive" features characteristic of "young" lakes, with rocky, stony or gravelly shores and bottoms. These "primitive" lakes occur in mountainous regions of hard rocks, often with steep marginal slopes undergoing relatively rapid surface erosion so far as the hardness of the rocks permits. The substrata are therefore rocky or consist of unstable stones or gravel. Such waters are as a rule themselves poor in dissolved mineral salts, but this is not the decisive factor or even necessarily an effective factor in these primitive aquatic plant habitats. Poverty in the finer inorganic silts and instability of the substratum are likely to be more important.

The characteristic aquatic plants of a primitive lake are therefore likely to be (1) those which can colonise rough and often unstable substrata, (2) those which can do without a copious supply of basic ions, and (3) those which flourish on preponderantly organic soils, since the lack of bases prevents the decomposition of humus, which tends to accumulate, just as it does on land in similar conditions.

Small mountain tarns remain in an extremely primitive condition and support a scanty flora of such specialised plants. The larger and deeper lakes of mountain districts may also remain largely primitive, but where there are important affluents bringing in silt which is laid down near their mouths, drifted by currents, and deposited on different parts of the lake floor, the vegetation becomes more various and includes many plants also found in the silted waters situated on the softer rocks of the lowlands. Nevertheless the lists from typical examples of the two kinds of water show very striking differences.

Life form. In any body of water with a well developed aquatic vegetation there is a great variety of life form among the species present. First of all there are a considerable number of species which are more or less amphibious—either in the sense that they live sometimes on land and sometimes in water, or that they can form leaves of different shape and structure, some adapted to an aerial and others to an aquatic life. The two types of leaf, submerged and aerial, may be borne on one and the same individual plant (as in water-lilies and some of the water crowfoots), or some individuals may lead an entirely submerged existence while others display their leaves in the air (as in other water crowfoots). Of amphibious species some are normally terrestrial but can exist in the water, some are aquatic but can exist on land, and others again inhabit a zone on the edge of a pond or lake which extends on both sides of the water's edge, and are apparently indifferent to the elements surrounding their shoots. The adjustments of form and structure in response to the different medium vary greatly in degree in different species. In some plants the change is comparatively slight, while others have air and water leaves so different in appearance and construction that they would not be suspected of belonging to the same species (Pl. 106, phots. 259–62). Most submerged water plants raise their flowers above the surface (phots. 259–60), but the reproductive processes of a few take place entirely under water.

Zonation of aquatic vegetation. Among water plants proper we can distinguish the three broad categories of "submerged", "floating leaf", and "reedswamp", and this division, correlated on the whole with depth of water and therefore showing a distinct zonation, can be recognised in most lakes, large ponds and slowly flowing rivers.[1] The completely submerged species occupy the deeper waters and descend as far as the decreasing light permits, the species with floating leaves occur in the shallower water near the margin, while the "reedswamp" species fringe the land, though they sometimes extend into water more than a metre in depth and also landwards into the wet soil where the water table is not far below the surface. The "floating leaf" and "reedswamp" plants frequently intermingle, e.g. waterlily and bulrush (Pl. 109, phot. 270), the former finding room (and protection) between the erect shoots of the latter where these are not too closely set ("open reedswamp").

There is a fourth category of species with largely submerged but partly emersed shoots, whose aerial leaves do not float, but rise freely into the air. These are intermediate between the species with floating leaves and the reedswamp plants, whose leaves are entirely or almost entirely aerial.

Besides these life forms there are the plants which are not rooted in the substratum but float freely in the water, and the vegetative shoots of these may be completely submerged or their leaves may float on the surface. The completely submerged forms include very many of the algae and a few flowering plants (*Ceratophyllum, Utricularia,*[2] *Lemna trisulca*), those with floating leaves including the other duckweeds (*L. minor, L. gibba*, etc., *Wolffia arrhiza*), the frogbit (*Hydrocharis morsus-ranae*), and the "water fern" *Azolla*.

Stages of the hydrosere. These different life forms characterise successive stages of the hydrosere. Submerged plants are the pioneers, and as the

---

[1] Pl. 107, phot. 266; Pl. 108.   [2] Pl. 106, phot. 259.

PLATE 106

Phot. 259. *Utricularia vulgaris* in aquarium. Free floating. Submerged leafy shoot with emersed inflorescence. *R. H. Yapp.*

Phot. 260. *Sagittaria sagittifolia* in aquarium. Submerged leaves band-shaped, emersed sagittate. *R. H. Yapp.*

Phot. 261. *Sagittaria sagittifolia*, submerged form (*vallisneriifolia*) with band-shaped leaves. River Cam. *J. Massart.*

Phot. 262. *Nuphar luteum, Potamogeton natans* (floating leaves). Emersed inflorescence of *Utricularia*, and shoots of *Scirpus lacustris*. Broad, Norfolk. *J. Massart.*

SUBMERGED, EMERSED AND FLOATING LEAVES

PLATE 107

Phot. 263. *Nymphaea alba.* *R. H. Yapp.*

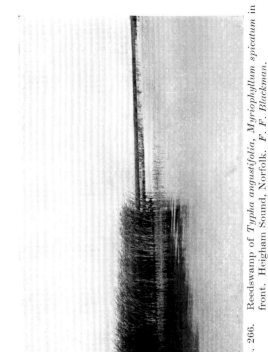

Phot. 264. *Nuphar luteum.* *R. H. Yapp.*

Phot. 265. *Nymphoides (Limnanthemum) peltatum* (floating leaf), *Oenanthe fluviatilis, Alisma plantago-aquatica, Sparganium erectum* (emersed). River Thames, Medley, Oxford. *A. H. Church.*

Phot. 266. Reedswamp of *Typha angustifolia, Myriophyllum spicatum* in front. Heigham Sound, Norfolk. *F. F. Blackman.*

soil level is gradually raised towards the water surface by the accumulation of organic debris resulting from the death of individual plants, or by inorganic silting, or by both together, they are succeeded by the plants with floating leaves, and these, in their turn, by the reedswamp dominants,[1] till the soil level reaches the water level and aquatic vegetation gives place to fen or marsh. But the particular communities entering into the sere depend upon a variety of conditions, such as the nature of the original substratum, the amount and nature of silting, exposure to current or wave action, and so on. As the water shallows it becomes quieter and warmer, increased protection is afforded and the conditions are more uniform, so that the floating leaf and reed swamp dominants are less various than those of the submerged communities.

Summary of life form groups. These life form groups may be summarised as follows:

(1) Wholly submerged plants:
    (*a*) Rooted or fixed.
    (*b*) Free floating (still waters).

(2) Plants with floating leaves (relatively still waters):
    (*a*) Rooted.
    (*b*) Free floating (still waters).

(3) Plants with partly emersed shoots.

(4) Plants (mostly tall) with functional leaves mainly or entirely emersed (reedswamp type).

Forms of submerged leaves. The submerged leaves of water plants are either entire and very thin, or "dissected", i.e. divided into filiform branches. All the living cells of the leaf are thus within a very short distance of the supply of gases dissolved in the surrounding water. Entire submerged leaves are frequently strap-shaped,[2] especially when growing in a river, and both the strap-shaped and the dissected type offer the least possible resistance to the water, the leaves streaming out with the flow of the current. This can be easily seen by looking over the parapet of a bridge at the submerged plants growing on the bed of a clear stream.

The aquatic plant communities of the British Isles have never been summarised, and it is still scarcely possible to write a general account of them. Several good descriptions of the vegetation of lakes and rivers have

---

[1] The plants with largely emersed shoots, such as the water plantain (*Alisma plantago-aquatica*, Pl. 107, phot. 265) and the arrowhead (*Sagittaria*, Pl. 106, phot. 260), mostly occur in silted waters just beyond or replacing the reedswamp, i.e. in fairly shallow water. Free floating vegetation, so far as vascular plants are concerned, is confined to waters rich in mineral salts, and a rich and varied plankton depends on the same conditions.

[2] Pl. 106, phots. 260, 261.

however appeared since the publication of *Types of British Vegetation*, and a selection from these will serve to illustrate the modes of occurrence of many of the communities and the general course of the hydroseres.

Examples illustrating the different habitats will be taken in the following order:—Ponds on the softer rocks; White Moss Loch; the Norfolk Broads; the Cumbrian Lakes; the most primitive lakes and tarns.

PONDS ON THE SOFTER ROCKS

Foul waters. Small stagnant ponds containing an excess of decaying organic matter, for example such as are constantly fouled by sewage or by the excrement of cattle, and also densely shaded pools overhung by thick trees or shrubs so that the bottom is thickly covered with decaying leaves and illumination is very low, may be entirely destitute of green plants. They are inhabited by anaerobic bacteria, saprophytic flagellata, and to some extent by blue-green algae, but the free oxygen necessary for respiration is present only in minimal quantities, and the substances produced by the anaerobic forms probably also inhibit the growth of algae or higher plants. The vegetation of such waters has not been studied ecologically in this country.

Lowland ponds. Ponds situated on strata relatively rich in mineral salts, especially if they receive rainwash from cultivated land, often become rapidly choked with vegetation unless they are constantly cleaned out. Flowering plants make up the bulk of the vegetation, and there are a few species of algae: bryophytes are generally absent, but two species of simple thalloid liverworts may occur—*Riccia fluitans* and *Ricciocarpus natans*. Both have a floating and also a terrestrial form, the former without rhizoids. The floating form of *Ricciocarpus natans* has long pendant scales on the ventral surface, and these give the floating thallus stability as well as increasing its photosynthetic and absorptive surface. The terrestrial forms bear rhizoids which penetrate the mud at the water's edge on which the liverwort grows. If the water level rises and the thallus is submerged, the plants continue to vegetate. Certain species of *Hypnum* sometimes occur in the same situations.

Trent valley ponds. A set of seven ponds described by Godwin (1923) in the alluvial gravels of the Trent valley near Trent Junction in southeast Derbyshire, including an oxbow cut off from the river Trent many years ago, possessed between them the following 32 species belonging to the aquatic communities proper.

Submerged and partially emersed vegetation

Alisma plantago-aquatica	Glyceria fluitans
Apium inundatum	Hottonia palustris
Butomus umbellatus	Lemna minor
Callitriche verna	L. trisulca
Elodea canadensis	Myosotis scorpioides (palustris)

Myriophyllum spicatum
Nasturtium officinale
Nuphar luteum
Oenanthe fistulosa
Oe. phellandrium

Polygonum amphibium
Potamogeton acutifolius
P. natans
Ranunculus circinatus

Reedswamp

Calamagrostis epigejos
Carex vulpina
Eleocharis palustris
Epilobium hirsutum
Equisetum fluviatile (limosum)
Glyceria maxima (aquatica)
Rumex hydrolapathum

Scirpus lacustris
Sium latifolium
Sparganium erectum
Thalictrum flavum
Typha angustifolia
T. latifolia

Besides the water plants proper, a number of marsh plants and other species of damp soil occurred round the edges of the ponds.

With the exception of the oxbow all these ponds were artificial, having been dug for ballast or other purposes at various dates during the nineteenth century. They afford therefore a fair sample, though by no means an exhaustive list, of the aquatic plants naturally colonising such habitats. Godwin points out that the most recently dug ponds were the poorest in species, and that most of the species occurred in one or more of the ponds, but not in all. Only one species indeed, *Alisma plantago-aquatica*, was found in all seven. Since the conditions were apparently fairly uniform in all the ponds this illustrates the largely chance distribution of the species constituting the actual flora of a small pond.

In some of Godwin's ponds zonation was very well shown. Thus in "Fletcher's Pond" there were three well-marked zones passing inshore from the open water: the first dominated by *Potamogeton natans*, the second by the yellow water-lily (*Nuphar luteum*), both rooted plants with floating leaves; the third by the common tall reedswamp dominant *Scirpus lacustris*, behind which, in shallow water, came a zone of the much lower growing *Eleocharis palustris*. In another pond with a considerable strip of very shallow water at its edge the reedswamp was represented by the horsetail *Equisetum fluviatile* (*limosum*), with *Eleocharis palustris* as an understorey and extending a little beyond the outer edge of the horsetail; *Scirpus lacustris* was here absent. These lower growing reed-swamp elements occur only in shallow water or in marsh, where the water level (at least in summer) is frequently below the soil surface.

Bramhope ponds. With these Trent valley ponds one may compare a set of eight ponds described by Norman Walker (1905), situated above the Bramhope railway tunnel about half-a-mile south of Bramhope, near Leeds in Yorkshire. These were small ponds lying close together, and owed their origin to surface drainage into excavations made in 1839 when the tunnel was built. Their floors sloped gently from the edge of the water to a maximum depth of 4 ft., and none had either inlet or outlet.

The total number of species of flowering plants was much smaller than in the Trent ponds, corresponding with the smaller area involved, and were largely different species. *Potamogeton natans* (which occurred in all the Bramhope ponds and in all but one of the Trent ponds), *Eleocharis palustris* and *Sparganium erectum* were however dominants of aquatic and reed-swamp communities in both sets.

Aquatic communities

Juncus bulbosus (clayey ponds)
Myriophyllum spicatum (few and dwarf)
Potamogeton natans (dominant in all ponds)
P. alpinus (clear water above shallow mud only)
Ranunculus sp. (sect. Batrachium)
Glyceria fluitans (almost alone in small clayey ponds)

Reedswamp

Eleocharis palustris	d	Oenanthe fistulosa	d
Juncus effusus	d	Sparganium erectum	d
Carex goodenowii	ld	Juncus articulatus	o

In the Juncetum effusi were also:

Galium palustre	Peplis portula
Glyceria fluitans	Ranunculus flammula
Myosotis caespitosa	Senecio aquaticus

Juncus effusus is not a typical reedswamp plant, but occurs (mainly on waterlogged soil rather than in water) in a great variety of situations, very abundantly in low-lying clay lands where the water table is close to the surface. In the case of the Bramhope ponds it throve best on the banks of the pond outside the water line, but extended into the water to a depth of 6 in. In the pond shown in Fig. 104 *Oenanthe fistulosa* began among the *Juncus* at a depth of 3 in., the slender stolons burrowing through the soft mud and stopping at the stiff clay soil. At a depth of about 6 in. the *Juncus* and *Oenanthe* gave way to *Eleocharis palustris* and *Sparganium erectum*. The last-named extended to a depth of more than 18 in., and gave way in its turn to *Potamogeton natans*, which occupied the centre of the pond.

It is interesting to note that this set of small ponds possessed together about the same number of species as the much larger Fletcher's Pond in the Trent valley which was made about the same time (1839 and 1836).

Hagley Pool. With the two sets of north midland ponds just described we may compare the vegetation of Hagley Pool near Oxford, an old detached backwater belonging to the network of waterways of the Thames system where it passes through the alluvial plain based on Oxford Clay.

Here development is on a larger scale and the vegetation much older. Correspondingly the reedswamp, which is dominated by *Sparganium erectum* (Pl. 107, phot. 265; Pl. 108), is considerably more varied.

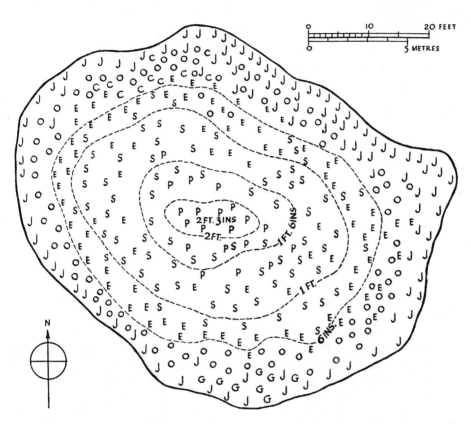

J = *Juncus effusus*
G = *Glyceria fluitans*
o = *Œnanthe fistulosa*
C = *Carex goodenowii*
E = *Eleocharis palustris*
S = *Sparganium erectum*
P = *Potamogeton natans*

FIG. 104. A SMALL POND AT BRAMHOPE, NEAR LEEDS

Juncus effusus is the main dominant of the marginal reedswamp also extending on to the bank. *Sparganium erectum*, with *Eleocharis palustris*, is dominant in deeper water, and *Potamogeton natans* forms a well marked community in the centre of the pond. After N. Walker, 1905.

Submerged community

Myriophyllum spicatum	Ranunculus circinatus
Oenanthe fluviatilis	Scirpus lacustris (submerged form)
Potamogeton lucens	
P. perfoliatus	Chara sp.

Floating leaf community

Hydrocharis morsus-ranae	Potamogeton natans
Lemna minor	Ranunculus peltatus
Nuphar luteum	

Reedswamp

Sparganium erectum	d	Iris pseudacorus	f
Carex riparia	a	Lythrum salicaria	f
Equisetum fluviatile (limosum)	a	Rorippa (Nasturtium) palustre	f
Glyceria maxima (aquatica)		Rumex hydrolapathum	f
Oenanthe fistulosa	a	Sium latifolium	f
Menyanthes trifoliata	la	Stachys palustris	f
Ranunculus lingua	la	Stellaria glauca	f
Carex acutiformis	f	Carex vesicaria	o
Galium uliginosum	f	Sium erectum	o

WHITE MOSS LOCH

A brief account may now be given of the vegetation of a very much larger sheet of water, the White Moss Loch (or simply "the White Moss") in Perthshire (Figs. 105–6) described by J. R. Matthews (1914). This is situated on the Old Red Sandstone in a natural depression to the north of the Ochil Hills in Fife and 175 ft. (*c.* 53 m.) above sea-level. The area of water was recorded by the Ordnance Survey in 1901 as 16 acres (*c.* 6·5 ha.), but in 1913 it did not exceed 10 acres (*c.* 4 ha.), so that the process of centripetal encroachment of the swamp and marsh vegetation was quite rapid. The greatest diameter was about 540 yards (*c.* 0·5 km.). The loch has no natural inflow or outflow stream, but ditches conduct into and out of it a considerable quantity of water during the winter months. There was nowhere more than a metre's depth of water in summer, though in winter the loch might be nearly twice as deep. In most winters the duration of ice does not exceed a fortnight, though in 1910 (a cold late winter all over the country) the loch was ice-bound for six weeks. Mud consisting of fine sand mixed with much organic material covered the bottom to a depth of at least two metres.

Two submerged communities were distinguished and two main consocies in the reedswamp.

Submerged communities. Elodetum occupied the centre of the loch where the water was more than 3 ft. deep and contained the following species:

Elodea canadensis	d	Potamogeton filiformis	f
Nitella translucens	f	P. perfoliatus	o

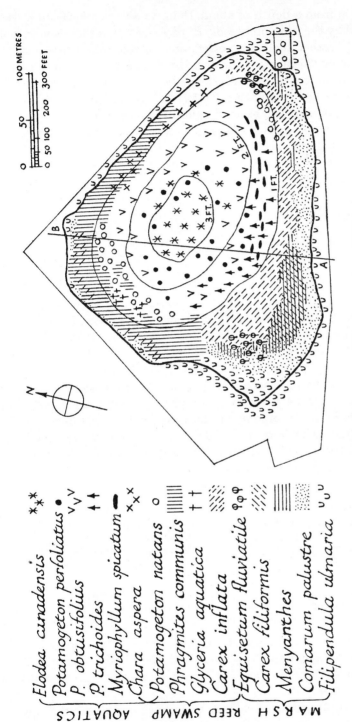

FIG. 105. WHITE MOSS LOCH (PERTHSHIRE) IN 1913

Willows and alders were planted about 1885 up to the sinuous line which was then close to the edge of the water. Zonation of marsh, reedswamp and aquatic vegetation is well shown. The reedswamp is mainly dominated by *Phragmites communis* (north) and *Carex inflata* (south). *Potamogeton natans* grows in about the same depth of water (30–60 cm.) as in the Bramhope Pond (Fig. 104). The deeper water in the centre of the loch is dominated by submerged species of *Potamogeton* and by *Elodea*. AB = transect line (Fig. 106). After Matthews, 1914.

AQUATICS {
Elodea canadensis ✱✱
Potamogeton perfoliatus ●
P. obtusifolius ᵛᵛ
P. trichoides ↟↟
Myriophyllum spicatum ━
Chara aspera ×ˣ

REED SWAMP {
Potamogeton natans ○
Phragmites communis ▥
Glyceria aquatica †✝
Carex inflata ⧄

MARSH {
Equisetum fluviatile φφ
Carex filiformis ⧄
Menyanthes ▥
Comarum palustre ⸳⸳⸳
Filipendula ulmaria ᵘᵘ

The next zone, from a depth of about 18 in. to 3 ft. (*c*. 45–90 cm.), was largely occupied by the two pondweeds, *P. perfoliatus* in rather deeper, and *P. obtusifolius* in rather shallower water; and outside this, bordering the reedswamp in about 30–40 cm. of water, three local societies were represented. Towards the south side *Myriophyllum spicatum* was dominant, with an understorey of *Potamogeton perfoliatus*, *P. trichoides* and *Nitella opaca*: along the east side and extending into the reedswamp the floor was carpeted with *Chara aspera* var. *inermis*; and on the north *Potamogeton natans* was dominant, with species of *Nitella* as undergrowth. Since species of pondweed were the predominant plants in the zone of water from 1–3 ft. deep, the community (Matthews's "Shallow water association") may fairly be called a *Potamogeton* associes. The list of species was as follows:

Potamogeton natans	ld	M. alterniflorum	f
P. perfoliatus	a	Chara aspera var.	
P. trichoides	a	subinermis	ld
P. obtusifolius	a	Nitella opaca	f
P. gramineus	f	N. flexilis	f
Myriophyllum spicatum	ld		

Reedswamp. In 1913 reedswamp dominated by *Phragmites* on the northern, and by *Carex inflata* on the southern side surrounded the loch, except for a space of about 80 yards at the eastern extremity which was probably kept clear by wave action. Since then however (1936) *Phragmites* has filled up this gap, joining the *Carex* on the southern side. There were and are no reeds on the southern side, but a little of the sedge associated with *Phragmites* on the northern. The contour line of 1 ft. (30 cm.) depth of water ran near the outer margin of the reedswamp, sometimes a little inside it, sometimes a little outside. The reeds on the northern side have also advanced towards the centre since 1913, so that the two reed "islands" on the north-west are no longer distinguishable.

Phragmitetum

Phragmites communis	d	Equisetum fluviatile	
Glyceria maxima (aquatica)	ld	(limosum)	la
Carex inflata	a	Carex flava	o
Littorella uniflora	a	Echinodorus ranunculoides	o

Caricetum inflatae

Carex inflata	d	Sparganium erectum	ld
Equisetum fluviatile		Potamogeton natans	l
(limosum)	lsd		

The reed persisted for a time, as it commonly does, after the level of the soil had been raised above the water, but both the Phragmitetum and the Caricetum were succeeded on the outer margin by the lower growing plants of the circumjacent marsh.

In all there were 21 species of charads and vascular plants in the aquatic communities and reedswamp of White Moss Loch, a substantially smaller

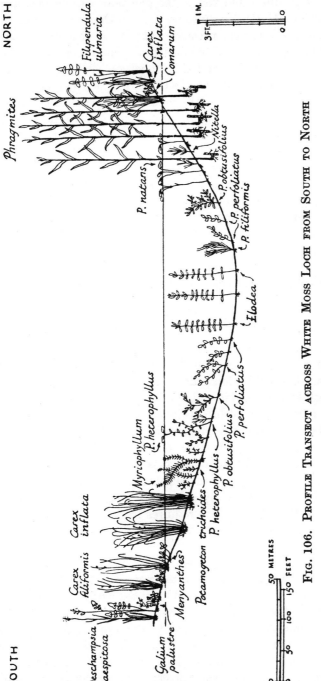

FIG. 106. PROFILE TRANSECT ACROSS WHITE MOSS LOCH FROM SOUTH TO NORTH

AB in Fig. 105. After Matthews, 1914.

number than in the Trent ponds. Conspicuous absentees were the reed-maces, the water-lilies and the water plantain, clearly indicating lack of inorganic silting. Matthews gives no data on the composition of the water or of the soil, but one may presume that these, situated on the Old Red Sandstone, were poorer in soluble basic salts than those of the Trent valley or of the Bramhope ponds, and this, together with the absence of silting, probably accounts for the difference (for example, the presence of a species like *Littorella uniflora*) and for the greater poverty of the flora.

Besides the advance of *Phragmites* referred to above the principal change between 1913 and 1936 is the colonisation (mainly of the *Carex inflata* swamp) along the south side of the loch, and also to the east and west, by young willows and alders derived from trees planted about 1885 between the fence and the black sinuous line in Fig. 105, which marked the inner limit of the plantation. Some of the young self-sown trees have now established themselves almost up to the level of the 1 ft. depth contour line as it was in 1913 (Matthews *in litt.*). Advance of the vegetation has evidently been much less rapid than at the beginning of the century.

We may now pass to larger areas of freshwater, where the ecological factors affecting aquatic vegetation are more various and operate on a bigger scale.

THE NORFOLK BROADS

These are shallow lakes situated in the upper parts of the old estuarine area of East Norfolk now entirely filled with alluvium. Most of the existing broads are surrounded by fen peat, and lie in the north of the region round the courses of the Bure and its tributaries the Ant and the Thurne (Fig. 107). There are a few remains of broads in the Yare valley (Fig. 108). The lower courses of the rivers run through mineral alluvium, the region repeating the geographical and geological features of the "Fenland", south of the Wash, on a smaller scale. Unlike the latter it is, however, only partially drained, and some of it still bears natural vegetation. There is a good preliminary description by Pallis of the aquatic and fen vegetation in *Types of British Vegetation* (1911), but the region requires fresh study on modern lines, and the aquatic vegetation can only be cursorily dealt with here.

The broads are not expansions of the rivers themselves but lie to one side of the stream beds, enclosed by fen peat which has formed around them. The process of "Verlandung" is often very rapid, so that several broads have greatly shrunk or even been obliterated within comparatively recent years (Fig. 108). The water of the broads and rivers has a pH value of more than 7, since, like that of the Fenland, it comes mainly from the chalk.

Pallis (1911) pointed out considerable differences between the aquatic vegetation of the broads of the various river valleys, differences which are much less marked between the corresponding reedswamps and less still between the fens. Thus in some of the broads of the Bure valley which are filled up close to the surface with organic mud, and have little or no free communication with the rivers, the aquatic vegetation proper is extremely poor or altogether absent: in those connected with the Ant, Thurne and Yare, on the other hand, where the circulation of water is much freer,

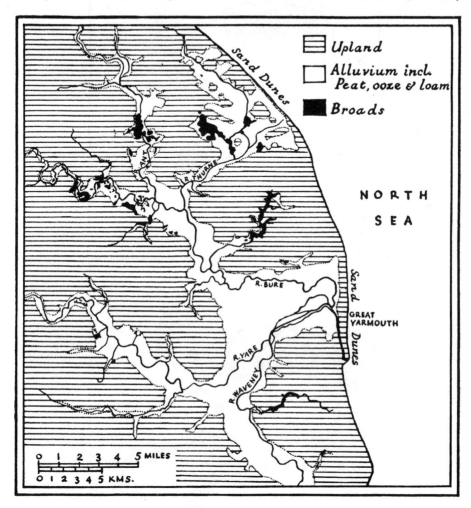

FIG. 107. THE FENLANDS OF EAST NORFOLK

The broads, shown in black, are confined to the upper parts of the old estuarine area where peat is formed. From Pallis, 1911.

aquatic vegetation is well developed, with *Potamogeton pectinatus* forma *interruptus* commonly dominant in the Thurne and Yare broads, and *Stratiotes aloides* in those of the Ant. Of abundant associated plants *Myriophyllum spicatum* and *Ranunculus circinatus* occur in the broads of the Ant and Thurne, *M. verticillatum* and *Ceratophyllum demersum* in those of the Ant and Yare. In the Thurne broads a series of charads and algae

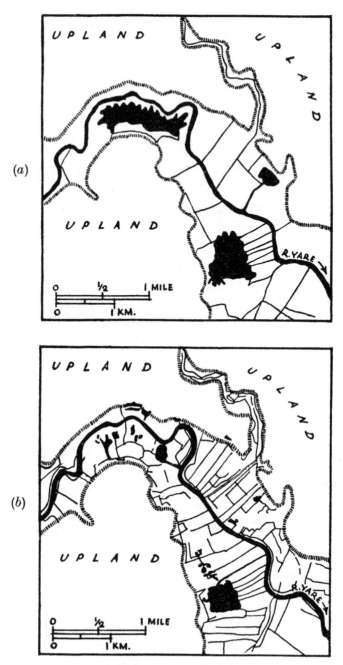

FIG. 108. SHRINKAGE OF BROADS

Part of the Yare valley, Norfolk. The broads (black) are entirely surrounded by peat.
The straight lines are dykes cut in the peat. (a) As shown in the Ordnance Survey of
1816–21. (b) The same area as surveyed between 1879 and 1886. The peat-forming plants
have encroached on the Broads and greatly reduced the areas of open water. Note the increased number of dykes.

PLATE 108

Phot. 267. Floating clump of *Glyceria maxima*: reedswamp and carr behind. Rockland Broad. *J. Massart.*

Phot. 268. Island of *Phragmites communis*: edge of reedswamp on left. Rockland Broad. *J. Massart.*

Phot. 269. Open reedswamp of *Scirpus lacustris* with under-storey of *Nymphaea alba*. Sutton Broad. *J. Massart.*

Phot. 270. Reedswamp of *Typha angustifolia, Sium erectum, Cicuta virosa, Epilobium hirsutum,* etc. *J. Massart.*

REEDSWAMP IN THE NORFOLK BROADS

may be dominant, while *Hippuris, Stratiotes,* and the very rare *Naias marina* occur. The duckweeds and water-lilies, the frog-bit and various pondweeds occur mainly in the broads of the Yare and Ant, in the latter associated with open reedswamp (Pl. 108, phot. 269).

The open reedswamps fringing the broads of all four rivers are dominated by *Typha angustifolia*[1] and *Phragmites communis*, but in the closed reedswamps of the Yare these species are replaced by *Glyceria maxima (aquatica)*[2] and *Phalaris arundinacea*. Some of these differences are probably correlated with the stronger current in the Yare which brings down more inorganic silt, favouring such plants as *Nymphaea*, and in the fens *Glyceria maxima* and *Phalaris arundinacea* (cf. Chapter XXXII, p. 643 and Chapter XXXIII, p. 653). In the Ant broads there is also a very wide open reedswamp dominated by *Scirpus lacustris*,[3] with which are associated the water-lilies, duckweeds and several pondweeds, including *Potamogeton natans*, recalling the Trent pond vegetation. In some of the Bure and Yare broads, too, there is a reedswamp composed of the great tussock-forming sedges, *Carex paniculata* and *C. acutiformis* (Pl. 88, phots. 214, 215).

REFERENCES

ARBER, AGNES. *Water plants: a study of aquatic Angiosperms.* Cambridge, 1920.

GODWIN, H. Dispersal of pond floras. *J. Ecol.* 11, 160–4. 1923.

MATTHEWS, J. R. The White Moss Loch: a study in biotic succession. *New Phytol.* 13, 134–48. 1914.

PALLIS, MARIETTA. "The River Valleys of East Norfolk." Chapter x. In *Types of British Vegetation*, pp. 214–45. 1911.

PEARSALL, W. H. On the classification of aquatic plant communities. *J. Ecol.* 6, 75–83. 1918.

Types of British Vegetation. Chapter VII, Aquatic Vegetation, pp. 187–96. 1911.

WALKER, NORMAN. Pond vegetation. *The Naturalist.* No. 585, pp. 305–11. October, 1905.

WATSON, W. The bryophytes and lichens of fresh water. *J. Ecol.* 7, 71–83. 1919.

[1] Pl. 107, phot. 266; Pl. 108, phot. 270. [2] Pl. 108, phot. 267.
[3] Pl. 108, phot. 269.

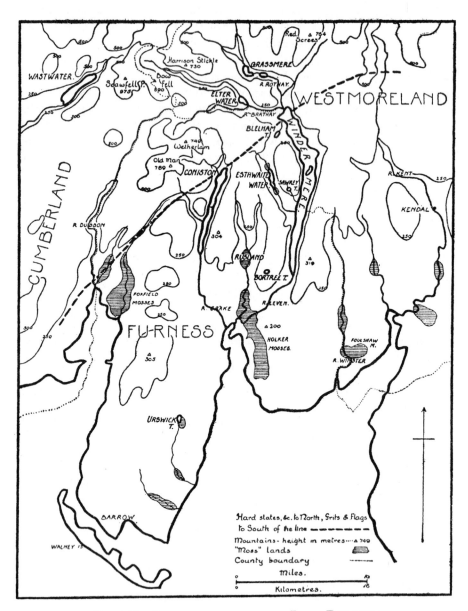

FIG. 109. SOUTHERN HALF OF THE LAKE DISTRICT

Showing the positions of Esthwaite Water and other lakes: also of the mosses (raised bogs) on the lower courses of the rivers (see p. 686). From Pearsall, 1917.

Chapter XXX

THE CUMBRIAN LAKES

Pearsall's very careful and thorough surveys of Esthwaite Water (1917–18) and of other lakes (1920) in Cumberland and Westmorland, i.e. in the region commonly known as the English Lake District (Fig. 109), provide a wide range of data relating to the conditions of life of aquatic, marsh, fen and bog vegetation in that region and by far the fullest account of the hydrosere that is available for any British waters.

In this chapter the aquatic communities of Esthwaite Water will first be briefly described with their characteristic habitats: then a general account of the aquatic vegetation of the lakes will be given, preceded by a notice of factors such as light and temperature and followed by a description of the observed successions.

Esthwaite Water. Esthwaite Water (Fig. 110) has only a quarter of the total residue of solid material found in the waters of the Norfolk broads (0·07 against 0·28 in parts per thousand), and this is distinctly higher than that of other lakes of the region (0·04–0·06). The "hardness" of the Esthwaite water is less than one-seventh of that of the Broads water (3·25 against 23·68). The organic residue of Esthwaite is relatively high and the water brown from peaty substances in solution. This is due to affluents coming from areas of acid peat and is correlated with relatively high acidity of the water.

In regard to the physiography of the aquatic habitats there is a marked difference between the eastern (lee) and the western (windward) shore, the former being far more subject to wave action, which cuts a sloping terrace, mainly below the water line (Fig. 111, AB), and continues it outwards into the lake (BC) by the accumulation of material washed back by the undertow of the waves. This gently graded terrace ends in a much steeper slope (CE). The terrace consists only of the coarser detritus (BCE), the finer sediments being carried down its steep slope and deposited farther out on the floor of the lake (EH). Besides this direct wave action a current is set up along the shore by the oblique incidence of the waves, and this scours the shallows clean and deposits silt in deeper bays. The affluents also bring quantities of silt into the lake and lay it down round their mouths. The kinds and degrees of silting are thus very various and they are of first importance in determining the vegetation.

Most of the leading aquatic communities of Esthwaite Water are well illustrated at Strickland Ees and the adjoining Fold Yeat Bay (Fig. 112).

Deep-water communities. (1) Of the entirely submerged species the charad *Nitella flexilis* (Figs. 112, 114) penetrates to the greatest depths, typically ranging from 1·8–3·6 m., with a light intensity of 0·05–0·102

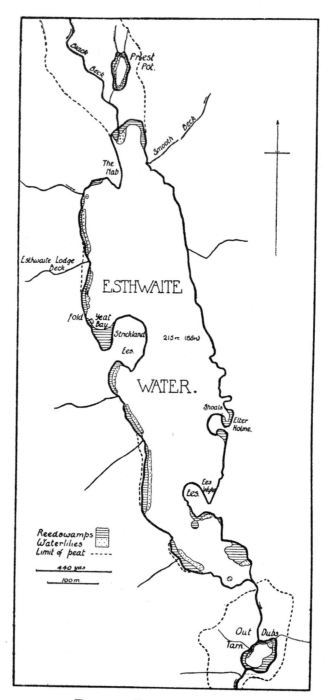

FIG. 110. ESTHWAITE WATER

Showing the positions of reedswamps and water lilies and the limits of peat formation: also
Strickland Ees (Fig. 112) and North Fen (Figs. 114, 115). From Pearsall, 1917.

(dysphotic), on blue-grey clayey inorganic mud with an average organic content of 16·9 per cent.

(2) A similar consocies is formed by the moss *Fontinalis antipyretica* (Figs. 112, 114), typically pure, but with *Elodea canadensis* and *Potamogeton obtusifolius* occasionally associated, and higher average organic content (18·9 per cent) in its substratum.

(3) *Sparganium minimum* forms another consocies (Figs. 112, 113, 114) on loose, grey or yellow-grey mud which is still more organic (22·4 per cent) outside the *Nymphaea alba* community ((7), p. 603) and at a depth of 2·4–3 m. (occasionally at 1·2–1·8 m.), frequently with the same two species that accompany *Fontinalis*.

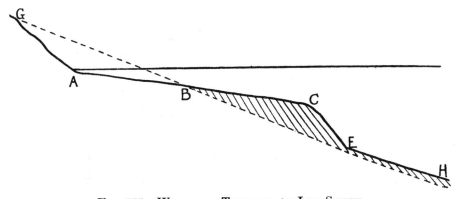

FIG. 111. WAVE-CUT TERRACE ON LEE SHORE

For description see text, p. 597. From Pearsall, 1917.

(4) A characteristic collection of species, which Pearsall calls the "linear-leaved associes" (Fig. 112), develops best at depths of 1·5–2·6 m. (light intensity 0·06–0·03—dysphotic) on blue-green clayey mud, like that of the *Nitella flexilis* community, but always with less than 15 per cent of organic matter (av. 13·7 per cent). This community shows two well-marked consocies dominated respectively by *Potamogeton pusillus* and *Naias flexilis* (average percentage of organic matter 7·3). The complete list of species is as follows:

Potamogeton pusillus	la	Callitriche autumnalis	la
P. pusillus subsp. lacustris	ld	Elodea canadensis	f
P. panormitanus	ld	Hydrilla verticillata var.	
P. obtusifolius	l	pomeranica	f
P. crispus var. serratus	o	Myriophyllum alterni-	
P. perfoliatus	r	florum	l
P. crispus	vr	Nitella flexilis	l
Naias flexilis	ld		

Pearsall remarks on the notable uniformity in life form of all the characteristic and more abundant species. "All have the pellucid linear leaves and delicate stems of the *Potamogeton pusillus* type. The elongate

subspecies *lacustris* has this character still better marked: the variety
serratus of *P. crispus* is bright green and translucent: *Hydrilla verticillata* is
represented by one of its slenderest varieties; and the peculiar elongate
form of *Elodea canadensis* also closely approximates to the type form of the
community. The uniform light conditions under which the associes develops have probably a direct influence on its growth form, but the constant character of the substratum is also a marked feature of the habitat.
The fragile growth form and the need for inorganic silt both indicate that
shelter is of great importance in the development of this associes. Exposure
to water movements would injure the plants and prevent the deposition of
silt."

Naias flexilis is the only member which fruits abundantly and is dependent on seed for dispersal. The other characteristic species rely mainly
on vegetative means of reproduction. The multitude of seeds produced by
Naias probably enables it to colonise newly deposited inorganic silts before
its associates.

(5) Of quite different life form—the rosette type—is the consocies of
Isoetes lacustris (Figs. 118, 119), developed on a "primitive" substratum of
rounded stones at depths of 1·5–2·7 m. The light intensity here varies from
0·03 to 0·06 (dysphotic). This consocies occurs only at the extreme southern
end of the lake, where no fluvial silt is left to be deposited, since the main
current in the lake is from north to south.

Along the foot of the steep slope bounding the wave-formed terrace
described above, where fine silt is deposited, a consocies of *Myriophyllum
alterniflorum* forms a narrow zone chiefly along the exposed eastern shore at
a depth of 1·25–1·5 m. (Fig. 112). This is the most favourable habitat on the
exposed shore, for below are bare stones devoid of silt, and shorewards is
the wave-swept gravel of the terrace. The well-developed root system of
Myriophyllum anchors the plant firmly and the finely cut leaves create the
minimum resistance to wave-wash. The water milfoil is usually alone, but
Potamogeton heterophyllus var. *longipedunculatus* may accompany it on the
exposed eastern bank, developing a thick carpet of short vegetative shoots
close to the substratum; and on the sheltered western shore *Myriophyllum*
is accompanied by *P. alpinus*.

Shallow-water communities. (6) Another well-marked community of the
rosette type of life form—a community found in relatively shallow water
round the shores of most lakes among the older rocks—is the *Littorella-
Lobelia* associes (Figs. 112, 114, 115), consisting only of the two species
Littorella uniflora and *Lobelia dortmanna*. This usually occurs at depths of
0·3–1·2 m. (euphotic), but occasionally to 2 m., on gravelly substrata,
sometimes overlaid by 3 or 4 cm. of black peaty mud, but never by
inorganic silt. At such depths there is vigorous wave action which keeps
the substratum unstable and prevents the deposition of silt. The mobile
gravel is successfully colonised by families of *Littorella* (single plants of

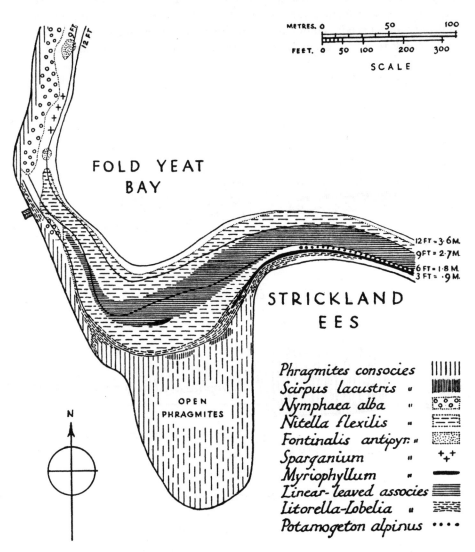

Legend:

Phragmites consocies	⦀⦀⦀
Scirpus lacustris "	▓▓▓
Nymphaea alba "	⦾⦾⦾
Nitella flexilis "	▭▭▭
Fontinalis antipyr. "	░░░
Sparganium "	+₊+
Myriophyllum "	▬▬▬
Linear-leaved associes	≡≡≡
Litorella-Lobelia "	▤▤▤
Potamogeton alpinus	••••

FIG. 112. STRICKLAND EES AND FOLD YEAT BAY, ESTHWAITE WATER

The greater part of the area of submerged vegetation is occupied by the *Nitella* consocies, which extends in places to a depth of 3·6 m. and the "linear-leaved associes", with *Sparganium minimum* and *Fontinalis* to the north. Next above come *Myriophyllum alterniflorum* and the *Littorella-Lobelia* associes, with *Nymphaea alba* on the terrace to the north (cf. Fig. 113). Behind is the Phragmitetum, occupying a large area in the shallow southward extension of Fold Yeat Bay, with patches of Scirpetum lacustris on its outer limit. From Pearsall, 1917.

which are uprooted by ducks and washed on to exposed shores), and the long-rooting shoots enable a single plant, once established, to extend vegetatively and form a family. *Lobelia*, with no vegetative reproduction but only small seeds, comes in when the *Littorella* colonies have advanced sufficiently to arrest seeds and silt among their leaves and roots, and the

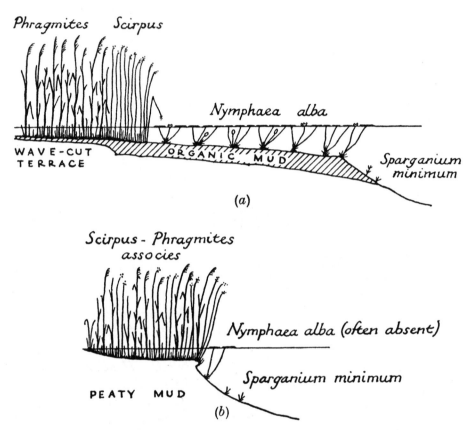

FIG. 113. PROFILE SECTIONS OF FOLD YEAT BAY

(*a*) *Phragmites* and *Scirpus* on the upper part of the wave cut terrace with *Nymphaea alba* below. *Sparganium minimum* on the steep slope to the floor of the lake.

(*b*) Terrace occupied entirely by reedswamp, which has almost entirely replaced *Nymphaea*. From Pearsall, 1917.

final sward of *Littorella-Lobelia* completely stabilises the gravel, in the interstices of which silt and plant remains are increasingly caught.

The flowers of *Littorella* are borne in scapes shorter than the leaves and are only produced when the plant is left above water: the inflorescences of *Lobelia*, borne on long scapes, are loose, few-flowered racemes of pale lilac flowers rising above the surface of the water.

Floating leaf communities. (7) *Nymphaeetum albae* (Figs. 112–115). The white water-lily consocies are the only floating leaf communities in this lake, and of these the one dominated by the common white water-lily is much the most widely distributed. It is characteristic of the western shore, where it can obtain the shelter that all floating leaf communities require, and may extend to an extreme depth of 2·5 m. The organic content of the substratum averages 23·5 per cent. Water-lilies are said to need abundant nitrates, but while the soils here are rich in nitrogen it appears to be all in the form of ammonia, no nitrates being determinable.

This community contains the following species:

Nymphaea alba	d	Ranunculus peltatus	
Nuphar luteum	lf	(not flowering)	r
(near the mouths of becks)		R. truncatus (not	
Lobelia dortmanna	l	flowering)	r

(8) *Nymphaeetum occidentalis.* This occurs in shallower water to a depth of 1·2 m. on black peaty mud, whose organic content is over 30 per cent, rarely more than 5 cm. thick, and lying on stones. The habitat is in the region of wave action so that no silt can accumulate, bases are deficient, and nitrates presumably in scanty supply. This is in accord with the distribution of the small white water-lily in upland tarns whose waters are acid.

Reedswamp communities. (9) *Phragmites-Scirpus* associes (Figs. 112–15). This is the characteristic reedswamp of Esthwaite, *Scirpus lacustris* being dominant towards the open water (to a depth of 1·3 m.) and *Phragmites communis* towards the land (to a depth of 0·92 m.). Reedswamp only occurs along stable sheltered stretches, not on loose gravel or on the primitive rounded stones of the lake shores. But if the gravel is stabilised, as by *Littorella, Phragmites* usually spreads and its substrata are generally organic, containing from 30 to 60 per cent of humus. Scirpetum often occurs alone in exposed situations, and its smooth elastic shoots probably resist wind and wave action better than those of *Phragmites*. The seeds of *Scirpus* can germinate under water.

(10) *Typhetum latifoliae* is confined to one situation at the mouth of a beck (see Figs. 114, 115), where abundant silt brought down by the stream favours rapid decay of organic matter (as shown by the quantities of marsh gas evolved) and thus probably the free production of nitrates.

(11) *Caricetum* occurs at the southern end of the lake in rather exposed places on peaty mud where silt is scanty. The organic content of the substratum probably always exceeds 60 per cent. The following species occur:

Carex lasiocarpa	ld	Equisetum fluviatile	
C. inflata	ld	(limosum)	f
C. vesicaria	a	Lobelia dortmanna	o
C. inflata × vesicaria	a	Nymphaea occidentalis	o

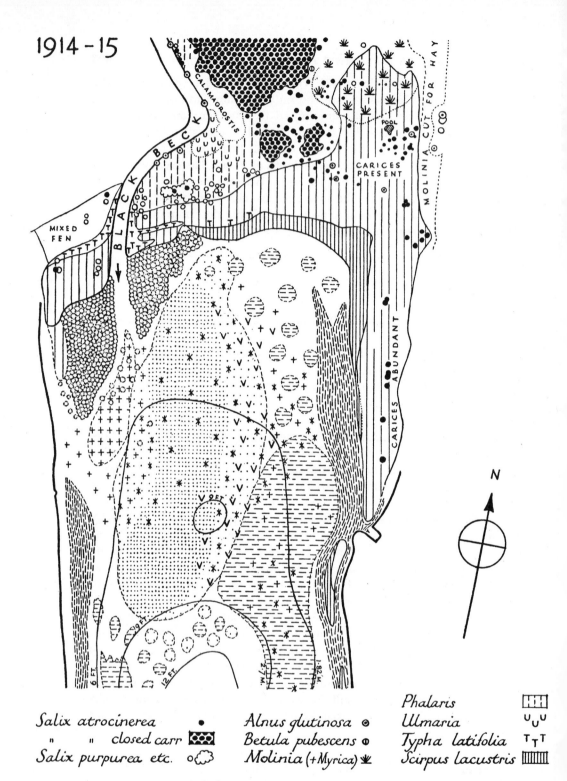

1914-15

MIXED FEN

CALAMAGROSTIS

BLACK BECK

CARICES PRESENT

MOLINIA CUT FOR HAY

POOL

CARICES ABUNDANT

6 FT.
2 FT.
2·7 M.
1·82 M.

N

Salix atrocinerea •
"   "   *closed carr* ▨
Salix purpurea etc. ○☁

Alnus glutinosa ⊚
Betula pubescens ⊕
Molinia (+*Myrica*) ☙

Phalaris ⊞⊞
Ulmaria ᵁᵁ
Typha latifolia ᵀᵀᵀ
Scirpus lacustris ▥

Figs. 114, 115. Communities surrounding the mouth of Black

The aquatic communities include *Nitella flexilis*, *Elodea* and *Sparganium minimum* in the
round the heavily silted mouth of the beck. Next come the reedswamps of Scirpetum and
(The fen communities in the upper parts of the maps are dealt with in Chapter XXXII.)

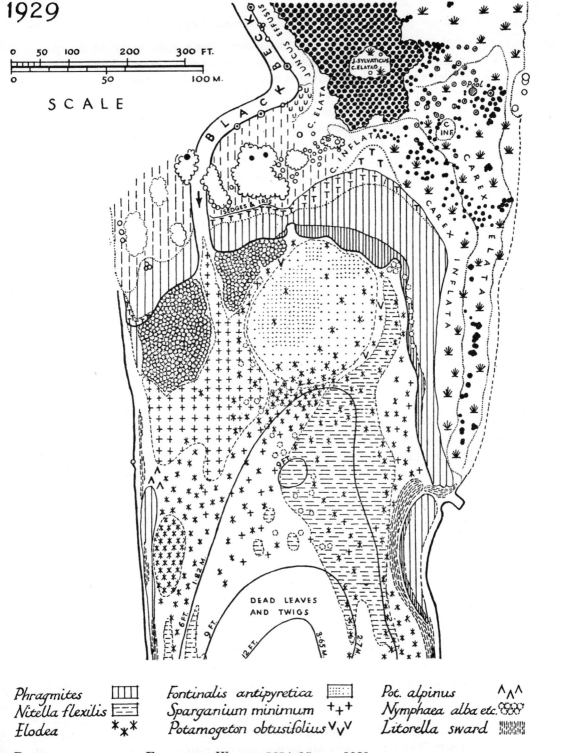

1929

SCALE

| 0 | 50 | 100 | 200 | 300 FT. |

| 0 | 50 | 100 M. |

Phragmites ▦   Fontinalis antipyretica ▨   Pot. alpinus ^^^

Nitella flexilis ▤   Sparganium minimum +++   Nymphaea alba etc. ◌◌◌

Elodea *×*   Potamogeton obtusifolius ᐯᐯᐯ   Litorella sward ▨▨▨

BECK AT THE HEAD OF ESTHWAITE WATER: 1914–15 AND 1929

deepest water, followed by *Fontinalis* and then consocies of *Nymphaea alba* richly developed Phragmitetum with Typhetum extending outwards from the mouth of the beck. See p. 606. From unpublished field maps kindly lent by Dr Pearsall.

Variations in habitat and succession. The aquatic and reedswamp communities have been described in the order of the average depths of water in which they grow: but it must not be supposed that the successions which occur necessarily follow that order. Besides the depth of water the habitats of the submerged communities depend closely on such factors as nature of substratum, wave action or shelter, silting, and amount of organic material in the mud. The diversified winding shore of a lake like Esthwaite, with promontories and deep bays, so that the shore line makes every possible angle with the direction of the prevailing wind, while silting is local and variable in rate, leads to the most various combinations of these factors in different places, and correspondingly various distribution of the communities. This is shown in Pearsall's detailed maps of the vegetation of various parts of the shore (Pearsall, 1917, pp. 194, 197, 200). One of these is reproduced in Fig. 112.

Figs. 114, 115 show the debouchment of the Black Beck into the head of the lake in 1914–15 and 1929. (The upper parts of these figures show the fen which succeeds the reedswamp in succession and are considered in Chapter XXXII, pp. 639–46.) The beck brings down a lot of silt, both organic and inorganic, and immediately around its mouth is the most richly silted area of the whole lake shore. The dysphotic consocies of *Nitella flexilis* and *Fontinalis antipyretica* extended in 1915 down to the 9 ft. contour and in patches to a depth of 12 ft. (3·65 m.), occupying most of the floor of the bay. By 1929 *Fontinalis* was largely restricted and displaced by *Sparganium minimum* advancing from the mouth of the beck on silt brought down by the stream, which had considerably raised this area of the floor. The patches of white waterlily on each side of the mouth advanced over this new silt, but maintained their relative positions. The *Littorella* consocies, which in 1915 formed broad zones on inorganic soil on both sides of the bay, was greatly reduced in 1929, while the *Scirpus* and *Phragmites* reedswamp advanced some 50 to 100 ft. into the water. *Typha latifolia*, flourishing on rich silt, progressed eastward in the *Phragmites* zone.

The other lakes. In his general paper on the aquatic vegetation of the English lakes (1920) Pearsall supplements the special study of Esthwaite by a wider range of data.

The numerous lakes of the English Lake District lie in a small area of mountain country: but all the larger ones are at comparatively low altitudes. They occupy the bottoms of long and narrow valleys which are glacial rock basins. Seven out of the twelve largest have more than half their area of greater depth than 20 m. The mountain rocks are of very uniform physical character—a series of very hard slates, flags and grits of Ordovician and Silurian age: they are all poor in lime and weather very slowly. The lake waters are consequently singularly pure, containing only

0·03–0·07 g. per litre of dissolved material. Only two of them (Bassen-
thwaite and Esthwaite) are coloured with peaty material. For
the twelve lakes the average content of the waters in the most
important substances in parts per million is as follows (Pearsall,
1920, 1930).

Dissolved substances

Residue	Mineral	45·3	Fe_2O_3	} 0·05
	Organic	8·3	Al_2O_3	
Na_2O	}	10·6	CO_2	5·2
K_2O			SO_3	11·2
CaO		5·7	Cl_2	7·6
MgO		2·8	SiO_2	2·4 to 0·1
			NO_3	0·2 to 0·02

The poverty in nitrates is very marked: phosphates as P vary from 0·02 to
0·001, the higher values occurring in early spring.

Light. In regard to the light factor, vegetation does not descend into
the water below the point at which the light intensity (as determined by
the rate of decomposition of potassium iodide) is reduced to 2 per cent of
full illumination on a bright day, and sometimes not so far. This point

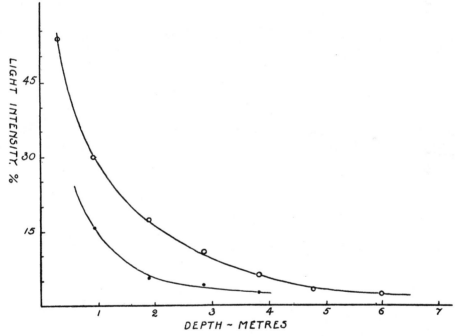

FIG. 116. DECREASE OF LIGHT INTENSITY WITH DEPTH OF WATER

The upper curve is based on data from Ullswater, Windermere and Derwentwater, the lower
from Esthwaite. From Pearsall, 1920.

varies in different lakes (Fig. 116) ranging from a depth of 10 m. in the very
clear Wastwater, where the vegetation descends as low as 7·7 m., to 4·1 m.
in the peat-coloured waters of Esthwaite, where the vegetation reaches to
4 m. only (Fig. 116). The failure of light is therefore one factor limiting the

downward penetration of the water plants and determining the depths at which certain species can exist; but variation of light intensity does not determine the general distribution of the vegetation.

Temperature. The temperatures of the lakes rise to maxima of about 17° C. in July and August, the means for these months being about 15·5° C. This compares with means of 20 to 22° C. in small ponds of the same district (Fig. 117). In April, May and June the ponds show means of 16, 20 and 21·5° C., while the lakes are very much colder with 4·9, 7·7 and 12·3° C.

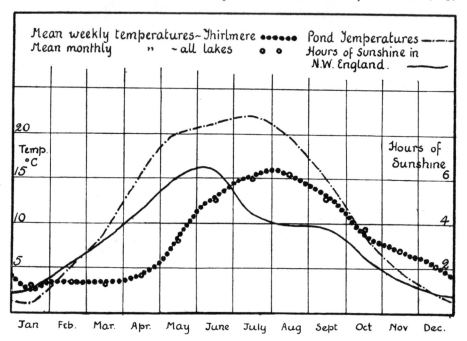

FIG. 117. TEMPERATURES OF LAKES AND PONDS THROUGHOUT THE YEAR

Note the earlier spring rise and much greater summer heights of the pond temperatures compared with those of the lakes, determining the earlier growth of pond vegetation. From Pearsall, 1920.

This marked difference in temperature in spring and early summer corresponds with earlier growth of vegetation in the ponds than in the lakes. The lake vegetation probably does not reach its maximum development before the end of August.

Aeration. The aeration of the lake waters is on the whole good. The oxygen content rises from July to August and September, but is highest in winter. The CO_2 content seems to be highest in August and September when the vegetation is most fully developed.

Silt. In these rocky lakes rooted vegetation is necessarily confined to places where the substratum is soft, i.e. in most cases where silt is present. Pearsall shows that different species react differently to silting, some succeeding best where it is rapid, others where it is slow or intermittent (Fig. 118). He concludes that "the silting factor underlies the distribution of all the types of aquatic vegetation found in this lake area, and that vegetation is only luxuriant in those places where silt is being deposited".

The decisive effect of silting on the vegetation, not only rooted but free floating, has been pointed out by Pearsall in another paper (1922), in which he has assembled an extensive series of data both from Great Britain and from other parts of the world, showing that abundance of free floating vegetation, both of flowering plants such as *Stratiotes*, *Hydrocharis*, *Ceratophyllum*, and Lemnaceae, and of plankton (Protococcaceae, Diatoms and Myxophyceae), is found only in waters rich in nitrates, silica, and calcium and magnesium salts. The waters poor in these salts,

The two types of lake and in which the ratio $\dfrac{K + Na}{Ca + Mg}$ is high, are characterised on the other hand by a Desmid plankton, and by the absence of free floating vascular plants. Lakes of the latter type occur on hard rocks not covered by soil so that silting is absent, and correspond, on the whole, to the so-called "Highland" type. Where silt is abundant all lakes *tend* to have waters of the so-called "calcareous" type, whatever the underlying rock.

By comparing soils on the lake floor where silt is abundant with similar samples at the same depth where silt is scanty (e.g. of boulder clay associated with very little silt) it was shown that the silted samples contain more potash and phosphates and less organic matter than those with little silt, and further that the potash is predominantly associated with the finer particles of the silt. In the absence of inorganic silting, with its

Potash and fine silt fresh supplies of bases, the organic matter derived from the decay of the plants which have colonised the substratum tends to accumulate, and to produce an acid peaty soil which is unfavourable to the

Subaqueous peats more luxuriant types of vegetation, just as are the similar terrestrial moor and heath soils.

An unmistakable correlation was shown between texture and potash content on the one hand and the type of vegetation present on the other. Thus the substrata of communities of *Potamogeton pusillus* have a fine silt fraction of 25·2 per cent and a mean potash content of 0·044, those of *Juncus fluitans* 7·5 and 0·025, and of *Isoetes lacustris* 11·8 and 0·021. Since the sediments become finer in passing to deeper water and the differences of light intensity are certainly not decisive, Pearsall holds that the zonation of vegetation is primarily the result of these differences in soil conditions, and that the variations in zonation between one lake and another and between different parts of the same lake are due to the effects of different degrees of wave action and of silting.

The following is a summary of the aquatic plant communities of the lakes of the English Lake District abstracted from Pearsall's account (1921).

Deep water communities. These are usually developed in light in-

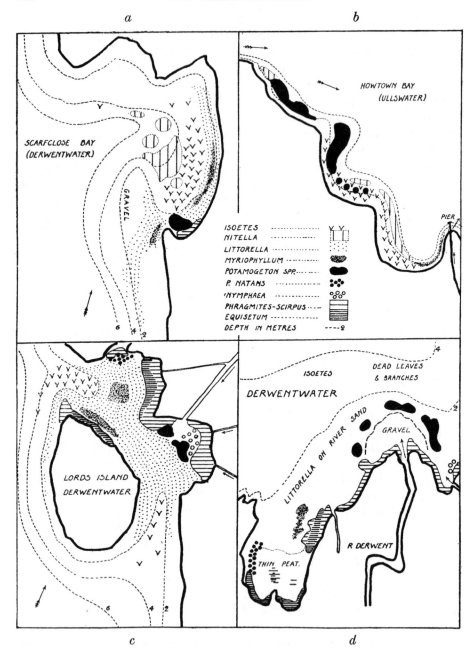

FIG. 118. RELATION OF DIFFERENT COMMUNITIES TO SILTING

On parts of the shore lines of different lakes the arrows show the silt-bearing currents (except in the south-west corners of *a* and *c*, where the barbed arrow represents wind and wave incidence across the lake). Submerged species of *Potamogeton* and *Nymphaea*, and *Scirpus-Phragmites* reedswamp occur near sources of silt; *Isoetes*, *Myriophyllum*, *Potamogeton natans*, and *Equisetum fluviatile*, away from them. In *c* wave action keeps the south-west shore of Lord's Island bare, and *Isoetes* develops on each side of the unsilted inlet behind the island. From Pearsall, 1920.

tensities of less than 15 per cent. The depth at which they occur depends on the degree of erosion and the abundance of fluvial silts. Thus in Wastwater, a rocky lake in which silts are scarce, they do not normally occur above 4 m., while in Esthwaite, where fluvial silt is abundant, they reach to within 1 m. of the surface.

(1) *Isoetes lacustris* consocies. This occurs normally on stones masked by a thin layer of silt, or on boulder clay: also on sand or occasionally thin peat. In most cases the soils are unsilted sterile glacial deposits with low potash content. With increase of silting associates begin to appear, the community becomes closed, and *Isoetes*, unable to alter its rooting level, dies. The following is the composition of this community:

Isoetes lacustris	d	Utricularia major	l
Nitella opaca	la	U. ochroleuca	l
Myriophyllum spicatum	la	Ranunculus spp.	r
Potamogeton perfoliatus	l	Sparganium minimum	r
Nitella flexilis	f	Fontinalis antipyretica	l
Chara fragilis	l		

Besides the above submerged plants, the following (normally terrestrial) mosses occur:

Climacium dendroides	r	Hypnum cuspidatum	r
Eurhynchium praelongum	r	Mnium cuspidatum	vr
E. rusciforme	r		

It is still uncertain whether they form integral parts of this deep water community or are surviving fragments of plants washed from the shore.

(2) *Juncus fluitans* consocies occurs in the more rocky lakes in well-defined areas near the mouths of streams where there is abundance of coarse silt or sand, relatively poor in potash but richer than the *Isoetes* substratum, and with only a moderate organic content. The following is the list of observed species:

Juncus bulbosus f. fluitans	d	Potamogeton pusillus	
Callitriche intermedia	la	subsp. lacustris	l
Myriophyllum spicatum	f	P. polygonifolius var.	
M. alterniflorum	r	pseudofluitans	r
Isoetes lacustris	f	Utricularia intermedia	l
Nitella opaca	f	Fontinalis antipyretica	l
N. flexilis	l	Chara fragilis	o

Juncus fluitans is locally subdominant in a very different habitat and with very different associates, such as *Lobelia dortmanna* and *Potamogeton natans*. This is in shallow water on a peaty soil with an average of 72 per cent of organic matter and a very low potash content. It seems probable that these unsilted peaty soils share an effective edaphic factor (scarcity of bases) with the coarser silts. A corresponding similarity is observed in terrestrial vegetation between dry peats and coarse sandy or gravelly soils, alike inhabited by heath plants.

(3) The *Nitella* associes is the only community typical of all the lakes, and it covers the most extensive areas on the deep water silts, developing in silted habitats where there is an accumulation of easily penetrable material ranging from fine sand to the finest muds, with an organic content which is never extreme. It is very variable, being in contact in different areas with most of the other deep water communities, both preceding and following (surrounding as it were) the *Potamogeton* and linear-leaved associes (see Figs. 115, 120). The following species are included:

Nitella opaca	d	Potamogeton praelongus	l
N. flexilis	ld	P. pusillus	r
Chara fragilis	o	subsp. lacustris	o
subsp. delicatula	la	P. perfoliatus	r
Myriophyllum spicatum	f	P. crispus	r
M. alterniflorum	l	Callitriche intermedia	r
Elodea canadensis	l	Eurhynchium rusciforme	r
Isoetes lacustris	l		

(4) The *Potamogeton pusillus* community, in which the subspecies *lacustris* is dominant, replaces the Nitelletum where fine sediments rich in potash are frequent and silting is rapid, as in Coniston, Windermere, Ullswater and Esthwaite. These fine sediments without much organic material are only deposited in deep water, so that the light intensity is always low (2–10 per cent), and this community is the only one which lives entirely in so weak a light. The community can range from fine rich silts to poor sands: its edaphic demands overlap those of *P. perfoliatus* and *P. praelongus*. The following species occur:

Potamogeton pusillus		Callitriche intermedia	o
subsp. lacustris	d	Myriophyllum spicatum	f
P. pusillus	o	M. alterniflorum	r
P. praelongus	l	M. verticillatum	r
P. perfoliatus	l	Isoetes lacustris	l
P. obtusifolius	r	Nitella opaca	f
P. crispus var. serratus	vr	N. flexilis	l
Elodea canadensis	l		

The next two communities (5) and (6) are found in Esthwaite only.

(5) The *Naias flexilis* consocies grows on very fine semiliquid mud, rich in potash and poor in organic matter. Here *Naias* plays the colonising role of *Nitella*, preceding *Potamogeton pusillus*.

Naias flexilis	d	Hydrilla verticillata var.	
Potamogeton pusillus	o	gracilis	o
P. panormitanus	o	Nitella flexilis	o

(6) The linear-leaved associes represents a development of the *P. pusillus* consocies resulting from the luxuriant silting conditions found in this lake, other linear-leaved species becoming associated. For a fuller description see pp. 599–600.

(7) The *Potamogeton obtusifolius* consocies occurs on fine silt, rich in

potash and with rather high organic content, especially in Esthwaite and Windermere, where it has replaced the *P. pusillus* consocies within a few years.

Potamogeton obtusifolius	d	Sparganium minimum	f
P. alpinus	o	Elodea canadensis	f
P. pusillus subsp. lacustris	r	Callitriche autumnalis	l
P. panormitanus	r	Myriophyllum spp.	o

(8) *P. perfoliatus* consocies:

P. perfoliatus (including vars. obtusifolius, lanceolatus and macrophyllus)	d	P. alpinus	r
		Myriophyllum spicatum	o
		Elodea canadensis	l
		Isoetes lacustris	l
P. praelongus	o	Nitella opaca	o
P. pusillus (agg.)	l	Chara fragilis	r
P. angustifolius	l		

(9) *P. praelongus* consocies:

P. praelongus	d	P. lucens	r
P. pusillus subsp. lacustris	o	Nitella opaca	f
P. perfoliatus	l	Myriophyllum spicatum	o
P. pusillus	r	Elodea canadensis	l
P. alpinus	r		

The *Potamogeton perfoliatus* and *P. praelongus* consocies are characteristic of well silted places in the lakes rich in silt (Coniston, Windermere, Ullswater and Esthwaite). *P. praelongus* grows on finer and more abundant silts in deeper water and therefore in lower light intensity than *P. perfoliatus*, within the habitat range of *P. pusillus*, and following that species or *Nitella*. *P. perfoliatus* on the other hand appears more often where silt is accumulating on the *Isoetes* habitat, though it may also succeed *Nitella*, *Juncus fluitans* or *P. pusillus* itself, but on coarser silts with a higher organic content. It appears that very rapid subaqueous silting favours the growth of the linear-leaved pondweeds, consolidation and stabilisation the establishment of the larger, broad-leaved species.

(10) The *P. alpinus* consocies succeeds *P. praelongus* or *P. pusillus* where silts are rich and very abundant, apparently only on silt which has been long stabilised (as in Windermere, Esthwaite and Derwentwater) and has accumulated a good deal of organic matter (mean 33·4 per cent), which takes a long time when silting is free. The following are characteristic species:

Potamogeton alpinus	d	P. obtusifolius	l
Elodea canadensis	f	Sparganium minimum	o
Myriophyllum spicatum	f	Nuphar luteum	o
M. verticillatum	l		

(11) The *Sparganium minimum* consocies grows in Derwentwater, Coniston, Esthwaite and Windermere in deep water, also on organic soils, and typically consists of scattered plants. Apparently no other submerged

plants are able to live in the habitat, though such places are nearly always being colonised by water-lilies.

(12) The *Fontinalis antipyretica* consocies shows no very definite relation to soil conditions or light intensity, but a decided preference for places near the mouths of streams. The soil is normally organic, but here and there the community may be sparsely developed on bare rock. The associates include:

Potamogeton obtusifolius   Sparganium minimum
Elodea canadensis   Juncus fluitans
Utricularia major   Eurhynchium rusciforme
U. ochroleuca   E. praelongum

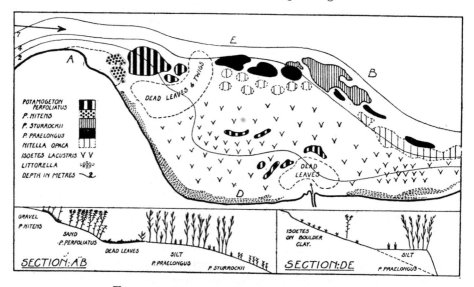

FIG. 119. THWAITE HILL BAY, ULLSWATER

Different species of *Potamogeton* occupy different parts of the silt terrace: *P. nitens* and *P. perfoliatus* on gravel and sand, *P. praelongus* on the finer silt at a greater depth. *Isoetes* occurs on unsilted boulder clay between the silt terrace and the shore line. *Littorella* lines the shore in shallow water ("*Potamogeton sturrockii*" is a synonym of *P. pusillus* subspecies *lacustris*). From Pearsall, 1920.

Succession. Among the deep-water communities three main lines of succession have been actually observed, or inferred on good grounds:

(1) On coarse soils in rocky lakes, poor in potash and ultimately organic:

Isoetes →Juncus fluitans →Potamogeton natans.

(2) On soils of intermediate character and most widespread in all the lakes:

Isoetes →Nitella →Potamogeton perfoliatus → $\left\{\begin{array}{l}\textit{Elodea-Sparganium minimum}\\ \textit{Potamogeton alpinus}\end{array}\right\}$
→P. natans and/or water-lilies.

(3) On the finest, richest and least organic silts, chiefly in Esthwaite Water:
Naias (or *Nitella*) →Linear-leaved species →*Nitella* →*Sparganium* →Water-lilies.

The proportion of potash decreases in (3) and the later stages of (2) and organic matter increases.

Changes due to silting factors appear to be usually reversible, while those due to increase of organic matter in the substratum are normally irreversible. The organic soils become more fibrous (less completely decayed) and yellow or brown rather than grey or black. In the later stages this organic matter accumulates more rapidly than the silt, and the parallel with raw humus formation on land is obvious. On the poorer and coarser subaqueous soils, the change (e.g. *Juncus fluitans* → *Potamogeton natans*) may in fact be completely identified with the formation of acid peat.

Shallow-water communities. In the Cumbrian lakes these normally occur in light of more than 15 per cent of full intensity, though they sometimes tolerate less. In the larger lakes they grow on soils that are, or have been, subject to wave erosion; or else on soils derived from those of deep water by the accumulation of silt and humus. The shallow water communities are not necessarily developed in succession to the deep water—they also colonise eroded shores *de novo*. The greater number of the shallow-water vascular plants in all freshwaters are partly emergent—either the inflorescence alone, as in *Myriophyllum*, or the upper surfaces of the floating leaves as well as the flowers, as in the water-lilies; or again, as in some of the ubiquitous plants of richer waters, such as *Alisma plantago-aquatica* and *Sagittaria*, the greater part of the photosynthetic shoots is developed above the surface of the water.

The shallow-water soils of the Cumbrian lakes are characterised by coarseness of sediment, poverty in potash, and a higher organic content than the soils of deep water, except where there is vigorous wave action or near the mouths of streams. There is a striking scarcity of pondweeds and water-lilies except in sheltered and silted areas.

(a) **Mainly submerged communities.** The first community is: (1) The *Littorella-Lobelia* associes. *Littorella uniflora* is the pioneer, spreading over and stabilising loose sand, gravel or morainic stones by means of its vegetative propagation. Here, on soil pulverised by wave action, it forms a continuous sward, collects sand and probably seeds, ultimately giving way to other communities. As the soil becomes more organic *Lobelia dortmanna* appears and tends to replace the *Littorella*.

Other dominants appear at a later stage, and the following communities may be distinguished, with progressive cessation of silting and increasing organic content.

(2) *Potamogeton* associes (average organic content 22·2 per cent):

P. gramineus (with var. longipedunculata)	P. angustifolius
	P. perfoliatus
P. nitens var. subgramineus	P. alpinus

Any one of these species may be dominant, and the last two persist when the organic content of the soil is over 40 per cent.

(3) *Myriophyllum* consocies (average organic content 42·7 per cent):

M. spicatum	d	M. verticillatum	
M. alterniflorum (Esthwaite)		Ranunculus peltatus var. truncatus	

(4) *Juncus fluitans* socies (average organic content 86·6 per cent) only occurring where the soil is more or less peaty. This is no more than a socies of the *Littorella-Lobelia* community, since the *Juncus* is not dominant over any considerable area. Associated are:

Sparganium minimum	f	Scirpus fluitans	o
Utricularia major forma		Nitella opaca	o
gigantea	la	Isoetes lacustris	r
Apium inundatum	l		

Myriophyllum (f–o) and *Ranunculus* sp. (o) are probably relict.

These communities merge into one another and are often indistinct. At one end of the scale is the *Littorella-Potamogeton* type, with inorganic soil relatively rich in potash, at the other the *Lobelia-Juncus fluitans* type, with organic soil poor in potash.

(b) **Floating leaf communities.** These are all limited to the most sheltered places, being excluded by wave action, though *Nymphaea occidentalis* and *Nuphar intermedium* are able to endure somewhat more vigorous wave action than the common white and yellow water-lilies. For the rest the following four communities show correlation with increasing organic content of the soil, parallel with that shown by the submerged communities.

(1) *Nymphaea alba* consocies, average organic content 25·7 per cent.

2) *Nuphar luteum* consocies, average organic content 38·6 per cent.

(3) *Nymphaea occidentalis* consocies, average organic content 48·5 per cent.

(4) *Potamogeton natans* consocies, average organic content 74·2 per cent.

Plants with floating leaves are usually very local in the Cumbrian lakes, particularly the large ones, because continual wave action along the exposed shores prevents the deposition of silt. Only Esthwaite, small and rich in silt, has any large areas of water-lilies: in Derwentwater, also small but with rather scanty silting, they are generally replaced by the consocies of *Potamogeton natans* except near the mouths of streams, where silt and water-lilies occur.

(c) **Reedswamp communities.** The reedswamp dominants again show the same general relation to the organic content of the substratum and to silting, but less definitely, for the vigorous growth of their rhizomes enables them to persist in spite of changing conditions. The following are the main communities:

(1) *Typha latifolia* consocies, average organic content 34·0 per cent.

(2) *Scirpus-Phragmites* associes, average organic content 40·7 per cent.

(3) *Equisetum fluviatile* (*limosum*) consocies, average organic content 60·9 per cent.

(4) *Carex* associes, average organic content 89·4 per cent.

Typha latifolia (the great reedmace or bulrush) occurs in these lakes only at the mouth of Black Beck in Esthwaite (see Figs. 114, 115) a place of abundant fine silting and rapid change of soil level.

Scirpus lacustris (the "true" bulrush) and the common reed (*Phragmites vulgaris*), the commonest European reedswamp dominants, form the characteristic reedswamp (see Figs. 112, 115, etc.), but they are nevertheless quite local in the Cumbrian lakes, because of the resistant nature and also of the instability of most of the shores. *Scirpus* extends into the deeper water and also prefers the more organic substrata, *Phragmites* forms the bulk of the reedswamp inshore and persists in the fens after the soil has been raised above the water level. The associes occurs in open bays where wave action and silting are both slight; also on sheltered western shores, in the mouths of streams, and in extensive shallows.

Equisetum fluviatile and *Carex* spp. form reedswamp in closed bays and on very sheltered shores, where both wave erosion and affluents are lacking. This reedswamp is therefore characteristic of the more rocky lakes and rarer in the more silted lakes such as Esthwaite. *Carex inflata* and *C. vesicaria* are the most frequent dominants of the Caricetum, with hybrids between them, and less commonly *C. lasiocarpa*, which is locally abundant.

The reedswamps are typically very pure communities in which the aerial shoots of the dominant plants are so closely set that there is little room for subordinate vegetation. But the fully developed community is usually fringed with open reedswamp where the dominants are beginning to establish themselves. The following species, among others, occur in this ecotone or mictium:

Sparganium natans	la	Scirpus fluitans	o
Apium inundatum	la	Callitriche intermedia	o
Utricularia major (neglecta)	la	C. stagnalis	r

These plants occur in their own habitat conditions in such a mictium. Open reedswamp of *Scirpus lacustris* with *Nymphaea alba* is a common type of mictium in other waters, e.g. in the Norfolk Broads (Pl. 109, phot. 270).

The two main hydroseres. From Pearsall's extensive series of data it is apparent that there are two main *types* of sere in the shallower waters of the Cumbrian lake shores.

Where silting is rapid and the substrata inorganic we have the *silted type*:

(A) *Littorella* → *Potamogeton* → *Nymphaea* → *Phragmites*.

Where silt is scarce and the soil is organic the *peaty type* occurs:

(B) *Lobelia* → *Myriophyllum* → *Potamogeton natans* → $\begin{cases} Equisetum \\ Carex. \end{cases}$

The actual successions are however very numerous, and the variations are brought about largely by *changes* in silting. Thus if silting is at first heavy and then stops altogether the general line of succession will shift from (A) to (B), and beginning with *Littorella* may end with *Carex*. On the other hand, if an increasingly organic soil becomes heavily silted the succession will change from (B) to (A) and end with *Phragmites*.

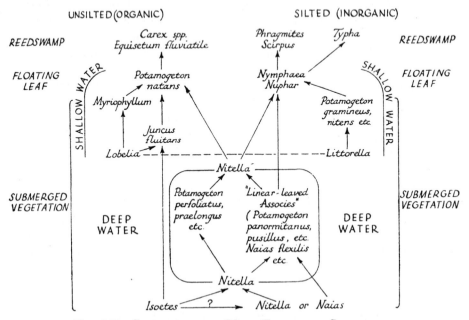

FIG. 120. SCHEME OF THE MAIN TRENDS OF SUCCESSION

In addition to somewhat excessive simplification, this diagram has one conspicuously un-natural feature in the separation on opposite sides of *Lobelia* and *Littorella*, which in fact commonly occur together in one associes, though the former increases as the soil becomes more organic, while the latter is the pioneer on inorganic silts. Such anomalies are inevitable when we attempt to exhibit the changes of three variants—succession in time, organic content of substratum, and depth of water—in a two-dimensional diagram.

The leading causal factors in the hydroseres are the increase of organic content in the substratum, and the raising of the soil level till it eventually reaches the surface of the water, so that the aquatic communities give way to fen. The succession brought about by these factors is checked or stopped by wave action, and altered by silting. Areas which are freely silted from the beginning bear a different set of communities from those which are not. Where silting is intermittent a "silted community" gives place, on cessation of silting, to a community characteristic of organic soil, while a fresh incidence of silting produces the reverse effect. Such changes are in fact reversible. But the changes in life form by which floating leaf and reedswamp dominants replace the submerged, occurring in the later stages of the sere, are irreversible, and so is the eventual predominance of accumulation of organic material.

The general direction of change is towards a more organic and acid soil with a decreased content of bases and especially potash, and is thus parallel to the formation of peaty and acidic moorland soils on land, a progression which as we shall see in Chapter xxxii is completed in the terrestrial "raised moss" of the Lake District lowlands.

By combining the observed seres in deep and shallow water we can construct a scheme such as that shown in Fig. 120, ignoring all but the main trends of succession.

THE MOST PRIMITIVE LAKES AND TARNS[1]

Pearsall has shown that the different lakes he studied form a series illustrating the evolution of glacial lake basins from a primitive type, with rocky floor, stony or peaty margins and little or no inorganic silt, to the richly silted types of which Esthwaite Water is an example.

The most primitive types are now to be found in the smaller mountain tarns where this development cannot take place, and of these we may distinguish (1) those with very little silt, what there is being inorganic and derived mainly from erosion of the shores; and (2) those in which the silt is peaty and derived from terrestrial sources, i.e. from the peat formed by adjoining land vegetation, while the accumulation of plant debris builds up a highly organic substratum.

(1) In this type silt is scarce, but what there is is inorganic. Any fluvial silt brought in by streams is coarse and sandy.

The vegetation is everywhere sparse. The species of deep water are:

Isoetes lacustris	la	Myriophyllum spicatum	l
Nitella opaca	la		

Where some inorganic silt is deposited we have:

Juncus fluitans (local, near streams)
Callitriche intermedia
Potamogeton polygonifolius var. pseudofluitans (entirely submerged)
P. pusillus subsp. lacustris (r)

In shallow water *Littorella uniflora* (lf) is the principal species. *Lobelia dortmanna* is very local.

(2) In this type the substratum becomes organic and inorganic silt is quite absent. When the lakes are very small (tarns) so that their shores are not eroded by wave action there is a good succession from the aquatic to the adjoining bog vegetation with gradual increase of acid peaty humus.

(a) Beginning in deeper water we have the *Juncus–Myriophyllum* open associes:

Juncus fluitans	l–la	Chara fragilis	l
Myriophyllum spicatum	la	Nitella opaca	l
Utricularia minor	f		

[1] Information kindly supplied by Dr W. H. Pearsall.

(*b*) The next community consists of *Sparganium minimum* (o–f), followed by a floating leaf community of

(*c*) *Nymphaea occidentalis* (a–d) and *Nuphar pumilum* (f–a). This is in turn followed by an open reed belt of

(*d*) *Equisetum fluviatile* (*limosum*), and this by

(*e*) Reedswamp of *Carex inflata* (d) with which are associated:

Eriophorum angustifolium	f	Potamogeton polygonifolius	l
Juncus fluitans	f	Utricularia minor	o
Menyanthes trifoliata	l		

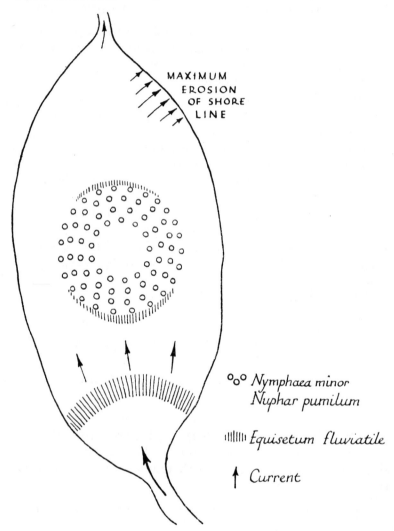

FIG. 121. DIAGRAM OF A "PRIMITIVE" TARN (LAKELET)

The shores are stony and barren with strong wave erosion on the leeward side. There is a development of characteristic reedswamp and small water-lily communities where slight silting occurs, i.e. beyond the mouth of the inlet stream and in the centre of the tarn just below the level of wave action. After a sketch supplied by Dr Pearsall.

In shallow water *Littorella* is rare or absent, *Subularia aquatica* local, while *Lobelia dortmanna* becomes locally frequent as the organic mud accumulates. This community often passes into open reedswamp of *Phragmites*, which is normally succeeded by open *Equisetum fluviatile*, often with *Scirpus lacustris* and *Potamogeton natans*. The closed reedswamp seems always to be formed by *Equisetum fluviatile* and *Carex inflata*. *Nymphaea occidentalis*, *Nuphar pumilum* and *Sparganium minimum* may occur in any of these reedswamps.

In tarns and lakelets of this type which are relatively advanced in development but subject to wave erosion of the shore, the water-lily and reed belt communities, (c) and (d), are often formed in a zone at some distance from the shore just below the zone of wave erosion, and also just beyond the mouth of a stream entering the tarn (Fig. 121). In both these areas a little silt is deposited and plant colonisation begins, while the shore zone is kept free from vegetation.

Transition to bog. As the peaty soil reaches the water level, bog plants such as *Sphagnum* spp., *Erica tetralix* and *Carex panicea* appear; and these are joined later by several other species of bog plants till the bog community is established. Among the bog species relicts of the reedswamp community (e) often linger. This is the most "organic" type of succession that occurs. Rather surprisingly the lowest *p*H value recorded is 5.

REFERENCES

PEARSALL, W. H. The aquatic and marsh vegetation of Esthwaite Water. *J. Ecol.* **5**, 180–202; **6**, 53–74. 1917–1918*a*.

PEARSALL, W. H. On the classification of aquatic plant communities. *J. Ecol.* **6**, 75–83. 1918*b*.

PEARSALL, W. H. The aquatic vegetation of the English Lakes. *J. Ecol.* **8**, 163–201. 1920 (1921).

PEARSALL, W. H. A suggestion as to factors influencing the distribution of free-floating vegetation. *J. Ecol.* **9**, 241–53. 1922.

PEARSALL, W. H. Phytoplankton in the English Lakes. I. The proportions in the waters of some dissolved substances of biological importance. *J. Ecol.* **18**, 306–20. 1930.

Chapter XXXI

THE VEGETATION OF RIVERS

The obvious difference between the conditions of life in a river and in a pond or lake is of course the existence in the former of a continuous current. Correspondingly the vegetation of a very sluggish stream is quite similar to that of a pond, both containing much the same zoned communities composed of the same species. But where the river current is considerable new factors come into play.

Current. In the first place the current tends to equalise conditions, such as temperature and content of dissolved salts and gases, in different parts of the river's course; but this action is far from being complete, because of the changing gradient and the various obstacles such as sharp bends which alter the rate of flow of the current, and also because of the varying nature of the strata through which the river flows. Secondly, the river current, like the waves and currents on a lake shore, erodes the banks, deposits silt here and removes it there—and not only silt, but also the vegetation itself, so that the river bed may be quite denuded of plants in some places and freshly colonised in others, and the vegetation of rivers continually subject to heavy floods remains sparse. Thirdly, the normal progressive process of "Verlandung", as the Germans call it, the "growing up" and ultimate obliteration of a pond or lake where there is no direct drainage through it, does not occur in a river. A river may become choked with vegetation so as to hold up the water and cause flooding: its course may be diverted, with the resulting formation of "oxbows", which if they are completely cut off from the stream, behave like ponds and ultimately fill up with vegetation; and extensive swamps and marshes, through which the river current winds its sluggish way, may be formed by the flooding of flat alluvial plains. But so long as water is supplied from its sources the river must continue to exist, and whenever any considerable body of water comes downstream the impact tends to clear the bed of vegetation which may have threatened to choke it. In rivers of medium or slow current the conditions in midstream are, however, very different from those near the banks, because the current is stronger and the depth of water usually greater. There is in fact a zonation of conditions, and consequently of vegetation, parallel to the direction of flow, i.e. to the banks, and therefore analogous to the zonation on the margin of a lake. Many such rivers are normally fringed with reed-swamp whose dominants have succeeded in colonising the shallower quieter water near the bank, though the current prevents its encroachment on the deeper water and thus holds up the progress of the succession.

Silting. In the last chapter we saw that silting had an all-important effect on the vegetation of lakes, and the same is true of rivers. Silt directly supplies most of the mineral nutrients for the rooted plants, as the salts dissolved in the whole body of water supply the free-floating plants and the algae, so that a richly silted river bed is the most prolific in vegetation. Silting is due to the slowing down of moving water which is carrying solid material in suspension, and thus depends on the velocity of the current, which is correlated with the nature of the river bed and **Speed of current** with the amount and texture of the silt carried or deposited. The general correspondence is shown in Table XVIII (Minnikin, 1920, Butcher 1933):

Table XVIII

Velocity of current per second	Nature of bed	Habitat
More than 4 ft. (1·21 m.)	Rock	Torrential
,, ,, 3 ft. (0·91 m.)	Heavy shingle	,,
,, ,, 2 ft. (0·60 m.)	Light shingle	Non-silted
,, ,, 1 ft. (0·30 m.)	Gravel	} Partly silted
,, ,, 8 in. (0·20 m.)	Sand	
,, ,, 5 in. (0·12 m.)	Silt	Silted
Less than 5 in. (12 cm.)	Mud	Pond-like

In a general way this change in the nature of a river bed with current velocity corresponds with the gradual decrease in average gradient of the bed of a river which rises in a mountainous region and flows through foothills and finally across plains towards the sea. But local topographical and geological features introduce many variations, and a river may increase and decrease the average rate of its current many times during its course. Bare rock and boulders are characteristic of the beds of mountain torrents, and as the gradient decreases and the rate of flow lessens the bed changes to gravel and sand, while the slowly moving waters of an alluvial plain have silty or muddy bottoms. But the currents of rivers are not constant even at any one spot and their beds are therefore unstable. Heavy precipitation in the mountains will add enormously in a few hours to the volume of water coming down and will correspondingly increase the rate of flow, shifting downstream the smaller particles that have been deposited. When the **Floods** flood subsides new particles brought down from upstream or flowing in from the banks will sink to the bottom as the water clears. Thus the condition of the bed at any spot is always changing in different directions, in broad contrast with the bottom of a pond or lake where change is on the whole slow and in the same direction, though lake bottoms also show fluctuating changes near the mouths of affluent streams and as the result of varying wave action and shore currents. Any obstacle to moving water carrying suspended material causes it to drop part of its load, and vegetation itself, once established, is an important agency in trapping silt. Silt does not consist solely of inorganic particles derived from eroded rocks, but may be partly or wholly organic, formed from debris of water plants themselves, from eroded peat or humus, and from the soil of manured agricultural land—not to mention sewage. A strong organic constituent normally increases the fertility of silt.

Turbidity. According to the amount of fine silt they carry rivers vary very much in clearness. Streams draining areas of limestone and of hard siliceous rock are in general clear, while those passing through agricultural country are more turbid. Flood water resulting from heavy rainfall will always bring soil into a stream and increase its turbidity. When a river is habitually very turbid it may be destitute of all submerged vegetation, owing to the diminution of light penetrating the waters, and perhaps partly to the unstable bed of silt and mud, as Butcher (1933) found in the river Tern in Shropshire.

Solutes and reaction of the water. The amount and the chemical nature of the solutes are of first importance to the plankton, to the algae, and also to the bryophytic vegetation of mountain torrents, which depend entirely on this source for their mineral nutrients; and the pH value of the water may also have a direct effect on the plants. On the whole, however, these factors, apart from the amount and nature of the silt with which they may be correlated, do not cause very marked differences in the rooted vegetation of vascular plants.

Thus Watson (1919) recognised that certain submerged mosses such as *Amblystegium filicinum*, *Fissidens crassipes*, *Hypnum commutatum*, *H. falcatum* var. *virescens*, *Orthotrichum rivulare*, *Philonotis calcarea*, and the thalloid liverwort *Pellia fabbroniana*, especially var. *lorea*, as well as the blue-green algae *Rivularia haematites* and *Scytonema* spp., are characteristic of calcareous streams. Butcher (1933) divides the rivers whose vegetation he investigated according to the proportion of lime dissolved in the water, but the rooted phanerogamic vegetation of a highly calcareous river like the Dove or the Itchen does not differ greatly from one which is moderately calcareous like the Tees, nor even from acid non-calcareous streams like those of the New Forest, providing that the silting factor is comparable. There are, however, a few species which are more characteristic of one or the other type. Thus *Myriophyllum spicatum*, *Potamogeton polygonifolius* and *P. alpinus* avoid the more calcareous waters, and *P. perfoliatus* and *P. crispus* the more acid.

The following data are mainly taken from Butcher's account (1933) of the vegetation of the different types of stream bed enumerated above and from Watson's paper on the Bryophytes (1919), supplemented by information about the Thames near Oxford kindly contributed by Mr H. Baker.

(1) **Very rapid current** (bed rocky, mountain torrent type). Vegetation consisting mainly of bryophytes, usually a mixture of species: leafy liverworts are often abundant and locally dominant. Flowering plants sparse or absent owing to lack of silt and of sufficiently secure rooting places.

Watson (1919) gives the following as the most characteristic liverworts, often occurring in pure patches and sometimes carpeting the rocky beds of mountain torrents:

Alicularia compressa	d or ld
Aneura sinuata	f
Aplozia riparia forma potamophila	lf
Chiloscyphus polyanthus var. rivularis	o
var. fragilis	f
Eucalyx obovatus var. rivularis	r
Marsupella aquatica	lf
Scapania dentata	a, ld
S. uliginosa and obliqua (alpine and subalpine)	lf

Among mosses the following are the most characteristic forms and are often locally dominant:

*Amblystegium filicinum	f	Fontinalis squamosa	f–a
A. fluviatile	o	Hyocomium flagellare	a
Brachythecium rivulare	a	*Hypnum commutatum	a
Bryum pseudotriquetrum	o, la	H. falcatum	o
Eurhynchium rusciforme	a, sd	H. riparium	f
Fontinalis antipyretica		Philonotis fontana	a
(and especially var.			
gracilis)	lf		

* Usually in calcareous streams.

Lichens are few, though they may be locally abundant. The commonest are crustaceous forms such as *Dermatocarpon aquaticum* and *Verrucaria submersa*. There are also various algae, of which *Sacheria mamillosa*, and the Myxophycean *Rivularia haematites* (calcareous), *Oscillatoria irrigua*, *Nostoc verrucosum* and *N. sphaericum* are frequent, *Cladophora glomerata* is abundant, and *Vaucheria sessilis* occasional (Watson, 1919).

Of flowering plants the following occur most commonly, but they are often absent altogether and never dominant:

Apium nodiflorum	f	Potamogeton pusillus	la
Callitriche spp. (commoner		Ranunculus fluitans	f
in slow rivers)	o–la	R. lenormandi	f
Glyceria fluitans	la	R. peltatus var. pseudo-	
Montia fontana var.		fluitans	o, f
rivularis	la	Stellaria uliginosa	o–la
Potamogeton crispus	la		

The upper course of the River Tees is a good example of this type. Its current is swift and its bed of solid rock, with some stones and sand, in which none of the higher plants can retain a foothold for long when the stream is in flood. The vegetation is therefore confined to mosses and the larger algae, which grow in firmly anchored cushions. The most characteristic species are:

Fontinalis antipyretica	a	Grimmia fontinaloides	r
Eurhynchium rusciforme	a		
Amblystegium fluviatile	o	Lemanea fluviatilis	a
Cinclidotus fontinaloides	o	L. mamillosa	o

(2) **Moderately swift current.** Bed composed of stones or boulders. Vegetation consisting partly of bryophytes but also of flowering plants where a little silt is trapped between the stones and the floods are not strong enough to tear out the plants.

The Dove and the Wharfe. Butcher found the following six species of flowering plants in the Dove (Derbyshire) and the first three, besides the mosses and algae, also in the Wharfe (Yorkshire), both calcareous streams draining the Mountain Limestone:

Ranunculus fluitans	d	Apium nodiflorum	r
Potamogeton densus	o–f	Myosotis scorpioides	
Sium erectum	o–a	(palustris)	r
Veronica anagallis-aquatica	o		

and of non-vascular plants:

Fontinalis antipyretica	f	Batrachospermum spp.	o
Eurhynchium rusciforme	f	Lemanea spp.	f
Cinclidotus fontinaloides	o		

The upper Thames. Even a river like the Thames, whose course is mainly "lowland", has quite a swift current in some of its upper reaches. The Thames has a course of 160 miles (*c.* 257 km.), and between its source in the oolites of the Cotswold Hills to the estuary below London it traverses a considerable variety of geological strata. The bed of the Thames has of course been largely artificialised by embanking, dredging, and the construction of locks, weirs, and "lashers".[1] In this way the river is regulated, made more easily navigable, and the flooding of adjacent meadowland is minimised. The vegetation is, of course, correspondingly affected, but most of the natural habitats are still provided, the swiftness of the current and the silting of the bed varying from place to place in relation to weirs, bridges, etc., as well as to the fall of the general river course. In its upper reaches, where it flows through the Oolites, the current is generally swift and the bed stony, and here the submerged vegetation includes five out of the six species of flowering plants found by Butcher in the Dove:

*Apium nodiflorum	*Ranunculus fluitans
*Myosotis scorpioides	*Veronica anagallis-aquatica
Nasturtium officinale	
Oenanthe fluviatilis	
*Potamogeton densus	*Fontinalis antipyretica
P. pectinatus forma interruptus	Cladophora sp.
Ranunculus circinatus	Enteromorpha intestinalis

* Also in the "moderately swift" reaches of the Dove.

The plants grow in longitudinal streaks of varying width with their shoots trailing downstream and sometimes reaching a length of 6 ft. (*c.* 2 m.).

[1] A "lasher" is an artificial waterfall taking that part of the stream which is not conducted through the accompanying lock.

PLATE 109

Phot. 271. *Nymphoides* (*Limnanthemum*) *peltatum*, *Nuphar luteum* (floating leaf), *Oenanthe fluviatilis* (emersed), *Sparganium erectum* (reedswamp). Medley, near Oxford. *A. H. Church.*

ZONATION IN THE THAMES

(3) **Moderate current.** Bed gravelly. The following species were observed by Butcher in various rivers in the north and north midlands:

Ranunculus fluitans[1]	f–d	Callitriche stagnalis	r–f
Sparganium simplex	f–d	Elodea canadensis	r–f
Potamogeton crispus	r–f	Myriophyllum spicatum	r–o
P. densus	r–f	Mimulus guttatus	o
P. pectinatus	o	Polygonum amphibium	f
P. pectinatus forma inter-		Sagittaria sagittifolia	f
ruptus	lsd	Scirpus lacustris	f
P. perfoliatus	r–d	(the last three in the Tern)	
P. pusillus	r		

In the Thames above the lock at Medley, just above Oxford, with a medium current and a gravelly bottom covered with thin silt in midstream, the following occurred:

(i) *Submerged community:*

Sagittaria sagittifolia, dominant, but producing submerged strap-shaped leaves only (f. *vallisneriifolia*).

Potamogeton lucens, subdominant.

P. perfoliatus, abundant.

Oenanthe fluviatilis,[2] abundant (a typical south English species).

Nuphar luteum with submerged leaves only, arising by extension of rhizomes from the zone of floating leaf plants (iii) nearer the bank where the yellow water-lily was dominant.

Hippuris vulgaris, submerged plumose shoots coming from rhizomes in (iv).

Next, across the mouth of a silted bay, was

(ii) a zone of *Scirpus lacustris*. This bulrush grew in a narrow band parallel with the axis of the river, its rhizomes embedded in silt and gravel below 4–6 ft. of water, and aerial shoots rising 7 ft. above the surface. Inside the line of bulrush was a thick bed of silt with only 2 ft. of water above it and inhabited by

(iii) a zone of *floating leaf species:*

Nuphar luteum, dominant.

Nymphoides (*Limnanthemum*) *peltatum*, sub-dominant, competing with the *Nuphar* by means of its long rhizomes (Pl. 109, phot. 271).

Potamogeton natans, *Polygonum amphibium*, and *Callitriche stagnalis* all frequent.

(iv) *Reedswamp*, in a few inches of water:

(*a*) Zone of *Hippuris vulgaris* as the pioneer dominant, forming broad expanses of aerial shoots up to 18 in. (45 cm.) high, and also with its rhizomes spreading out beyond (ii) into deep water, where they bore sub-

---

[1] Pl. 110, phot. 272.　　　　　　　　　[2] Pl. 109, phot. 271.

merged plumose shoots; and with *Sagittaria sagittifolia* frequent in isolated patches among the *Hippuris*. The mare's-tail is not a very common plant in the Thames, and this position as pioneer of the reedswamp is commonly occupied by *Sagittaria*.

(*b*) The next zone is not here dominated by the typical tall reedswamp species, but by a number of smaller plants which ordinarily accompany the reed dominants where they can find room:

Mentha aquatica	a	Myosotis scorpioides	
Nasturtium officinale	a	(palustris)	f
Veronica beccabunga	a	Rorippa islandica	
Apium nodiflorum	f	(Nasturtium palustre)	o
Veronica anagallis-aquatica	f		

The taller species colonised this zone in patches, starting from transported rhizomes swept downstream in winter floods and rapidly spreading:

Sparganium erectum	f	Alisma plantago-aquatica	o
Iris pseudacorus	f	Glyceria maxima (aquatica)	o
Rumex hydrolapathum	o		

(4) **Medium to slow current.** Bed sandy and silted.

The examples described by Butcher are mostly from lowland streams rising from low hills, springs or marshy ground—the common type of river in the midlands and the east and south of England. The phanerogamic vegetation is very much richer owing to the better nutritive conditions and the lessened danger of detachment of rooted plants by the current, though the vegetation of any given stretch of river bed is more or less transitory (Fig. 122).

In the River Lark in East Anglia the fastest stretches contain several of the same species as in (2) and (3) above, with *Ranunculus fluitans* again dominant, but with the addition of *Oenanthe fluviatilis*, a species confined to the south. Where the current is slow Butcher found the composition of the vegetation to be as follows (Pl. 110, phot. 273):

Sparganium simplex	d	Potamogeton perfoliatus	f
Sagittaria sagittifolia	sd	P. praelongus	r
Elodea canadensis	f	P. pusillus	r
Potamogeton crispus	f	Callitriche stagnalis	o

In the medium and slow-flowing stretches of the Itchen (Hampshire) there were the following:

Elodea canadensis	f–d	Ranunculus fluitans	r
Hippuris vulgaris	a–sd	R. pseudo-fluitans	r
Sparganium simplex	a–sd	Sium erectum	r
Callitriche stagnalis	f–sd	Apium nodiflorum	r
Oenanthe fluviatilis	f–a	Lemna trisulca	r
Potamogeton densus	f	*Reedswamp*	
P. pusillus	f	Scirpus lacustris	a
P. crispus	r	Typha latifolia	r

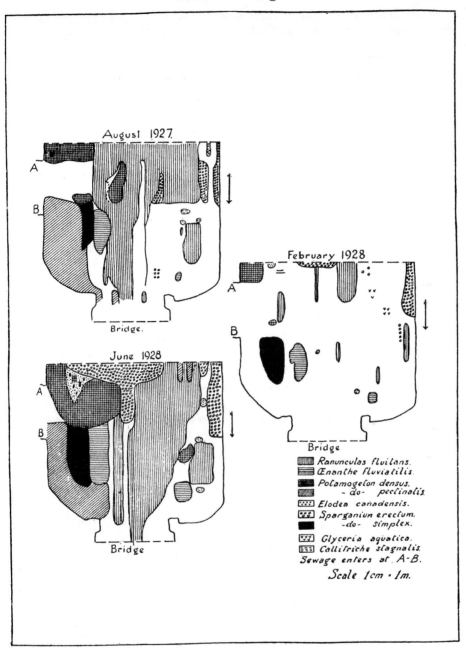

FIG. 122. AQUATIC VEGETATION IN THE RIVER LARK (SUFFOLK)

Showing the changes in the areas occupied by different species from season to season.
From Butcher, 1933.

In stretches of the upper Thames with medium current several of the same species occur, with *Sagittaria* dominant and *Nuphar* occasional:

Sagittaria sagittifolia	d	Sparganium simplex	o
Potamogeton perfoliatus	a	Nuphar luteum	o
Oenanthe fluviatilis	a	Hippuris vulgaris	r
Callitriche stagnalis	f		
Elodea canadensis	f	Cladophora glomerata	o
Scirpus lacustris	f	Vaucheria sessilis	o

In the River Cam above Cambridge, with moderate to slow current, the submerged vegetation is made up of the following species (*Types of British Vegetation*, p. 191):

Apium nodiflorum (submerged form)	Potamogeton lucens (and f. acuminatus)
Callitriche stagnalis	P. perfoliatus
Elodea canadensis	P. praelongus
Epilobium hirsutum (submerged form)	Ranunculus circinatus
Myosotis scorpioides (submerged form)	Sagittaria sagittifolia (f. vallisneriifolia)[1]
Oenanthe fluviatilis (and submerged form)	Sparganium simplex (f. longissima)
	Scirpus lacustris (submerged form)
Potamogeton crispus (and f. serratus)	Veronica beccabunga (submerged form)
P. densus	

Glyceria reedswamp. *Glyceria maxima* (*aquatica*) is one of the common dominants of the reedswamp of the upper Thames, as it is in many of the East Anglian rivers (Pl. 111). The Glycerietum commonly starts as a fringe of a few plants in shallow water close to the bank, and grows rapidly by vigorous extension of the rhizomes. The stream may be several feet deep close to this fringe and contain the submerged aquatic community characteristic of a moderate current. At the outer edge of the fringe the rhizomes and procumbent leafy shoots of *Glyceria* are free-floating and extend 1–2 ft. (30–60 cm.) into the stream. During winter the aerial shoots die, but do not dry out like those of *Phragmites*, and become compacted into a black slimy humus, the rotting mass from previous years forming the substratum for the current year's growth. Thus as growth proceeds outwards the portion of the mass of humus nearest the bank is gradually raised and the growth of the pioneer becomes stunted. The substratum is then invaded by other reedswamp species, such as *Lysimachia vulgaris* and *Lythrum salicaria*, but mainly by subaquatic, marsh and terrestrial species, among which the following are common:

Alopecurus geniculatus	Plantago lanceolata
Filipendula ulmaria	Poa trivialis
Mentha aquatica	Ranunculus repens
Myosotis scorpioides	Rorippa (Nasturtium) amphibia
Nasturtium officinale	R. islandica (Nasturtium palustre)

Thus the growth of Glycerietum maximae depends much more on the accumulation of black humus produced by itself than on the trapping of silt, and at least the outer edge floats on watery organic ooze. The closely

[1] Pl. 106, phot. 261, p. 582.

PLATE 110

Phot. 272. Non-silted community in the River Tees at Neasham, Yorkshire. *Ranunculus fluitans* dominant. *R. W. Butcher* (1933).

Phot. 273. Silted community in the River Lark, Suffolk. *Sagittaria sagittifolia* (f. *vallisneriifolia*) and *Sparganium simplex* (submerged form) dominant. *R. W. Butcher* (1933).

SILTED AND NON-SILTED RIVER COMMUNITIES

PLATE 111

Phot. 274. Reedswamp of *Glyceria maxima*, summer. River Lark, Suffolk. *R. W. Butcher* (1933).

Phot. 275. The same view, winter condition. *R. W. Butcher* (1933).

GLYCERIETUM MAXIMAE IN THE LARK

matted rhizomes and decaying debris of *Glyceria* appear to be more re-sistant to the scouring of floods than the plants of *Sparganium erectum*, which, in the Thames, not uncommonly appears as isolated patches in a Glycerietum and may compete successfully with *Glyceria*.

(5) **Current very slow or negligible.** River bed of fine mud.

In the Lark, under nearly stagnant and muddy conditions, Butcher (1933) observed the following community:

Potamogeton lucens	d	Nuphar luteum	f
P. pectinatus	sd	Zannichellia palustris	o
Sparganium simplex	a	Callitriche stagnalis	o
Potamogeton natans	f	Scirpus lacustris	r
P. crispus	f	Sparganium erectum	r

With the appearance of the yellow water-lily and the floating-leaved pondweed this community is obviously beginning to approximate to that of a pond.

In very slow-moving or almost stationary water in the Thames, over fine silt and mud, we have the same dominants with different associated species:

Potamogeton lucens	d	Potamogeton compressus	o
P. pectinatus	sd	P. pectinatus forma	
Elodea canadensis	a	interruptus	o
*Sagittaria sagittifolia	a	Hippuris vulgaris (sub-	
*Oenanthe fluviatilis	a	merged)	o
Potamogeton pusillus	o	Ceratophyllum demersum	r

* Flowering.

Reedswamp, too, is well developed, and the following species occur in canalised portions with no current and a heavy deposit of fine mud:

Sparganium erectum[1]	d	Typha latifolia	f
Glyceria maxima[2]	d	Scirpus lacustris	f
Phalaris arundinacea	f	Alisma plantago-aquatica	f

Turbid rivers. When the water is turbid, cutting down the light in-tensity, and the substratum is unstable under flooding, the submerged vegetation may be almost absent, as in the River Tern in Shropshire. In such rivers there is however often a littoral reedswamp, which Butcher found in the Tern to consist of the following species:

Sparganium erectum	d	Myosotis scorpioides	f
Scirpus lacustris	sd	Typha latifolia	o
Sagittaria sagittifolia	f	Alisma plantago-aquatica	o
Butomus umbellatus	f	Rorippa amphibia	o

Riparian community. Parts of the river bank of the Thames (and of many other rivers) particularly in the concavities of bends subject to the scour of the stream in flood, commonly rise vertically 2–3 ft. (*c.* 60–90 cm.) above the average water level. This exposed wall, generally of clay and overlying alluvium, affords a station for a characteristic open community,

[1] Pl. 109.   [2] Pl. 111.

which is transitory, since it is cut away at intervals by the river. The following species occur:

Achillea ptarmica	Lycopus europaeus
Barbarea vulgaris	Lysimachia nummularia
Brassica rapa	Lythrum salicaria
Cochlearia armoracia	Polygonum amphibium*
Hypericum acutum (tetrapterum)	Potentilla reptans
Juncus articulatus (lamprocarpus)	Scutellaria galericulata

* Terrestrial state with rhizomes extending into the water as the floating aquatic form.

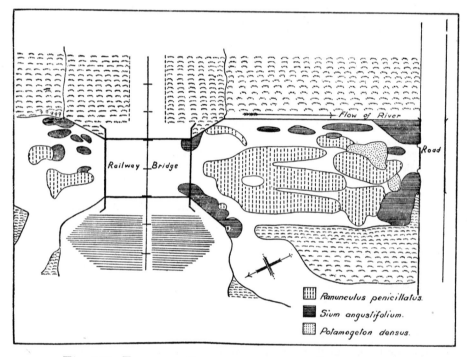

FIG. 123. EFFECT OF SHADE ON DIFFERENT SPECIES IN THE RIVER ITCHEN (HAMPSHIRE)

Sium erectum (angustifolium) grows farthest under the bridge, while *Ranunculus pseudo-fluitans (penicillatus)* diminishes in quantity as it approaches the bridge. From Butcher, 1927.

Shading effect. Differentiation of habitat by degrees of light intensity does not seem to be so conspicuous among aquatic as among terrestrial plants, but Fig. 123 illustrates the tolerance of shade by *Sium erectum* in contrast with other species.

Summary. Butcher summarises the river communities he has studied and of which examples have been given, as *torrential, non-silted, silted* and *"littoral"*, the last-named being so-called because it approximates to the vegetation near the sides of ponds. The first is dominated by bryophytes, the second by *Ranunculus fluitans* (Pl. 110, phot. 272), *R. pseudo-fluitans* or *Myriophyllum spicatum*, while the dominants of the last two are much more varied (Pls. 109, 111).

As Butcher (1933) points out, the "ideal" river, which is richest in plant and animal life, is one which is silted, but not excessively, particularly not by silt which is preponderantly organic, since the layer of water immediately above thick organic silt becomes de-oxygenated, so that the fauna is neither varied nor plentiful.

REFERENCES

BAKER, H. Unpublished memoranda on aquatic vegetation of the Thames near Oxford.

BUTCHER, R. W. A preliminary account of the vegetation of the River Itchen. *J. Ecol.* **15**, 55–65. 1927.

BUTCHER, R. W., PENTELOW, F. T. K. and WOODLEY, J. W. A biological investigation of the River Lark. *Fisheries Investigation Series*, No. 3. Ministry of Agriculture and Fisheries, 1931.

BUTCHER, R. W. Studies on the ecology of rivers. I. On the distribution of macrophytic vegetation in the rivers of Britain. *J. Ecol.* **21**, 58–91. 1933.

MINNIKIN, R. C. *Practical River and Canal Engineering.* London, 1920.

WATSON, W. The bryophytes and lichens of fresh water. *J. Ecol.* **7**, 71–83. 1919.

Chapter XXXII

MARSH AND FEN VEGETATION

MARSH. THE ESTHWAITE "FENS"

Definitions. Words like *marsh, fen, bog* and *swamp* are rather loosely employed in ordinary language; and when words in common use are defined in a technical sense with the necessary limitations and exclusions, there is always some danger of misunderstanding. Nevertheless in the ordinary usage of these four terms there is a certain rough uniformity which corresponds well enough with the nature of the vegetation and soil, and invites an attempt to stabilise their meanings in an ecological sense.

In this book (as in *Types of British Vegetation*) the term *marsh* is applied to the "soil-vegetation type" in which the soil is waterlogged, the summer water level being close to, or conforming with, but not normally much above, the ground level, and in which the soil has an inorganic (mineral) basis; *fen* to a corresponding type (whose vegetation is closely similar) in which the soil is organic (peat) but is somewhat or decidedly alkaline, nearly neutral, or somewhat, but not extremely, acid. *Bog*, on the other hand (bearing a radically different vegetation), forms peat which is extremely acid. *Swamp* is used for the type in which the normal summer water level is above the soil surface: it is usually dominated by reeds (*Phragmites*) or by other tall grasses, sedges or rushes, often accompanied by dicotyledonous species of similar habit: the commonest kind of swamp is therefore called *reedswamp*.

In marsh, fen and swamp the water is telluric in origin, in bog it may or may not be.

Swamp or reed-swamp has already been described (Chapters XXIX–XXXI) as the last term of aquatic vegetation. It also forms a transition to the land vegetation of marsh or fen. When the soil level rises to or above the water level, marsh plants begin to colonise the swamp, and the reed-plants begin to give way to them, though certain species, notably the common reed (*Phragmites*), may maintain themselves in the marsh or fen and even spread within it by means of their strong horizontal rhizomes.

Marsh. Marsh commonly occurs on low river banks or on the shores of lakes and on the undrained flood plains of rivers, and its mineral soil often consists of alluvial silt; but marsh exists wherever mineral soil is waterlogged, irrespective of its origin. Where the relation of water level to soil level remains approximately stable, marsh (or fen) represents an edaphic climax. But if from any cause the soil surface is progressively built up above the water level (or the water table is lowered) so that the root systems of the plants are better aerated, the marsh (or fen) vegetation gives way to a

more completely terrestrial type—grassland, scrub or forest—and ultimately to the climatic climax. This certainly happens quite normally as a
continuation of the hydrosere wherever the soil is progressively and effect-
ively raised, but the progress to climax forest is difficult to demonstrate in
this country, because of the regulation of rivers checking the silting of
alluvial plains, the draining of marshy ground disturbing the natural seres,
the destruction of forest, and the resulting paucity of seed parents for the
later stages of the succession, and especially for the climax dominants.

Marsh vegetation is commonly zoned round or along the edge of any
permanent body of water, unless the bank is very steep so that the soil
is not waterlogged at all. It is very various in detail, and the factors
determining the different communities are not at all thoroughly understood.

Esthwaite marshes. Pearsall (1918) describes various zoned marsh
communities on the shores of Esthwaite Lake with a substratum of rounded
morainic stones, very little eroded or silted.

(1) On slightly eroded tracts of this morainic shore line are consocies of
Eleocharis palustris and *Phalaris arundinacea*, accompanied by the purple
loosestrife (*Lythrum salicaria*), extending below the summer water level
and about 10–15 cm. above it. These are said to give way to *Phragmites*
where peat accumulates or silting sets in; and all three communities may
equally well be called reedswamp (cf. p. 585). The dominants and *Lythrum*
all have tough rhizomes able to penetrate the hard substratum.

(2) Where pasture land abuts on the lake, and there is presumably
softer soil, a mixed community occurs, extending about 10–30 cm. above
the summer water-level, with no dominants and containing the following
species:

Achillea ptarmica	f	Lysimachia nummularia	la
Caltha palustris	l	Lythrum salicaria	o
Carex diversicolor (flacca)	f	Myosotis caespitosa	f
C. flava	f	Polygonum hydropiper	o
C. goodenowii	o	Prunella vulgaris	f
Comarum palustre	l	Ranunculus flammula	f
Hydrocotyle vulgaris	f	Senecio aquaticus	l
Hypochaeris radicata	o		
Juncus articulatus	a	Climacium dendroides	l
J. effusus	f	Hypnum cuspidatum	r
J. conglomeratus	l	Rhacomitrium heterosti-	
Littorella uniflora	o	chum	f

Apart from the bryophytes these are all genuine marsh plants except
Littorella, which, as we have seen (Chapter xxx), is primarily a shallow-
water aquatic, *Prunella*, which is a pasture plant, and *Hypochaeris*, a deep-
rooted pasture or wayside species commonly found on drier ground.

(3) Where woods of *Quercus sessiliflora*, the climax forest of the hillsides
of this region, approach the lake, the water's edge is fringed with a narrow

belt of alderwood with ash and grey sallow, and a ground vegetation which is partly marsh and partly woodland:

Alnus glutinosa	a	Juncus acutiflorus	ld
Fraxinus excelsior	f	J. articulatus	o
Salix atrocinerea	f	Lychnis flos-cuculi	f
		Lythrum salicaria	f
Achillea ptarmica	f	Meconopsis cambrica	r
Angelica sylvestris	la	Melandrium dioicum	f
Agrostis stolonifera	l	Molinia caerulea	ld
Brachypodium sylvaticum	la	Phalaris arundinacea	ld
Caltha palustris	f	Poterium officinale	o
Centaurea "nigra"	r	Ranunculus flammula	f
Circaea lutetiana	o	R. repens	o
Cirsium palustre	la	Rumex acetosa	l
Deschampsia caespitosa	la	R. obtusifolius	o
Epilobium montanum	o	Scrophularia nodosa	f
Filipendula ulmaria	f	Senecio aquaticus	o
Galium aparine	o	Succisa pratensis	a
Holcus lanatus	o	Valeriana officinalis	f
		V. sambucifolia	l

Phalaris and *Juncus acutiflorus* are locally dominant nearer the lake where the light intensity is 50 per cent, *Molinia, Filipendula* and *Deschampsia,* and finally *Circaea* and *Mercurialis perennis,* in the edge of the oakwood itself with a light intensity of 5 per cent.

(4) Behind the slope of unstable gravel, 1 m. wide, on the outward face of an open bank of morainic material which was formerly submerged and superficially eroded, but which now has a maximum elevation of 2·75 m. and a slope of 1 in 15, the following zoned communities occur:

(a) A line of *Carex hudsoni* (water level 7–14 cm. below the soil surface) which acts as a breakwater for the zones behind it:

(b)

Caltha palustris	Succisa pratensis
Lythrum salicaria	Valeriana sambucifolia

(c) A dense sward of

Carex panicea	a	Lythrum salicaria	f
C. stellulata	f	Molinia caerulea	f
Comarum palustre	f	Prunella vulgaris	o
Filipendula ulmaria	lf	Ranunculus flammula	a
Hydrocotyle vulgaris	f	R. repens	f
Juncus articulatus	la	Viola palustris	f
J. effusus	o		

(d) With accumulation of peat ("acid fen")

Molinia caerulea	Myrica gale

(e) A thicket of small trees and shrubs:

Alnus glutinosa	a	Fraxinus excelsior	l
Betula pubescens	a	Quercus sessiliflora	o
Corylus avellana	l	Salix atrocinerea	a
Frangula alnus	f		

The thicket is extending to the *Myrica-Molinia* zone, and as the water level falls (which it will do owing to the cutting back of the lake outlet) the successive zones will presumably move forward towards the lake. Here then we have acid fen succeeding marsh, but quickly overtaken by forest— Alnetum followed by Quercetum sessiliflorae.

Cornish marshes. In north Cornwall various types of alluvial marsh have been recognised by Magor (unpublished). The first two, situated near the sea over soil derived from Devonian slates, but not subject to flooding by salt or brackish water, may fairly be called reedswamp.

(1) The first is marked by the prevalence of tall rushes, grasses and dicotyledonous herbs, the dominants often forming almost pure local communities:

Epilobium hirsutum	co–d	Festuca elatior	f
Sparganium erectum	co–d	Filipendula ulmaria	f
Phalaris arundinacea	ld	Juncus articulatus	f
Juncus conglomeratus	va	Lotus uliginosus	f
Alopecurus geniculatus	a	Lycopus europaeus	f
Apium nodiflorum	a	Nasturtium officinale	f
Myosotis scorpioides		Poa pratensis	f
(palustris)	a	Potentilla anserina	f
Ranunculus acris	a	Rumex conglomeratus	f
Solanum dulcamara	a	Veronica beccabunga	f
Caltha palustris	f	Carex hirta	o
Calystegia sepium	f	Juncus acutiflorus	o
Carex remota	f	Oenanthe crocata	o
C. vulpina	f	Orchis praetermissa	o
Equisetum fluviatile		Vicia cracca	o
(limosum)	f		

(2) Another type, also dominated by similar tall plants but with moving water, has a distinctly different list of species, only about half being common to the two types: it is very local, occurring at Polzeath and Mennic Bay. Here the rare galingale (*Cyperus longus*) is co-dominant with *Sparganium erectum*:

Cyperus longus	co–d	Holcus lanatus	f
Sparganium erectum	co–d	Lotus uliginosus	f
Carex riparia	ld	Lychnis flos-cuculi	f
Galium palustre	va	Menyanthes trifoliata	f
Juncus inflexus (glaucus)	va	Oenanthe crocata	f
Mentha aquatica	va	Phalaris arundinacea	f
Polygonum persicaria	va	Poa pratensis	f
Epilobium hirsutum	a	Potentilla anserina	f
Equisetum fluviatile		Pulicaria dysenterica	f
(limosum)	a	Ranunculus acris	f
Myosotis scorpioides		R. repens	f
(palustris)	a	Rumex hydrolapathum	f
Rumex conglomeratus	a	Senecio aquaticus	f
Iris pseudacorus	la	Solanum dulcamara	f
Apium nodiflorum	f	Plantago lanceolata	o
Bellis perennis	f	Scrophularia aquatica	o
Caltha palustris var. minor	f	Plantago major	r
Filipendula ulmaria	f	Scirpus pauciflorus	r

(3) A third type is developed higher up the rivers and farther from the maritime influence. Here *Juncus articulatus* is dominant and woody plants (alders and sallows) have begun to colonise the marsh:

Juncus articulatus	d	Polygonum lapathifolium	f
Mentha aquatica	va	Potentilla procumbens	f
Cirsium palustre	a	Pulicaria dysenterica	f
Filipendula ulmaria	a	Rumex acetosa	f
Iris pseudacorus	a	Senecio aquaticus	f
Juncus effusus	a	Succisa pratensis	f
Lotus uliginosus	a	Veronica chamaedrys	f
Lychnis flos-cuculi	a	Athyrium filix-femina	o
Mentha aquatica × arvensis	a	Callitriche stagnalis	o
Poa pratensis	a	Galium palustre	o
Ranunculus acris	a	Hypericum acutum	
R. flammula	a	(tetrapterum)	o
Alnus glutinosa	la	Montia fontana	o
Sparganium erectum	la	Potentilla anserina	o
Epilobium hirsutum	la	Rubus sp.	o
Hydrocotyle vulgaris	f	Rumex conglomeratus	o
Juncus conglomeratus	f	Salix atrocinerea	o
Lythrum salicaria	f	S. aurita	o
Pedicularis sylvatica	f	Dryopteris spinulosa	o
Plantago lanceolata	f		

Of these thirty-eight species only fifteen occur in (2) and only nine in (1), eight species being common to all three types.

Magor comments on the different habitat requirements of the different species of rush: *Juncus conglomeratus*, a very abundant species in type (1), is wholly replaced by *J. inflexus* (*glaucus*) in the Cyperetum longi, and largely by *J. articulatus* with *J. effusus* in the riverside freshwater marsh farther from the sea, while yet other species of the genus occur in salt marsh and in peat bog.

Silting and peat formation. Although marsh has been defined as the wet soil-vegetation complex in which the soil is mainly mineral it must be understood, of course, that humus is constantly formed in marsh soil as a necessary result of the continual growth and decay of the vegetation. Under waterlogged conditions, indeed, this increase in humus will always tend to transform the soil from preponderantly mineral to preponderantly organic. We have seen (Chapter IV, p. 88) that "meadow soil", with impeded drainage, contains the largest proportion of humus among the terrestrial types. Increase in the proportion of humus can only be checked by an increased rate of disintegration following better aeration and concomitant increase in the population of soil organisms which carry out the disintegrative processes; or by the fresh accretion of mineral silt. On a river flood-plain periodically overspread by fresh silt we may thus have the soil continually raised, and a terrestrial succession initiated, leading to the climatic climax. In the earlier stages the soil surface is raised largely by the deposition of inorganic silt; in the later, when it is above the level of any but exceptional floods, mainly by the accumulation of plant debris, which, at such levels, is less waterlogged and consequently better aerated. In such cases marsh will form the section of the hydrosere between purely aquatic and purely terrestrial vegetation.

Fen. But where periodic mineral silting is a less important factor or is absent altogether the soil is mainly or entirely organic from the outset, and where the supply of basic ions is adequate the type of vegetation succeeding reedswamp is typical *fen*.[1] The organic soil is formed by the decay of the plant debris under relatively anaerobic conditions, and is therefore *peat*. It is however irrigated by water relatively rich in basic ions and often alkaline in reaction (see Chapter IV).

The East Anglian Fenland (Fig. 126, p. 148) lying between Cambridge and the Wash is the largest and best known area of fen in the country, and since its waters come mainly from calcareous rocks they are alkaline, and the fen vegetation (of which nothing more than fragments now remain in the East Anglian Fenland) is developed in a medium of relatively high pH value. The effect of this general alkaline "buffer" is mainly seen in the general absence of markedly oxyphilous species, which only occur locally where the peat is built up above the level of the alkaline ground water. The bulk of the fen vegetation consists of plants of waterlogged soil which are however in no way tied to alkaline conditions, but flourish equally well in marsh or fen with neutral or somewhat acid waters, though some "calcicolous" species are present. Smaller areas of fenland exist in many other parts of the country.

Esthwaite fens. The East Anglian fens are very little or not at all silted by river flooding, so that their soil is pure peat, but much vegetation which we cannot naturally separate from fens is silted to a greater or less degree. Thus Pearsall (1918) described fen at the N. end of Esthwaite Water (Figs. 124, 125), parts of which are heavily silted and whose waters are at the same time somewhat acid. Here we have waterlogged soils varying from mainly inorganic to mainly organic, i.e. transitions between marsh and fen according to our definition, but which, since their development and their vegetation are closely similar, are here treated with the fens.[2] The building up of such fens, while contributed to by inorganic silting, is largely brought about by the growth of peat. Only in the "reedgrass fen" does the in-

[1] Where the ground water is poor in basic ions and very acid in reaction, "bog" or "moss" is formed, characterised by quite a different type of vegetation and of peat (see Chapter XXXIV).

[2] Pearsall separates the apparently more or less static communities of waterlogged mineral soil surrounding the lake as "marsh" from those which are undergoing active development, which he calls "fens". It is usually better to make the distinction depend on the preponderance of mineral or inorganic material in the soil, irrespective of whether it is being raised by silting or not; but it must be recognised that no sharp line can be drawn between the two, which bear very similar vegetation so long as the supply of bases is adequate. The distinction is not in fact, so far as we know at present, of any great *vegetational* importance.

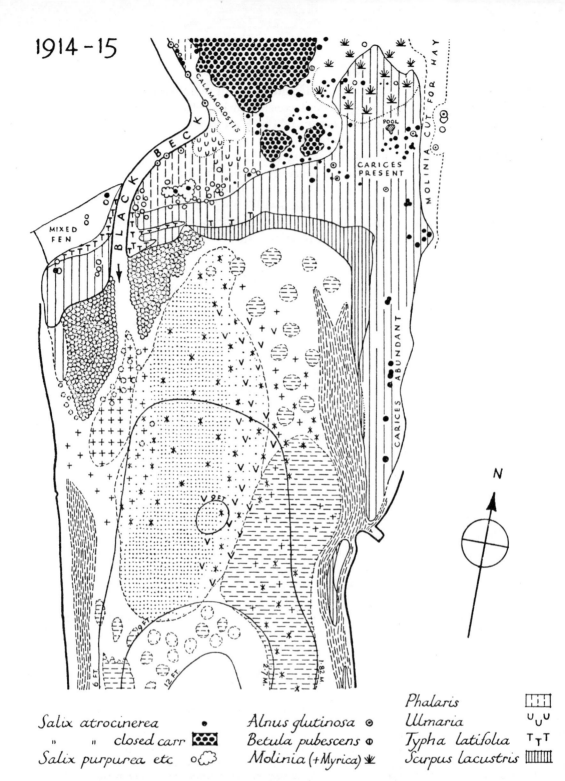

1914–15

Salix atrocinerea •   Alnus glutinosa ⊙   Phalaris

     "   "   closed carr   Betula pubescens ⊕   Ulmaria

Salix purpurea etc ⃝   Molinia (+Myrica) ⋇   Typha latifolia

                                                        Scirpus lacustris

FIG. 124. NORTH FEN AT THE HEAD OF ESTHWAITE WATER, IN 1914–15

The aquatic communities occupying nearly three-quarters of the figure are described in connexion with Fig. 114 (p. 604) of which this is a reprint. The fen to the north is divisible into (1) a strip bordering the Black Beck, with rapid sedimentation, *Calamagrostis canescens*, *Phalaris arundinacea*, local *Filipendula ulmaria*, *Salix purpurea* and *S. decipiens*; (2) a central strip of moderate sedimentation with carr of *Salix atrocinerea*; and (3) an eastern strip of very slow sedimentation with *Molinia caerulea* and *Myrica gale*.

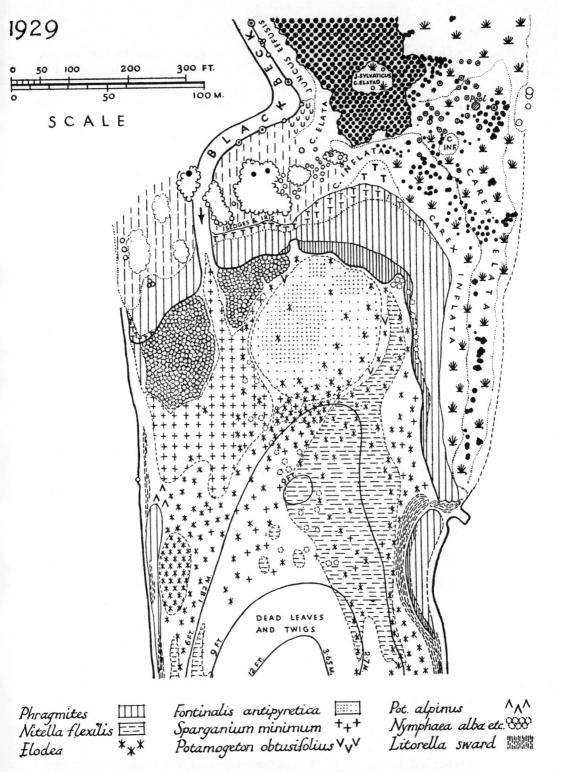

1929

SCALE

Phragmites | Fontinalis antipyretica | Pot. alpinus
Nitella flexilis | Sparganium minimum | Nymphaea alba etc.
Elodea | Potamogeton obtusifolius | Litorella sward

FIG. 125. NORTH FEN, ESTHWAITE WATER, IN 1929

Compare with Fig. 124. During the 15 years' interval between the two surveys the Carices have increased and invaded the central and western strips, the Salices have increased in all parts, and Molinietum has superseded Phragmitetum over a wide zone in the north-eastern and eastern parts. For symbols of fen species see Fig. 124.

organic largely preponderate over the organic fraction, so that on our definition it ought to be called "marsh". But such a separation would be quite artificial. The following figures (Pearsall 1918) show the relative proportions of organic and inorganic material in the Esthwaite fens and the preceding reedswamps:

Table XX. *Esthwaite fen soils: ratio of organic to inorganic constituents and pH values**

		Sedimentation		
		Rapid	Moderate	Slow
Reedswamp:				
Ratio O./I.		1·3 (*Typha*)	1·51 (*Phragmites*)	1·76 (*Phragmites*)
pH		6·4–6·0		5·5–5·4
Fen:				
"Mixed fen"	Ratio ⎰1·06		?1·5–2·0	2·98 (*Molinia*)
"Reed grass"	O./I. ⎱0·26			
pH		5·7–5·4		5·0–4·5
				(4·9–4·4 in 1936)
Carr (*Salix atrocinerea*):				
Ratio O./I.			2·2	
pH			Young 5·5–5·2	
			Middle 5·3–5·1	
			Old 5·0–4·8	

* The inorganic fraction necessarily includes "ash" derived from the plant debris, as well as inorganic silt. The pH values were taken in 1929. All the figures relate to the area shown in Figs. 124 and 125, sedimentation being rapid on the left near the beck, moderate in the middle, and slow on the right.

From the figures in this table, which are calculated from the averages of Pearsall's data, it is apparent that all the soils of the Esthwaite fens except that of the "reed grass" associes have a preponderance of organic material (in "mixed fen" the fractions are practically equal), that this is relatively high in the reedswamps, increasing with diminished sedimentation very greatly in the *Molinia* fen, where silting is at a minimum or absent and the organic is practically three times the inorganic fraction. In the richly silted fens it is much lower. All the pH values are on the acid side of neutrality, and there has been a general tendency for them to fall since 1929.

Succession on the Esthwaite fens. (1) *Area of rapid sedimentation*. The North Fen at Esthwaite is strongly influenced by the Black Beck, a rapid stream round whose exit into the head of the lake the fen has grown up, and which regularly floods and deposits silt upon the parts adjacent to the course of the stream (Figs. 124, 125). The soil in this part of the fen is relatively high and well drained, consisting largely of inorganic silt (O./I.=0·26, 1·06), and the water level depends on the rise and fall of the beck. When the bed of the beck has been cleared above the outfall much more silt is brought down into the lake, and part is also deposited on the adjacent fen during floods. A rapid development of the vegetation ensues.

(a) The reedswamp at the mouth of the beck is dominated by *Typha latifolia* (*p*H 6–6·4), this being the only part of the lake where the great reedmace occurs. *Phragmites* accompanies it, especially towards the open water, and the chief associates of the two dominants are *Caltha palustris*, *Menyanthes trifoliata*, *Ranunculus lingua* and *Scutellaria galericulata*.

(b) The "mixed fen" associes which follows this reedswamp in the succession has a black muddy soil with organic and inorganic fractions nearly equal and numerous abundant species, of which the following are typical.

Locally dominant or subdominant

Carex hudsonii (elata)   Menyanthes trifoliata
Galium palustre   Phalaris arundinacea
Iris pseudacorus   Typha latifolia

Abundant or locally abundant

Agrostis gigantea   Menyanthes trifoliata
Carex inflata   Myosotis scorpioides
Comarum palustre   Phragmites communis
Filipendula ulmaria   Ranunculus lingua
Lotus uliginosus   Salix atrocinerea subsp. aquatica
Lysimachia vulgaris   × S. decipiens
Lythrum salicaria   S. purpurea

(c) Adjoining the beck is a belt of what Pearsall calls the "reed grass associes" because relatively pure consocies of *Phalaris arundinacea* and *Calamagrostis canescens* are the conspicuous features, the former in the lower lying parts towards the lake and the latter in the older and drier parts of this zone of rapid silting. The soil is a grey-brown clayey silt with only about 20 per cent of humus. In summer the water level may be as much as a foot (30 cm.) below the surface. Besides the reed grasses, consocies of *Filipendula ulmaria* occur, as well as the following associates:

Caltha palustris	lf	× Salix decipiens	l
Galium palustre	lf	Scutellaria galericulata	o
Juncus effusus	o	Urtica dioica	o
Salix atrocinerea	l	Valeriana officinalis	l
S. purpurea	f		

Salis purpurea and × *S. decipiens* are particularly characteristic of rapidly silted areas. A few alders fringe the stream. If developed extensively enough this type of habitat might ultimately produce woodland of alder and birch. It is probably only on well aerated gravelly soils that oakwood would eventually appear.

(2) *Area of moderate sedimentation.* This is the central vertical strip, containing "closed carr", in Figs. 124 and 125. (a) The reedswamp is here composed of *Scirpus lacustris* towards the open water and *Phragmites communis* inshore. The soil is peaty, and closer and tougher than in (1a). Herbaceous associates are sparse, the bog bean (*Menyanthes*) being the only common species.

41-2

(*b*) A zone in which *Carex hudsonii* (*elata*) or *C. inflata* is dominant succeeds the *Phragmites* reedswamp with but little change of soil, and the water level is close to the surface (5–10 cm.). The following species occur:

Carex hudsonii (elata)	d	Lythrum salicaria	f
C. inflata	l	Phalaris arundinacea	l
C. vesicaria	lsd	Phragmites communis	f
Comarum palustre	f	Scutellaria galericulata	f
Eriophorum angustifolium	r	Senecio aquaticus	l
Galium palustre	f	Typha latifolia	r
Hydrocotyle vulgaris	o		

The appearance of *Eriophorum angustifolium*, which is more commonly associated with acid bogs though not extremely oxyphilous, is of interest: it also occurs locally in the East Anglian fens.

(*c*) The next zone, called by Pearsall "open carr", is evidently a transitional zone in which woody plants are colonising the fen (Fig. 124). The soil is very variable in organic content and also in water level, which ranges from the surface to 15 cm. below it. Together with the woody plants there is also a great variety of herbaceous fen species, of which it is worth while to cite the full list given by Pearsall, for comparison with the lists from the East Anglian and Irish fens given on pp. 656, 664–6. It will be noted that a good many of the species, though by no means all, are the same:

Achillea ptarmica	Juncus articulatus
Agrostis stolonifera	J. effusus
A. tenuis	J. acutiflorus
Angelica sylvestris	Lathyrus pratensis
Anthoxanthum odoratum	Lotus uliginosus
Caltha palustris	Lychnis flos-cuculi
Cardamine pratensis	Lycopus europaeus
Carex canescens	Lysimachia vulgaris
C. goodenowii var. juncella	Lythrum salicaria
C. inflata	Mentha aquatica
C. hudsonii (elata)	M. arvensis
C. panicea	Menyanthes trifoliata
C. paniculata	Molinia caerulea
C. vesicaria	Orchis maculata
"Centaurea nigra"	Parnassia palustris
Cirsium palustre	Phalaris arundinacea
Comarum palustre	Phleum pratense
Crepis paludosa	Phragmites communis
Deschampsia caespitosa	Poterium officinale
Dryopteris spinulosa	Prunella vulgaris
Eleocharis palustris	Ranunculus repens
Epilobium parviflorum	Rumex acetosa
Equisetum fluviatile	R. crispus
Filipendula ulmaria	R. obtusifolius
Galeopsis tetrahit	Scrophularia nodosa
Galium palustre	Scutellaria galericulata
Holcus lanatus	Senecio aquaticus
Hydrocotyle vulgaris	Stachys palustris and var. canescens
Iris pseudacorus	Stellaria palustris

Succisa pratensis
Urtica dioica
Valeriana officinalis
V. sambucifolia
Veronica scutellata
Viola canina
V. palustris

Woody species

Alnus glutinosa
Betula pubescens
Rosa caesia (coriifolia)
Rubus idaeus
Salix atrocinerea
S. purpurea
Solanum dulcamara
Viburnum opulus

Of the herbaceous species *Filipendula ulmaria* and *Molinia* are the most abundant. Of the woody plants *Salix atrocinerea* is the most abundant species: drier, more inorganic soils bear *Salix purpurea*;[1] and *Betula pubescens* occurs where the organic content is very high.

(*d*) In the closed carr (Figs. 124, 125) the soil is a brown clayey mud, variable in the ratio of organic to inorganic constituents, but with a high average (2·2). The water level is also variable, frequently reaching the surface, and the drainage is very poor. *Salix atrocinerea* is dominant in close dense canopy, with *S. purpurea* and *S. aurita* rare, and *Alnus glutinosa*, *Betula pubescens* and *Frangula alnus* local. The light intensity beneath the sallows is low and obviously limits the ground vegetation, which is very sparse. In old carr the values range from 1 to 5 per cent, and there are practically no plants below the trees and shrubs. Where a scattered vegetation is present the light values are from 3 to 14 per cent. Where it rises to 70 per cent there is a dense vegetation of flowering *Filipendula ulmaria*, *Carex hudsonii* and *Lythrum salicaria*. Under the deeper shade of the carr, the Carices, *Phragmites* and *Iris* very seldom flower. The ground vegetation consists of the following species:

Agrostis gigantea	l	Iris pseudacorus	l
Caltha palustris	l	Lythrum salicaria	f
Carex hudsonii	la	Mentha arvensis	l
C. vesicaria	f	Molinia caerulea var.	
Comarum palustre	o	viridiflora	la
Filipendula ulmaria	la	Phragmites communis	la
Galium palustre	o	Valeriana sambucifolia	o

The mosses and liverworts are characteristic:

Amblystegium serpens	la	Plagiochila asplenioides	l
Fissidens taxifolius	f	Plagiothecium denticu-	
Hypnum patientiae	o	latum	f
Lophocolea bidentata	f	Pterygophyllum lucens	o
Mnium punctatum	f		

Between 1914 and 1929 the thickets of *Salix purpurea* and the closed carr of *S. atrocinerea* extended considerably (Figs. 124, 125), the latter filling up the interval between the main area of carr and the two outlying patches present in 1914.

[1] In 1929 *Salix atrocinerea* only occurred where the *p*H value was above 5 and *S. purpurea* where it was above 5·4.

No trace of regeneration is found in the carr, which appears to be transitory, the *Salix* bushes ultimately dying and disappearing. The transience of the grey sallow is in accord with what happens in the East Anglian fens. At Esthwaite, however, there is no evidence that any kind of woodland follows this stage. When the carr disappears *Molinia* is still present and it is probable that a bog with *Myrica gale*, *Erica tetralix*, etc., develops, as in (3) below. Here we encounter the influence of the cool wet climate which tends to produce bog as climax vegetation on all undrained soils with deficiency of oxygen.

(3) *Area of slow sedimentation*. This is the vertical strip on the right of Figs. 124 and 125.

(*a*) *Reedswamp*. As in (2*a*) this consists of an outer zone of *Scirpus lacustris* and an inner zone of *Phragmites communis*. The peaty soil has an organic/inorganic ratio of 1·76 and the peat is closer and less decayed. In 1914 *Phragmites* was still dominant for a distance of 45–50 m. beyond where the water level was at or just below the soil surface. In 1929, i.e. in the course of the 15 years since 1914, the outer edge of the *Phragmites* reedswamp advanced about 30 m. into the water and its inner portion was replaced by the Molinietum to a depth of about 100 metres (Figs. 124 and 125). Here then, this part of the succession is actually demonstrated as having occurred within a comparatively short term of years.

(*b*) *Molinietum*. As the *Phragmites* thins out on the inner edge of the reedswamp *Comarum palustre* becomes abundant, *Carex inflata* and *Lysimachia nummularia* frequent, and *Molinia caerulea* colonises the peat, gradually replacing *Phragmites* as the dominant when the water level is about 10 cm. below the peat surface. Here the organic/inorganic ratio is about 3. In striking contrast to the "mixed fen" and "open carr" of (2) the Molinietum is very poor in species, containing only the following:

Molinia caerulea	d	Myrica gale	lsd
Galium saxatile	f	Dryopteris dilatata	r
Potentilla erecta	l		
Succisa pratensis	l	Sphagnum acutifolium	l

Transition to bog. This Molinietum can barely be classed as fen, but is rather a stage in the development of bog. Later in the succession the *Molinia* dies out while *Myrica gale* persists. Thus under the conditions of this north-western climate, and with waters draining from mainly acidic rocks, the communities resembling those of "true fen" are only developed where there is abundant inorganic silting, and are very transitory.

The limits of fen. Though the alkaline fens of East Anglia have here been spoken of as "true fen", it is clear that no line based on soil reaction can be drawn between them and fens which are quite acid such as the "mixed fen" at Esthwaite (*p*H *c*. 5·5). Many or most of the "true fen" species can and do flourish in peat of this degree of acidity. Probably more important is the base status of the peat, irrespective of soil reaction (cf. the

exacting woodland flora of the acid clay-with-flints on the Chiltern plateau,
Chapter XIX). What is wanted in order to understand the whole range of
these fen communities is a close investigation of the conditions under which
they can develop in different localities. As the ground water becomes
poorer in bases and acidity increases a point seems to be reached at which
an impoverished reedswamp of *Carex inflata* or *Equisetum fluviatile* (see
Chapter XXX, p. 617) takes the place of *Phragmites* (itself able to tolerate a
wide range of conditions in respect of acidity and base status). In poorer
conditions still Sphagnetum probably follows directly on the acidic water
vegetation. It is important to keep the terms *fen* and *moss* (or *bog*) for the
two well-marked types of vegetation, whatever the particular combination
of edaphic and climatic factors involved in producing them.

It is clear however that Molinietum is intermediate between luxuriant
fen on the one hand, and bog or moss on the other (cf. Chapters XXXIII and
XXXIV), and a re-investigation of the requirements of *Molinia*, its possible
associates, and the range of conditions under which Molinietum develops is
the immediate desideratum.

REFERENCES

MAGOR, E. W. Geographical distribution of vegetation in Cornwall: Camelford and
    Wadebridge district. (Unpublished.)
PEARSALL, W. H. The aquatic and marsh vegetation of Esthwaite Water. *J. Ecol.* **6**,
    53–74. 1918.
Supplementary information about the later development of the North Fen at
Esthwaite has been kindly supplied by Dr Pearsall.

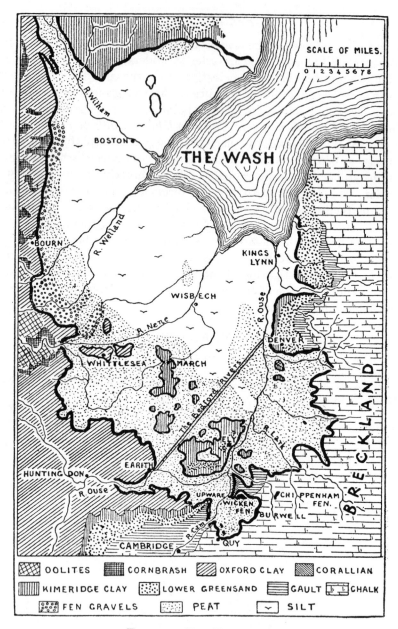

FIG. 126. THE FENLAND

The area of fenland is bounded by a thick black line, and consists of the "Marshland" (marine silt—unshaded) nearest the sea, and the "Fenland" in the narrower sense (peat—dotted) bordering the upland. The whole is drained and cultivated except a few small areas still bearing natural vegetation. Two of these, Wicken and Chippenham Fens, in the south are shown in black, the latter a "valley fen" outside the general peat area. The Cretaceous rocks (east and south-east) and the Jurassic rocks (south-west and west) bordering the Fenland and also forming "islands" in the southern part, are distinctively shaded, but the glacial deposits which cover much of their surface (Fig. 31, p. 108) are not shown. From Yapp, 1908.

Chapter XXXIII

THE EAST ANGLIAN FENS. NORTH IRISH FENS

THE EAST ANGLIAN FENS

The English "Fenland", *par excellence*, is the great tract of nearly flat peat and silt land about 1500 square miles in extent, scarcely raised above sea-level and occupying the northern portion of Cambridgeshire, with parts of the adjacent counties of Lincoln, Huntingdon, Norfolk and Suffolk (Fig. 126). It lies around the Wash, which is now the common estuary of the rivers Witham, Welland, Nen and Ouse. This great, very shallow basin, bounded to east and west respectively by Cretaceous and Jurassic rocks, was originally formed in pre-glacial times by an ancient river flowing north-east to join the confluent of the Rhine and the Thames in a plain which is now occupied by the North Sea. The whole region was invaded by ice-sheets during the Glacial Period, and since the withdrawal of the ice has suffered a depression of as much as 200 ft., broken by periods of stability or slight elevation, with corresponding transgressions and retreats of the sea. The result is that the floor of the Fenland consists of interdigitating wedges of marine silt and freshwater peat—the silt deposited by the invading tides, the peat built up by the fen plants which grew in marshes formed by the holding up of fresh water descending from the surrounding uplands. The surface soil is silt ("Marshland") towards the Wash, peat ("Fenland" in the narrower sense) towards the landward edge of the basin (see Fig. 126).

Practically the whole area is now cultivated. Drainage and reclamation began in Roman—perhaps in pre-Roman—times, but was not thoroughly carried out till the seventeenth century (see Chapter VII, pp. 186–7). The drained peat and silt (especially the latter) is very fertile, and is famous for its potatoes and locally for its fruit and its bulbs. Latterly, too, much sugar beet has been grown, since the relatively dry, sunny and somewhat continental climate is one of the most suitable in England for this crop. Before drainage the area must have been a waste of saltmarsh on the seaward, of fen and reedswamp on the landward side, the whole intersected by lagoons ("meres") and sluggish streams, tidal for considerable, though variable, distances from the sea. But when the land rose, or the peat grew up above the water level, the fen was extensively invaded by trees—oak, yew, pine and birch—the remains of which are still constantly being dug out of the peat. Recent evidence shows that raised bog also was formed in places towards the edges of the basin.

The few remaining fragments of peat fen bearing natural vegetation, which are all close to the edge of the basin, often owe their preservation to their use as "washes" or "catch-waters" into which flood water from the uplands could be turned when the cultivated areas were threatened with inundation. The surface of these surviving fens is at a higher level than that of the cultivated fenland, whose peat has wasted down, as the result of drying and ploughing, to a level markedly lower than that of the embanked

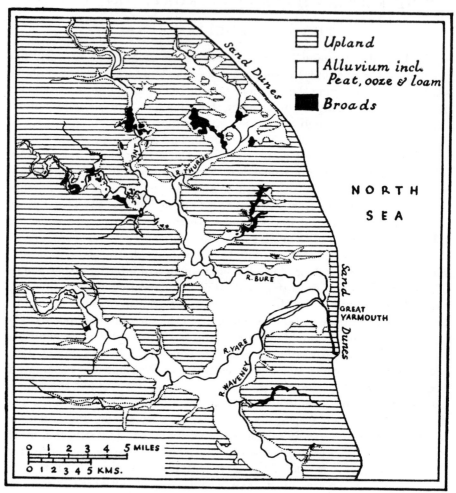

FIG. 127. THE FENLANDS OF EAST NORFOLK

The broads, shown in black, are confined to the upper parts of the old estuarine area where peat is formed. From Pallis, 1911.

rivers, into which the drainage water of the ditches ("lodes" or "dykes") is pumped—formerly by windmills, now by steam or oil (Diesel) engines. The whole drainage system is regulated by pumps and sluices. The present condition of the existing fragments of fen—of which Wicken, Chippenham, Woodwalton and Holme Fens are the largest and best known (the two first shown in Fig. 126)—is the result of human activity. The natural vegetation

of Wicken Fen, for example, has been constantly cut, and since it has been
occupied in strips by different owners, each of whom may have treated his
strip according to a different regime varying from time to time, the fen
shows a patchwork of different kinds of vegetation—dominated by
Cladium, by *Molinia*, by mixed herbs, sedges and grasses, or by bushes.
The disentangling of the detailed causes which have led to these varied
results has not been an easy task, but prolonged observation and experiment, aided by comparison with fen regions elsewhere which have not
been so closely under human control, have been successful in elucidating the

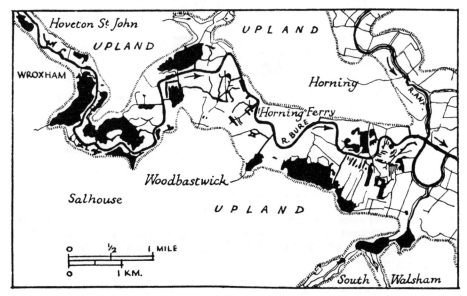

FIG. 128. PART OF THE BURE VALLEY (EAST NORFOLK)

The broads (black) lie just off the existing course of the river.

main outlines of the story together with many of the detailed mechanisms
(Godwin *et al.*, 1929–1936).

East Norfolk fens. A fen region in which the vegetation has been in large
part freer to develop on natural lines is found in East Norfolk. Here three
rivers—the Bure, Yare and Waveney and their tributaries—flow through an
alluvial plain, converging near Great Yarmouth, where their combined
waters enter the North Sea (Fig. 127). This region of the Norfolk "Broads"
—shallow lagoons which mostly lie surrounded by peat just off the courses
of the rivers (Figs. 128–9)—though much smaller than the Fenland, is quite
similar in its physical conditions, its soils and its natural vegetation; and
there is the same contrast of peat in the upper parts of the river valleys
with silt nearer the sea (Fig. 130). Before reclamation, indeed, the Fenland

---

[1] Fig. 129 and Pl. 88, phot. 213, show the situation of two broads to right and left of
the River Bure from which they are separated by fen peat.

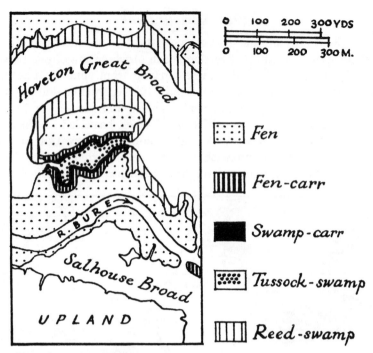

FIG. 129. HOVETON AND SALHOUSE BROADS (RIVER BURE)

These broads lie opposite one another on either side of the river. They are surrounded by reedswamp of *Phragmites communis* and *Typha angustifolia* which is succeeded by fen. The small piece of water between the fen island in Hoveton Broad and the fen bordering the river is however fringed with "tussock swamp" of *Carex paniculata* and *C. acutiformis*, and this is followed by "swamp carr" (cf. Pl. 88, phots. 214–15). From Pallis, 1911.

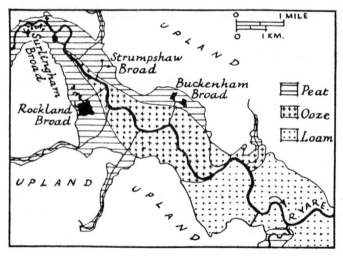

FIG. 130. PART OF THE YARE VALLEY

The peat, "ooze" and "loam" (silt) regions of "alluvium" are shown succeeding one another down the valley. The remains of the broads lie exclusively in the peat region (cf. Figs. 107, 108, pp. 593–4). From Pallis, 1911.

must have been much like parts of the Broads region to-day. The rivers of both areas drain calcareous soils so that their waters are alkaline, and in both the peat is correspondingly basic in reaction and bears an almost identical natural vegetation. But in East Norfolk the primary succession can be much more easily traced, and by its aid the different semi-natural plant communities seen at such a locality as Wicken Fen can be interpreted. The process of "Verlandung" has made considerable progress during the last century: several open broads of some size having "grown up" so that the open water is now restricted to quite small areas (Fig. 108, p. 594; Fig. 130).

Soil of Wicken Fen. The soil of Wicken Fen is a black, almost structure-less peat. The water content of the surface peat (6 cm.) expressed as a percentage of the volume of the fresh sample is about 80, at greater depths often higher. The air content expressed in the same way is about 5 per cent, in one sample from "litter" (Molinietum) as high as 11 per cent.

The ratio of the weight of organic to inorganic substance in the surface peat (6 cm.) varies from 1·2 to 4·0: at greater depths similar values are generally obtained, but they may be higher, reaching as much as 11·5 in one case under carr, at a depth of 20 cm. Thus much higher ratios occur than in the silted Esthwaite fens (Chapter XXXII, p. 642).

The percentage of oxygen in the soil air at a depth of 20 cm. is about 18, of CO_2 about 4 at the end of the summer. These values are distinctly lower for oxygen and higher for CO_2 than at the same depth under grass turf growing on ordinary garden loam. The percentage of CO_2 dissolved in the fen water is much higher still, and of oxygen very low indeed, the highest values being less than 0·5 per cent of the atmospheric concentration. This suggests very clearly the lack of aeration that must be suffered by the root systems during flooding.

"Under former lacustrine conditions fresh-water shell marl was formed over much of the Cambridgeshire fenlands, and this bed, 5–10 cm. in thickness, is to be found about 30 or 40 cm. below the peat surface over large parts of Wicken Sedge Fen" (Godwin, Mobbs and Bharucha, 1932). The surface peat above this bed shows an alkaline reaction almost everywhere. The top centimetre or so is usually neutral or slightly acid: at somewhat greater depths the *p*H value varies from 7 to 8·3, with an average of 7·5.

Vegetation. In Chapter XXVIII a brief notice was given to the aquatic vegetation of the Broads, and the reedswamps were seen to consist generally of *Typha angustifolia* and *Phragmites communis*, with an open community of *Scirpus lacustris* on the side towards open water in Barton Broad on the Ant, and a closed swamp, passing to fen, of *Glyceria maxima* (*aquatica*) and *Phalaris arundinacea* in the Yare valley. These last dominants are probably associated with heavier silting and a higher *p*H value. The climax fen carr, dominated by alder, was considered in Chapter XXIII.

Cladietum. In the Bure valley and those of its tributaries the Ant and the Thurne the beginning of fen vegetation is marked by the incidence of the saw-sedge, *Cladium mariscus*, which colonises the very shallow water of closed reedswamp, and rapidly builds up the peat to the water surface. This also happens at Wicken Fen, where open water is absent except in the artificial and constantly cleared lodes and dykes. *Cladium* dominates the areas of very shallow water and maintains itself where it has formed peat just above the summer water level, i.e. in the lowest lying areas of fen. *Cladium* has long strap-shaped evergreen leaves, which may be as much as 3 m. long, with sharp serrated edges. These bend over at a height of about 1·5 m., so that the upper surface of the community is remarkably level. The leaves live for two or three years and when they die decay very slowly, remaining propped among the living leaves and forming a continuous elastic mass or "mattress". Owing to this habit *Cladium* excludes most other species, and at Wicken there are, within the Cladietum, only scattered individuals of tall yellow loosestrife (*Lysimachia vulgaris*), dwarf willow (*Salix repens* var. *fusca*), and *Phragmites*, which maintains itself for a long time in the fen by means of its vigorous and extensive underground rhizomes, pushing horizontally below the surface peat and sending up aerial shoots at intervals.

The Cladietum at Wicken Fen is invaded by bushes down to a level almost as low as the saw sedge itself extends,[1] the lower limit of bush growth being probably fixed by the depth of winter flooding (Godwin and Bharucha, 1932). Thus *Cladium* fen is almost immediately followed by carr (fen scrub or wood). All the other herbaceous fen communities at Wicken are the result of mowing or of removal of carr. But in East Norfolk, if the slope of the basin is slight and there is a deficiency of seed parents in the immediate neighbourhood the bush colonisation is slow and sparse, and the extent of Cladietum and of the other fen communities described by Pallis (1911) may be considerable and maintained for a long time before it is superseded by carr, though some of them are doubtless maintained by mowing as at Wicken. It may be presumed that the consocies of **Glycerietum, Phalaretum,** *Glyceria maxima* and *Phalaris arundinacea* characteristic of Yare valley fen, where *Cladium* is said to be rare, are parts of a prisere corresponding with the Cladietum in the other East Norfolk valleys and at Wicken.

The following lists of typical species are given by Pallis (1911) for the East Norfolk fens:

Bure Valley Fen

Dominants:

    Cladium mariscus
    Juncus subnodulosus
    Phragmites communis

Local dominants:

    Molinia caerulea
    Carex lasiocarpa

Characteristic associated species:

Liparis loeselii	r		
Pyrola rotundifolia	r	Myrica gale	f

[1] Fig. 131.

Yare Valley Fen

Dominants:

  Glyceria maxima (forming consocies)
  Phalaris arundinacea (forming con-
    socies)

Local dominant:

  Poa trivialis

Associated species:

Cladium mariscus	r	Myrica gale	o
Filipendula ulmaria	va	Thalictrum flavum	f
Galium palustre	va	Valeriana officinalis	va
Lychnis flos-cuculi	va		
Myosotis scorpioides			
(palustris)	va		

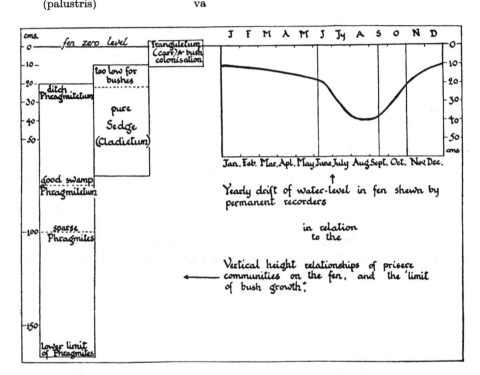

FIG. 131. LEVEL OF BUSH COLONISATION AT WICKEN FEN

The lowest level at which bushes can colonise the fen is about the winter water level,
i.e. 30 cm. below the summer water level. From Godwin, 1936.

Certain oxyphilous plants have been omitted from these lists, because
they are all quite local species which indicate the development of acidity
and are referred to below, but the occurrence of *Pyrola* and the greater
frequency of *Myrica* in the Bure valley fen already indicate a tendency
to develop acidity. The woody species in Pallis's lists (other than the dwarf
Myrica), which have also been omitted, are obviously the fore-runners of
carr.

General list. The general list for all the East Norfolk fens includes:

Angelica sylvestris
Caltha palustris
Calamagrostis canescens
Carex disticha
C. hudsonii (elata)
C. flava
C. hornschuchiana (fulva)
C. lasiocarpa
C. panicea
C. paradoxa
Cladium mariscus
Comarum palustre
Dryopteris thelypteris
Epipactis palustris
Filipendula ulmaria
Galium uliginosum
G. palustre
Glyceria maxima
Hydrocotyle vulgaris
Hypericum elodes
H. acutum (tetrapterum)

Juncus subnodulosus
Lathyrus palustris
Lychnis flos-cuculi
Lysimachia vulgaris
Lythrum salicaria
Menyanthes trifoliata
Molinia caerulea
Myosotis scorpioides
Oenanthe fistulosa
O. lachenalii
Ophioglossum vulgatum
Orchis incarnata
Peucedanum palustre
Phalaris arundinacea
Phragmites communis
Potentilla erecta
Thalictrum flavum
Utricularia intermedia
U. minor
Valeriana dioica
V. officinalis

Five of these species are seen in Pl. 112, phot. 276. The list is far from being exhaustive, a great many other species occurring here and there, but those cited are the most characteristic and widespread.

Unevenness of surface. "The surface of the fen is uneven because the growth of the peat is unequal, owing to the differences in habit and mode of growth of the different plants.... This uneven growth of the fen peat reacts on the plants in various ways, for instance through the initiation of local differences in the relation of soil surface and water level; thus a locally varying habitat is provided..." (Pallis, 1911). These inequalities of surface account for the presence of species of different requirements, some growing in shallow pools between the tussocks of the larger plants such as *Cladium* or *Molinia*, others several inches above the water level in the humus formed on the tussocks themselves.

Oxyphilous communities. In this last situation and round the roots of colonising shrubs the natural acidity of the humus is not neutralised by the alkaline ground water, and here the Sphagna and other oxyphilous species such as sundew and cotton-grass occur locally (Pl. 112, phot. 278). The following are recorded by Pallis:

Drosera anglica
Eriophorum angustifolium
Sphagnum cymbifolium

Sphagnum intermedium
   (? = S. recurvum)
S. squarrosum

Godwin and Turner (1933), from the immediate neighbourhood of Calthorpe Broad, whose water has a pH value between 7 and 8, recorded the following species growing on peat which showed pH values below 7, falling

PLATE 112

Phot. 276. Fen community: *Phragmites communis, Juncus subnodulosus, Dryopteris thelypteris, Epipactis palustris, Lysimachia vulgaris. M. Pallis.*

. 277. "Mixed Fen." *Sium erectum, S. latifolium, Phragmites, Carex* sp., etc. *C. G. P. Laidlaw.*

Phot. 278. Oxyphilous society in fen: *Drosera anglica, Myrica gale,* with *Hydrocotyle, Carex,* etc. *J. Massart.*

FEN VEGETATION

PLATE 113

Phot. 279. Developing carr in the Norfolk Broads region: *Fraxinus excelsior*, *Betula pubescens*, *Salix atrocinerea*. The fen is dominated by *Phragmites*, with some *Juncus subnodulosus* and *Myrica gale*. *M. Pallis*.

Phot. 280. Young carr of *Frangula alnus* at Wicken Fen. Dead leaves of *Cladium* from the preceding Cladietum caught in the crotches. *G. E. Briggs*.

FEN CARR

to 5 (noted several times), while water squeezed from a *Sphagnum* tussock showed *p*H 4·2:

Polytrichum commune L.	Sphagnum fimbriatum Wils.
	S. plumulosum Röll.
	S. squarrosum Pers.
	S. subsecundum Nees

These species, with the exception of *S. plumulosum*, are known from their habitats to have high mineral requirements. *Eriophorum angustifolium* also occurred (in one place a large area of nearly pure cotton-grass) in soil of *p*H values varying from about 7 down to as low as 4·3.

These local societies of Sphagna and associated species may well indicate the beginning of the tendency to develop raised bog on the basis of fen (see Chapter XXXIV, pp. 675–6), a tendency which in this region is overtaken and nullified by the colonisation of woody plants and the development of scrub and woodland.

On the edges of the fenland bordering the upland rather similar communities of oxyphilous plants may occur. One such, at Potter Heigham, mentioned by Pallis (1911) included the following:

Calluna vulgaris	Potentilla erecta
Drosera rotundifolia	Polytrichum commune
Erica tetralix	Sphagnum spp.
Eriophorum angustifolium	Aulacomnium palustre

These marginal oxyphilous communities may be due to the higher level of the peat built up over the edge of the upland, so that it is raised above the alkaline fen water, or alternatively to acid water drainage from neighbouring sandy upland soil; and they may correspond with the local tendency to form acidic bog on the edges of fenland, which seems to have found extensive expression in the past (p. 649).

Development of carr. Godwin and Bharucha (1932) have shown that the critical level for the successful establishment of woody plants is situated at about the height of the water table in winter (Fig. 131). Abundant shrub colonisation of the Cladietum at Wicken Fen can be seen at this level, and on old trenched ground where water stands in the furrows during summer this colonisation is confined to the intervening ridges.

Carr at Wicken Fen. The colonising woody species at Wicken are the alder buckthorn (*Frangula alnus*), which is far the most abundant, the grey sallow (*Salix atrocinerea*), the common buckthorn (*Rhamnus catharticus*) and the guelder rose (*Viburnum opulus*) (Figs. 132, 133). Privet (*Ligustrum vulgare*) is locally plentiful, while hawthorn (*Crataegus monogyna*) and blackthorn (*Prunus spinosa*) occur occasionally. Birch and alder have been planted here and there about the fen, and seedlings of birch are not uncommon. Curiously enough the alder, which is undoubtedly a natural dominant of carr in Norfolk, makes no headway at Wicken, though its pollen is abundant throughout the Wicken peat, and planted alders thrive and shed viable seeds. In a neighbouring fen there is some ash and *Populus canescens*, and oak seedlings are occasionally found.

Developing carr at Wicken is dominated by *Frangula alnus*, though why this shrub should be at first so much more abundant than *Rhamnus cath-*

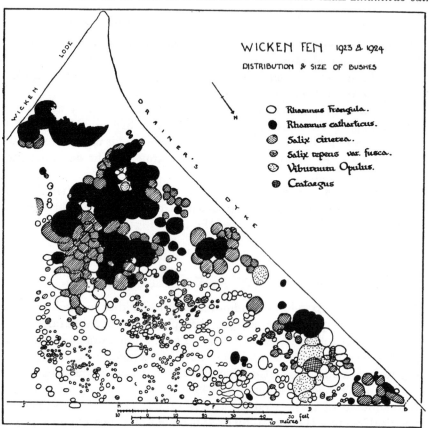

FIG. 132. BUSH COLONISATION IN A CORNER OF WICKEN FEN, 1923–24

An area of "mixed sedge" (*Cladium* and *Molinia*) colonised by bushes. To the south (upper part of the map) nearly closed carr of *Rhamnus catharticus* and *Salix atrocinerea* has become established. To the north (lower part of map) the sedge is being colonised by very numerous small bushes, mainly of *Frangula alnus*. All bushes more than 3 ft. high and 1 ft. in diameter are recorded. From Godwin, 1936. *Salix* "*cinerea*" = *S. atrocinerea*, "*Rhamnus Frangula*" = *Frangula alnus*.

articus is not clear. Abundance of *Salix atrocinerea* is characteristic of young carr, while older carr is commonly dominated by *Rhamnus catharticus* and *Viburnum opulus* (Figs. 132, 133). *Frangula* suffers from "die-back" caused by the fungi *Nectria cinnabarina* and *Fusarium* sp. and this is an important cause of its diminution and loss of dominance in adult carr (Godwin, 1936).

At first *Cladium* and other fen plants maintain themselves between the bushes, but as the canopy closes they are gradually suppressed. Dead

Cladium may often be found below the shrubs of closed carr, the leaves sometimes caught up in the crotches of the branches (Pl. 113, phot. 280).

In time the ground becomes bare, and is then recolonised by a sparse but

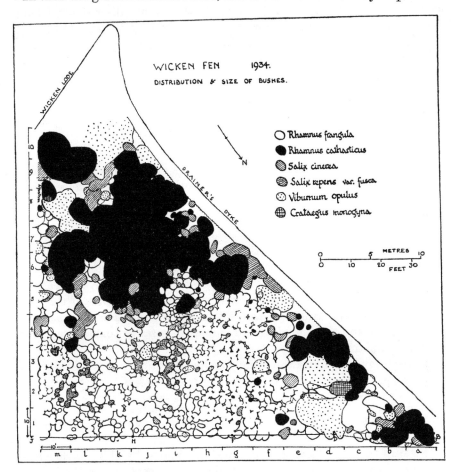

FIG. 133. PROGRESS OF BUSH COLONISATION IN 1934

During the 10 years' interval the clump of carr has extended and become quite closed, *Salix atrocinerea* giving way to *Rhamnus catharticus*. Small clumps of this dominant are established in the north corner. *Viburnum opulus* has also increased in this region and at the southern extremity of the plot. *Frangula alnus* has almost covered the main area of mixed sedge, though *Cladium* still survives between the bushes. From Godwin, 1936.

very characteristic vegetation consisting of the following species, none of which flowers under these conditions (Pl. 114, phot. 281):

Agrostis stolonifera	lva	Rubus caesius	va
Calystegia sepium	a	Symphytum officinale	f
Dryopteris thelypteris	va	Urtica dioica	f
Iris pseudacorus	a	Hypnum cuspidatum	a
Lysimachia vulgaris	a	Mnium affine	f

Carr in East Norfolk. The carr developed on the fens of East Norfolk is far more varied (Pl. 113, phot. 279), as is only to be expected from its much greater extent, and includes practically all the Wicken species, as well as alder. Some account of the alderwoods in the Broads district has already been given in Chapter XXIII, but the main facts may be repeated here to complete the story. Pallis (1911) gives the following list:

Developing carr

Alnus glutinosa	a	Quercus robur	o
Betula pubescens	a	Rhamnus catharticus	a
Frangula alnus	a	Ribes nigrum	o
Fraxinus excelsior	f	Salix atrocinerea	a
Ligustrum vulgare	o	Viburnum opulus	a

The undergrowth consists of the fen vegetation.

In the adult carr the same trees and shrubs occur, with alder typically dominant and a corresponding reduction of other species. *Betula pubescens* is sometimes dominant in fen carr.

The undergrowth characteristically includes the three shrubs *Ribes nigrum* (f), *R. rubrum* (f) and *R. grossularia* (o), the black and red currants and the gooseberry. It is sometimes said that these are bird-sown from cottage gardens, and such dispersal cannot of course be excluded, but from a comparison of fen and marsh woods as a whole there is every reason to suppose that these bushes are perfectly natural constituents of the carrs. Of scramblers and climbers the bittersweet (*Solanum dulcamara*) is characteristic, and the hop (*Humulus lupulus*) is fairly constant and may be very abundant.

The herb layer includes the following species:

Caltha palustris	f	Iris pseudacorus	a
Carex acutiformis	o	Osmunda regalis	r
C. paniculata	a	Urtica dioica	a
Dryopteris thelypteris	a	Mnium hornum	a
Filipendula ulmaria	f		

"Swamp carr." "*Swamp carr*", also dominated by alder, is developed according to Pallis (1911) not on consolidated fen peat but on the tussocks or stools of reedswamp plants such as the large Carices (*C. paniculata, C. riparia, C. acutiformis*), on the edge of open water, and is a swampy wood often partially floating and with much open water between the peat islands (Pl. 88, phots. 214–15). It is commonly dominated by alder with *Salix atrocinerea* abundant and a ground vegetation of fen fern and the great Carices. The details of its origin and fate have not been sufficiently studied, but Godwin and Turner (1933) think that swamp carr may arise from young fen carr through depression of floating peat by the weight of the developing trees and shrubs (Chapter XXIII, p. 463) in the absence of consolidation by oak, etc.

There is some evidence (see Chapter XXIII) that alderwood may develop into oakwood and from the subfossil remains in the fen peat it seems that

PLATE 114

Phot. 281. Mature carr of *Rhamnus catharticus* at Wicken Fen. *Eupatorium cannabinum, Iris pseudacorus, Rubus caesius, Urtica dioica*: *Crataegus monogyna* on the extreme left. Cf. Pl. 90, phot. 220. *H. Godwin.*

FEN CARR

oakwoods have established themselves here in the past, possibly sometimes through an intermediate stage of birch and pinewood. But how far these various types of forest on peat are stages of the prisere and how far responses to fluctuations of climate is still uncertain. A drowned river basin so near the sea may never have allowed full expression to the development of later stages of the prisere.

Anthropogenic fen communities. Most of the existing fen, indeed all of it at Wicken and probably most in East Norfolk, is the result of human activity in mowing the sedge and clearing carr. At Wicken Fen Godwin and Tansley (1929) distinguished four such anthropogenic but semi-natural herbaceous fen communities, apart from the lodes and their artificial banks, and apart from the Cladietum which is a stage in the prisere: (1) "Mixed sedge" (Cladio-Molinietum), (2) "litter" (Molinietum), (3) the fen droves, and (4) "mixed fen" with an abundance of conspicuous flowers.

(1) *"Mixed sedge" (Cladio-Molinietum).* Cladietum, as we have seen, is in the first place a natural stage of the prisere following reedswamp and giving way in its turn to carr as the result of colonisation by bushes. But *Cladium* (the "sedge" *par excellence* of the fenmen) is useful for thatching or kindling and was at one time regularly mown for those purposes, as the *Phragmites* was for thatching. If Cladietum is mown once in three or four years the colonising bushes cannot make headway, but the *Cladium* survives and is maintained as a semi-natural community. Its vigour is however somewhat impaired, because *Cladium* is an evergreen plant and its vegetative capacity is reduced by the frequent removal of the living leaves. The plants consequently grow less luxuriantly, lose their exclusive dominance, and *Molinia* enters the community and comes to share dominance with the *Cladium*. *Phragmites*, less tall than in the reedswamp, is frequent, and a number of other species occur in local patches including:

Angelica sylvestris	Lysimachia vulgaris
Eupatorium cannabinum	Peucedanum palustre
Hydrocotyle vulgaris	Salix repens var. fusca

These are abundantly evident the first year after mowing, when they flower freely, but at the end of the 4-year interval they are hardly noticeable. The effect of the "coppicing cycle" on the field layer of coppiced oak-hazel wood (Chapter XIII, pp. 277–8) may be compared with this "mowing cycle". The mattress of dead *Cladium* leaves is still formed, and the peat of course continues to grow, raising the ground level of the mixed sedge several inches above the critical level at which bushes can first invade, and in fact above the soil level of many examples of mature carr. If cutting is stopped the community is quickly colonised by bushes.

(2) *"Litter" (Molinietum).* If the sedge or mixed sedge is cut every year the *Cladium* suffers so severely by the continuous removal of the living leaves that it cannot maintain itself, and the fen becomes dominated by *Molinia* which does not suffer from mowing, since only the dead leaves are

removed by the autumn cutting. In this Molinietum *Phragmites* is still frequent, while *Carex panicea* and *Juncus subnodulosus* become subdominant. Dicotyledonous plants are more abundant than in mixed sedge, and include:

Angelica sylvestris	Succisa pratensis
Cirsium anglicum	Thalictrum flavum
Filipendula ulmaria	Valeriana dioica
Hydrocotyle vulgaris	

This community often occupies ground with the peat surface at precisely the same level as that of mixed sedge, and the difference is simply the result of the different treatment. On the cessation of mowing the ground is immediately colonised by bushes. The cut "litter" was used for cattle bedding and from this use the fenmen's name is taken.

(3) *The fen droves* (wide tracks kept clear to facilitate access to different parts of the fen) which are cut twice or thrice a year, are an extreme example of the effect of repeated cutting and traffic. Here there is quite a different flora, including many species of rush, sedge, grass and dicotyledonous plants, several of which are "dry land plants", not found elsewhere on the fen. Tall species are eliminated and many dwarf species are able to survive which do not exist in the communities of the fen proper.

(4) *"Mixed fen"* results from the clearance of carr. A host of plants, many of which have conspicuous flowers, is rapidly established with the access of full illumination. Such areas can be readily recognised by the presence of several relicts from the undergrowth of the carr. The following are among the species which occur:

*Agrostis stolonifera	Lathyrus palustris
Angelica sylvestris	Lysimachia vulgaris
Calamagrostis epigejos	Lythrum salicaria
C. canescens	Molinia caerulea
Cirsium anglicum	Peucedanum palustre
*Calystegia sepium	*Phalaris arundinacea
*Dryopteris thelypteris	Phragmites communis
Epilobium hirsutum	*Rubus caesius
Eupatorium cannabinum	Sium erectum[1]
*Filipendula ulmaria	S. latifolium[2]
*Iris pseudacorus	Symphytum officinale
Juncus subnodulosus	Thalictrum flavum

* Species relict from the undergrowth of carr.

The "mixed fen", if left alone, rapidly reverts to carr, most rapidly of course if the stools of the cut carr shrubs have not been removed.

THE NORTH-EAST IRISH FENS

Around Lough Neagh in northern Ireland, but more especially round and to the south of its southern end lies an extensive area of fenland,

[1] Pl. 112, phot. 277 shows a small portion of "mixed fen" in which species of *Sium* are locally dominant.

probably the most extensive in the British Isles still remaining comparatively unspoiled. The water of Lough Neagh itself is approximately neutral in reaction, and in the drains cut through the fen peat it may have a *p*H value of 6–6·5, but is never more acid than that (Small, 1931). This relative alkalinity of the ground water seems to be the factor which determines the existence of fen in a region whose climate favours the production of acid bog and "moor". The rivers draining into Lough Neagh come partly from the basalts of Antrim, Derry, and Tyrone, and the basic material (largely calcium bicarbonate) which they bring, though much less than is dissolved from the chalk by the East Anglian streams, is sufficient to provide, as Small (1931) points out, an effective alkaline buffer, keeping the peat water from becoming at all markedly acid. This buffer action is the more efficient because of the deep winter flooding in these fens and the great reservoir of neutral water in Lough Neagh. The level of Lough Neagh itself normally rises a metre and sometimes nearly 2 m. when the rivers are in flood. The extreme range at Ballybay Bridge on the Ballybay River several miles south of Lough Neagh, recorded between November 1930 and October 1931, was about 120 cm. (White, 1932). About 50 cm. of water covered the communities of the lower fen during more than half the year 1930–1, though it was an abnormal year, since the usual winter level was never reached and unusual floods occurred during the summer. It seems probable that the zonation of the fen communities is largely determined in these fens by the winter water levels, as indeed it is at Wicken (Godwin, 1931), where the vertical range of both water level and plant communities is very much less, mainly because of the close regulation of the water level by the pumps and sluices.

The flora of the Lough Neagh fens is very much like that of the East Anglian fens, though a certain number of species, some of which are absent altogether from Ireland, are wanting: others, such as *Cladium* and *Rhamnus catharticus*, abundant in East Anglia, are local or rare.

Among biotic factors no mention is made of regular periodic mowing of the fens, as at Wicken, and there is no evidence that the cutting which is done has any decisive effect in altering the vegetation. Peat cutting is extensive, and the black fen peat (moulded into turves with mud) is of much better fuel quality than the yellow and brown bog and moor peats. Many of the "drains" which are met with are simply trenches dug for peat and have no effect on the general drainage.

Zonation. The zonation of the fen communities is very well marked and the different zones are dominated by a number of species forming local societies. The vertical range of most of these species is considerable, so that there is much overlapping, but their regions of dominance are rather sharply limited. The reedswamp, which occurs abundantly in "drains" showing open water throughout the year, as well as along the shores of the lakes, consists of the ordinary dominants *Scirpus lacustris, Phragmites* *communis* and *Typha latifolia,* as well as *Sparganium* and various other local dominants and associated species.

The vegetation of the fen itself appears to segregate into a great number of more or less zoned societies, and in this respect to differ from the East Anglian fens, which have fewer dominants. Miss White (1932) gives a list of the following fen societies, some of whose dominants are equally conspicuous in the reedswamps.

Lower fen. (1) Society of *Hippuris vulgaris.* This is often the initial society of the fen succession, growing as luxuriantly in the lower fen, where the water level is often an inch or two below the soil surface, as in the reedswamp. Associated may be *Bidens* spp., *Ranunculus* spp., *Alisma ranunculoides,* etc.

(2) Society of *Equisetum fluviatile (limosum)* and *E. palustre.* This also may be an initial society of the fen, or it may invade (1).

(3) The society of *Menyanthes trifoliata,* again, may initiate the fen succession, becoming established on the edge of a reedswamp (in which for example *Butomus* is dominant) or it may replace the *Equisetum* society.

(4) The *Lythrum salicaria* society occupies the next zone of dominance, and is frequently invaded by

(5) The society of *Lysimachia vulgaris,* with which *Lythrum* is also often co-dominant.

(6) A society of *Iris pseudacorus* may be initial, replacing reedswamp, or it may replace (1) and (2).

(7) *Phragmites* not only appears as a dominant of the reedswamp but also as a fen dominant, though it does not grow so tall in the latter situation. It ranges slightly higher than *Iris.*

(8) *Carex inflata,* too, with *C. hudsonii,* is often an important species in the reedswamp, occurring in the adjacent fen as a local dominant. It often colonises peat which has been cut to deep levels. *Cladium mariscus* and *Typha angustifolia* are confined to this zone.

(9) The society of *Comarum palustre* often invades and replaces the *Equisetum* society. It may be regarded as the uppermost society of the Lower Fen. The marsh cinquefoil ranges decidedly higher than the preceding species, extending as a subordinate species even into the Upper Fen.

Middle fen. None of the following societies is ever initial in the fen sere:

(10) The society of marsh marigold (*Caltha palustris*) frequently succeeds the marsh cinquefoil (9), or it may follow the iris (6) or the horsetails (2), but more often with the bog bean (3) intervening.

(11) The cuckoo-flower (*Cardamine pratensis*) society often follows (10), *Caltha* and *Cardamine* frequently maintaining co-dominance over the greater part of the zone they occupy. Again, *Cardamine* may succeed the marsh cinquefoil or the bog bean.

Society dominants		I	II	III	IV	V	VI	VII	VIII	IX	X	XI	XII	XIII	XIV	XV	XVI	XVII	XVIII	XIX	XX
		\<Upper Fen.\>						\<Mid Fen.\>							\<Lower Fen.\>						
Molinia caerulea	I	D																			
Juncus articulatus	II		D																		
Juncus conglomeratus	III			D																	
Filipendula ulmaria	IV				D																
Rhinanthus sp. and grasses	V					D															
Grasses	VI						D														
Hydrocotyle vulgaris	VII							D													
Carex panicea and grasses	VIII								D												
Carex spp.	IX									D											
Cardamine pratensis	X										D										
Caltha palustris	XI											D									
Comarum palustre	XII												D								
Carex inflata	XIII													D							
Phragmites communis	XIV														D						
Iris pseudacorus	XV															D					
Lysimachia vulgaris	XVI																D				
Lythrum salicaria	XVII																	D			
Menyanthes trifoliata	XVIII																		D		
Equisetum fluviatile} E. palustre}	XIX																			D	
Hippuris vulgaris	XX																				D

FIG. 134. TABLE OF FEN SOCIETY DOMINANTS IN THE LOUGH NEAGH FENS (NORTH-EAST IRELAND)

Each of the dominant species is marked D opposite the number of the society it dominates. Its extension through other societies is shown by horizontal lines, continuous where the species is abundant, frequent or occasional, dotted where it is rare. From White, 1932.

(12) The next society is said to be dominated by species of *Carex*, but no details are given.

(13) Then comes a society dominated by *Carex panicea* and various grasses: it succeeds either (10) or (11), or even (9). The grasses range upwards into the upper fen and generally also dominate a higher zone (see 15 below). They form grassy meadows along several of the rivers.

(14) The marsh pennywort (*Hydrocotyle vulgaris*) society has the greatest range of any fen species—extending from the upper limit of the upper fen nearly to the bottom of the lower—but it is only dominant in the middle fen where it successfully invades (9) and may itself be invaded by (19) or (20) of the Upper Fen, being often co-dominant with *Juncus articulatus* over considerable areas.

Upper fen. (15) and (16) are dominated by grasses and the yellow rattle (*Rhinanthus crista-galli*) which may succeed (10) or (11). The following grasses are listed as occurring in the upper fen and the upper part of the middle fen, but no details of the frequency of the different species are given. Presumably several are co-dominant in these societies.

Agropyron repens	D. flexuosa
Agrostis canina	Festuca elatior
Aira praecox	F. ovina
Alopecurus geniculatus	Glyceria fluitans
A. pratensis	Holcus lanatus
Anthoxanthum odoratum	H. mollis
Briza media	Lolium perenne
Calamagrostis neglecta var. hookeri	Poa annua
Cynosurus cristatus	P. pratensis
Dactylis glomerata	P. trivialis
Deschampsia caespitosa	

Nothing is said of the effect of mowing or pasturing on these communities, nor of the relations of the different species in the vegetation. It is clear that the list includes moisture-loving species with others of drier situations, and some belonging to neutral with others of acid grassland.

(17) The meadowsweet (*Filipendula ulmaria*) dominates a society at this level and is said to succeed (12) and (14).

(18) The common rush (said to be *Juncus conglomeratus*) replaces (14) at several places, and in one instance is invading (9).

(19) *Juncus articulatus* may also invade (14) and the two are co-dominant in some places.

(20) Molinietum is the uppermost society of the fen. It may also invade (14).

Eriophoretum angustifolii does not occupy any definite place in the series given above. It occurs in waterlogged areas at various levels, and is said to contain a definite set of associated species.

Nature of the succession. It will be obvious from the foregoing not only that there is much overlapping between communities and frequent co-dominance in the ecotones, but that almost any society of the "middle fen" may invade and replace various societies (though usually the higher members) of the lower fen, and that the societies of the upper fen similarly succeed, not one another, but some society (very commonly the *Hydrocotyle* society) of the middle fen. In other words there is never anything like a complete linear series of the whole 20 societies on the ground, but very various shorter series consisting of perhaps half a dozen communities in any one local sere. Nevertheless Miss White insists that the dominant of each of the 20 societies has its own "niche" in the entire series. This is shown, together with the range of each dominant as an associated species, in Fig. 134 (p. 665), reproduced from her paper. The large number of these societies suggests that many differences of level have been brought about by peat cutting, and that this may account for the extreme fragmentation of the vegetation, in contrast to the greater extent of pure communities in the East Anglian fens.

Relation to "moor". The four uppermost societies (17–20) of the upper fen, taken together, "may be regarded as the climax fen community and as the pioneer moor community, because in each of these societies the substratum is more acidic than [in the] mid or lower fen societies. When *Sphagnum* becomes established in the fen there is soon a well-marked increase in the number of moorland species, and typical fen species become fewer in number." "This change in vegetation occurs when the level of the soil has been raised almost above the level of the winter floods. When the buffering of the flood waters ceases to control the acidity of the vegetation, the substratum changes to relative acidity" (White, 1932).

Vegetation of "ramparts". "Ramparts", i.e. ridges whose surface is above the general level, the peat between them having been cut away, are of frequent occurrence in many areas of the north Armagh fens. These are "really relicts of the moor formation, which, beyond doubt, was formerly much more extensive in this district than it is at present". They "serve as a place upon which to dry and stack the mud peat which is cut from the fen". "They are built up of yellowish brown peat composed of decayed *Sphagnum, Eriophorum* or *Calluna*...always acid in reaction." In the existing vegetation of the "ramparts" sometimes *Sphagnum*, sometimes *Calluna* is dominant, and accompanying these are sundews, rushes, *Erica tetralix, Narthecium ossifragum, Viola palustris, Molinia, Nardus,* and on well-drained summits *Blechnum spicant, Pteridium, Ulex europaeus* and *Galium saxatile.* Seedlings of fen plants are sometimes found on the "ramparts", having germinated in peat debris thrown up from neighbouring "drains", but they do not survive the summer. Inversely the seedlings of *Calluna, Erica tetralix,* etc., occasionally occur in the fen near the ramparts, but very rarely reach maturity. The small colonies of these species sometimes found in the fen are restricted to large sods of the "rampart" peat which have been isolated by cutting round and whose surface is well above the fen level.

Fen scrub or carr. The undershrubs *Myrica gale* and *Salix repens* occur here and there in the north Armagh fens, but their distribution is very limited and apparently sporadic. Where they are succeeded by taller bushes they do not persist, and are evidently unable to endure shading.

In carr proper *Salix atrocinerea* and *S. caprea* are most frequently the pioneers and continue as local dominants of the most advanced carr that is developed in the region. If we may judge from the analogy of the East Anglian fens this is probably evidence that the carr is young. The woody canopy never becomes dense enough to exclude the herbaceous fen vegetation, as it does in the East Anglian fens. It is not clear whether this is due to a difference of climate. "In very many places...it would appear that intensive grazing or the cutting of the fen vegetation is sufficient to keep the bush colonisation—the establishment of carr—at a standstill" (White, 1932, p. 278). But there seems no doubt that fen passes into an oxyphilous vegetation (*Sphagnum, Calluna,* etc.) far more frequently than in eastern England, and this difference is presumably correlated with the different climate.

The following woody species are recorded:

Salix atrocinerea and S. caprea		a, ld or co-d	
Alnus glutinosa	la	Rhamnus catharticus	vr
Betula pubescens		Salix alba	
Crataegus monogyna		S. pentandra	
Fraxinus excelsior		S. purpurea	ld
Hedera helix		S. viminalis	ld
Ilex aquifolium		Ulex europaeus	la
Ligustrum vulgare	r	Ulmus sp.	
Lonicera periclymenum		Viburnum opulus	r

The presence of these trees and shrubs is reported as without effect on the herbaceous fen vegetation, which is the same whether woody plants are present or not.

"Swamp carr" is formed in places on the shores of Lough Neagh by direct invasion of the reedswamp by the sallows: in other places it arises by extension of the adjacent fen carr over the reedswamp where this is passing into fen.

SUMMARY OF THE LATER HYDROSERE

By way of a link between the account of marsh and fen given in this chapter and the last and the following chapters dealing with bog or moss it will be convenient here to summarise the hydroseres from the point at which "land vegetation" begins, i.e. where the soil level is approximately the same as the summer water level. We have seen (p. 638) that it is theoretically possible for silting to continue (during floods) after **Marsh** this level has been attained, so that the hydrosere progresses through a stage of *marsh* (preponderantly mineral soil) to wet forest, and ultimately, by the accumulation of humus (which is then well aerated) to climax forest. But no such successions have been described for the British Isles, few of whose alluvial plains are both ungrazed and also subject to silt-carrying floods; and thus there is little opportunity for this sequence to occur undisturbed by man. The commoner case is the development of *fen*

Fen
at or above the summer water level, succeeded either by the invasion of woody plants (carr) or by the colonisation of *Sphagnum* and its associates.

A common fen pioneer and dominant in the calcareous basins of East Anglia is *Cladium mariscus*, and this is accompanied by a few other species of which the most abundant are *Carex panicea* and *Juncus subnodulosus*. Other fens are dominated by *Glyceria maxima*, *Phalaris arundinacea*, etc. In the small local calcareous fens of the south midlands, *J. subnodulosus* sometimes forms reedswamp and is succeeded by *Schoenus nigricans* as the first fen dominant. In other places *Phragmites communis* may dominate the fen as well as the reedswamp. In northern Ireland also the common reed may dominate the first stage of fen but the dominants of the "lower fen" are much more various, perhaps owing to the variety of water level and habitat introduced by extensive peat cutting.

It is only at Wicken Fen in East Anglia that exact observations have been made on the critical level for the colonisation of woody plants, which

Carr
initiates the formation of *carr*. Godwin and Bharucha (1932) have shown that seedlings of *Frangula alnus* can establish themselves in Cladietum at about the average winter water level (i.e. about 30 cm. above the average summer water level), but not below this level, which suggests that the possible level of establishment is determined by winter flooding. Thus if we limit the term fen to the herbaceous peat communities whose soil surface is not below the average summer water level, it can develop here only over a vertical range of about 30 cm. (more at Lough Neagh). Most of the fen actually existing is above this level and is maintained by mowing, which prevents the development of woody vegetation.

At Wicken the most abundant woody colonist on fen is *Frangula alnus*, followed by *Salix atrocinerea* (which is relatively short lived), *Rhamnus catharticus* and *Viburnum opulus*—the two last usually dominating the tallest and oldest carr. In the more extensive natural fens and carrs of east Norfolk *Fraxinus excelsior*, *Betula pubescens* and *Alnus glutinosa* also enter into the structure of carr, both birch and alder, but especially alder, becoming dominant in the fen wood. At Calthorpe Broad two generations of

Oakwood
Quercus robur, derived from planted parents, have established themselves freely in developing carr and can suppress *Alnus* (Godwin and Turner, 1933). It seems probable that oak forest would ultimately succeed fen wood, at least in East Anglia and probably over the greater part of England.

In the north-west, however, it is otherwise. Where a fen is heavily silted with inorganic silt, as on the North Fen at Esthwaite, "mixed fen" develops, often with *Calamagrostis* or *Phalaris arundinacea* dominant, and this is colonised by *Salix purpurea* and *S. fragilis*. *Alnus* follows, and probably *Betula pubescens* if the ground level is raised and sufficiently dry. Probably only on gravel would genuine woodland (Quercetum sessiliflorae) develop, and this would really be a marsh rather than a fen sere; but no examples have been described, though *zonations* of the kind are known (see Chapter XXXII, pp. 635–7). Where the silting is less and the proportion of organic material higher, *Molinia* enters the community (still

Molinietum and transitory carr containing many fen species), which is also colonised by *Salix atrocinerea*. The sallow is however transitory and the carr formed (Figs. 124, 125) does not regenerate. The soil is waterlogged, is freely colonised by *Sphagnum*, and passes into bog with *Molinia* and *Myrica gale, Erica tetralix* and some *Calluna*. Only when it is drained does *Betula* appear. Where sedimentation is very slight or absent the organic ratio is still higher, the reaction more acid and Molinietum immediately follows reedswamp. *Myrica gale* is characteristic of this Molinietum, which

Raised bog ultimately gives way to typical *raised bog*, with *Sphagnum* abundant or dominant, and the associated acid bog plants.

These are the conditions on flat undrained soil in the Lake District and apparently also in south-west Scotland. In north-east Ireland, also, woody vegetation does not follow the *Salix* carr, which never becomes closed, and there is a general tendency to the development of oxyphilous vegetation.

This is in contrast on the one hand to the East Anglian fens where it seems that oakwood would normally succeed carr in spite of the local tendency to develop *Sphagnum*, etc., and on the other to the blanket bogs

Blanket bog of western Scotland and Ireland where fen and carr do not appear in the hydrosere at all, but blanket bog normally develops from the acidic aquatic vegetation of pools and on the general surface of undrained, flat or gently sloping ground. These conclusions are tentatively summarised in the accompanying diagram (Fig. 135).

Correlation with climate. It is difficult to avoid the conclusion that these differences in the hydrosere are determined by the differences of climate—that the wetter and cooler the climate the greater the tendency to bog or moss development. In the East Anglian climate Sphagneta still appear locally and there is evidence that in some post-glacial periods raised bog was locally formed in this region, but oak forest would now seem to be the natural culmination of the sere. In the more oceanic climate of north-west England, south-west Scotland and north-east Ireland (which lie relatively close together) the climax on flat undrained ground is raised moss or bog, whether this develops from fen formed in a calcareous basin, as in the Armagh fens, or almost straight from aquatic vegetation (Lake District). Differences of inorganic silting or of the reaction of the ground waters suffice to modify the succession. Finally, in the extreme oceanic climate, moss or bog develops immediately following the aquatic stages of the hydrosere or directly on wet soil.

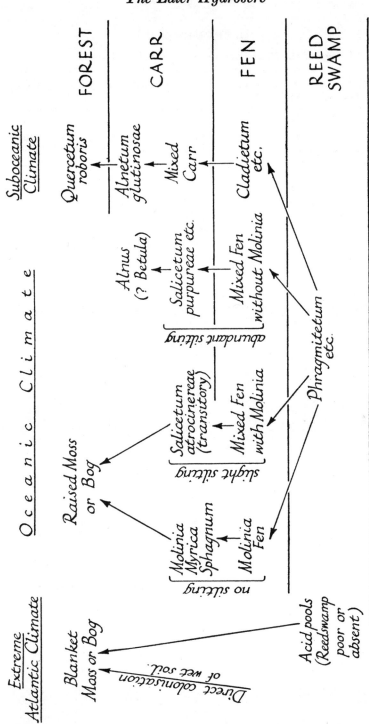

FIG. 185. CLIMATIC RELATIONS OF HYDROSERES

REFERENCES

DUFF, M. Ecology of the Moss Lane region. *Proc. Roy. Irish Acad.* **39**. 1930.

GODWIN, H. The "sedge" and "litter" of Wicken Fen. *J. Ecol.* **17**, 148–60. 1929.

GODWIN, H. Botany of Cambridgeshire in the *Victoria County History of Cambridgeshire*. Oxford, 1938.

GODWIN, H. Studies in the ecology of Wicken Fen. I. The ground water level of the fen. *J. Ecol.* **19**, 449–73. 1931.

GODWIN, H. and BHARUCHA, F. R. Studies in the ecology of Wicken Fen. II. The fen water-table and its control of plant communities. *J. Ecol.* **20**, 157–91. 1932.

GODWIN, H. and BHARUCHA, F. R. Studies in the ecology of Wicken Fen. III. The establishment and development of fen scrub. *J. Ecol.* **24**, 82–116. 1936.

GODWIN, H., MOBBS, R. H. and BHARUCHA, F. R. Soil factors in Wicken Sedge Fen. *The Natural History of Wicken Fen*, Part VI, 601–14. 1932.

GODWIN, H. and TANSLEY, A. G. The vegetation of Wicken Fen. *The Natural History of Wicken Fen*, Part V. 1929.

GODWIN, H., and TURNER, J. S. Soil acidity in relation to vegetational succession in Calthorpe Broad, Norfolk. *J. Ecol.* **21**, 235–62. 1933.

PALLIS, M. In *Types of British Vegetation*, Chapter X. 1911.

SMALL, J. The fenlands of Lough Neagh. *J. Ecol.* **19**, 383–8. 1931.

WHITE, J. M. The Fens of North Armagh. *Proc. Roy. Irish Acad.* **40**, 233–83. 1932.

YAPP, R. H. Wicken Fen [Sketches of vegetation at home and abroad, IV]. *New Phyt.* **7**, 61–81. 1908.

Chapter XXXIV

THE MOSS OR BOG FORMATION

TERMINOLOGY. VALLEY BOG, RAISED BOG, AND BLANKET BOG

BOG COMMUNITIES—(1) SPHAGNETUM

Terminology. The plant communities forming and continuing to grow upon constantly wet acid peat are almost everywhere called *mosses* in northern England and southern Scotland, and *bogs* in Ireland. In southern England, except on the high lands of the south-western peninsula, they are now almost confined to local depressions in sandy soil where the drainage is impeded, and there is no general common name for this type of vegetation.

The word *moor* (or *moorland*) applies in ordinary English speech primarily to (usually) high-lying country covered with heather and other Ericaceous dwarf shrubs, mainly Vaccinia; though it is often used more widely to refer to land bearing the whole series of oxyphilous communities and not regarded primarily as pastureland, from bog or moss to acidic grassland such as Nardetum or Molinietum. For example the elevated moorlands of the south-western peninsula, the largest areas of which are Dartmoor, Exmoor and Bodmin Moor, include both drier types on shallow sandy and peaty soil and wetter types on deep peat.

In German the word *Moor* applies to *any* area of *deep* peat, whether acid or alkaline, and the different types of peat were distinguished by Weber as *oligotrophic*, i.e. poor in nutritive (basic) salts, which we call in English "moor peat", "moss peat" or "bog peat", and *eutrophic*, i.e. rich in nutritive salts, which we call "fen peat". Under suitable climatic conditions oligotrophic peat may be built up on the top of eutrophic peat, i.e. fen (German *Niedermoor*) may be succeeded by moss or bog (German *Hochmoor*), often through a stage of *mesotrophic* peat, or as the Germans call it *Uebergangsmoor* ("transition moor") with vegetation intermediate between the two extreme types. The historical records of this change in the vegetation of a peat area may be preserved in the peat in the shape of remains of characteristic fen plants in the lower layers of peat, and of bog plants in the upper. Many examples of this succession have been described on the continent and some in England (see Fig. 47, p. 168 and Tables on pp. 692–5).

We cannot however use the English word "moor" as an ecological term in the German sense of an area covered with deep peat. According to the Oxford Dictionary moor means in modern English "a tract of unenclosed waste land" "usually covered with heather". The older meanings of "moor", which include waste marshland where peat might or might not be formed, are now obsolete. The root of the word is generally held by scholars to be connected with a word meaning "to die", and thus applied primarily to "dead" or barren, i.e. infertile, land. Its very wide original application is thus easily understood.

In fact the word is now generally used, as has been said, for *any* tract of unenclosed land (generally elevated) with acid peaty soil and not used *primarily* as pasture, so that it would not be applied to the bent-fescue hill pastures or to the limestone grasslands. Such "moorlands" may be wet or dry, and their soil may or may not be deep peat, but it is always acid and bears a related set of plant communities. These may be dominated by heather (*Calluna*), bilberry, etc. (*Vaccinium*), matgrass (*Nardus*), purple moorgrass (*Molinia*), cotton-grass (*Eriophorum*) or deer sedge (*Scirpus caespitosus*). In common speech the most typical "moors" are undoubtedly the "heather moors" or "grouse moors", dominated by *Calluna* and related vegetation and ordinarily used for the preservation and shooting of grouse (*Lagopus scoticus*). The soil may be of deep peat, but with a relatively dry surface, or the peat may be shallow and much mixed with mineral matter.

"Moss" or "bog". In the search for a suitable English word that can be used as a technical term for the natural group of *wet* peat-forming and peat-inhabiting communities we are thus limited to "moss" and "bog". Both have a strong claim founded on widespread common use for precisely the kind of soil-vegetation complex for which a designation is required. "Moss" has the additional advantage of linguistic correspondence with the Scandinavian words *Mosse* (Swedish), *Mose* (Danish and Norwegian) and the German *Moos*, which are applied to just the same vegetation. On the other hand it has the drawback—not perhaps very serious—of possible confusion with the taxonomic group of *Musci*, among which the Sphagna (the characteristic bog mosses) may be important, and are often primary, dominants or constituents of the formation in question. The disadvantage of the word "bog" is that it is sometimes loosely used in common language for *any* wet soil into which the foot sinks, but this again is not a very serious drawback.

In the present work, therefore, "moss" and "bog" are used synonymously for the wet acid peat vegetation; and "moor" and "moorland" are not used as technical terms, but only in the wide popular sense. The terms "low moor" and "high moor" (literal translations of the German *Niedermoor* and *Hochmoor*) are avoided altogether, both because "moor" cannot properly be used in English to include fen (*Niedermoor*) while it is always used for upland heath (which is neither *Niedermoor* nor *Hochmoor*), and because the adjectives inevitably carry misleading implications of altitude.

The plant communities which form and inhabit wet acid peat have often been divided into "lowland" and "upland", but they are more naturally classified as *valley bog, raised bog* and *blanket bog*—names which refer to real differences in habitat, structure and mode of development.[1]

[1] It is likely that with fuller knowledge additional types of bog will be recognised, but for the present the classification given will suffice.

Valley bog and Raised bog. Valley bog is developed where water, draining from relatively acidic rocks, stagnates in a flat bottomed valley or depression, so as to keep the soil constantly wet. In such situations species of *Sphagnum* and associated plants appear and produce a bog limited to the area of wet soil. Such bogs are common in the mountainous regions of Palaeogenic rocks in the north and west of the British Isles, but they have not been investigated ecologically, and very little is known of their structure and development. The "wet heath" communities (see pp. 734–41) occurring in depressions of the English lowland heaths, where the ground water is nearly level with the surface, are essentially of this type.

In a sufficiently moist climate the characteristic *raised bog* (German *Hochmoor*) may develop on the top of a valley bog. The valley bog itself, however, being fed by drainage water, is never so poor in soluble mineral constituents as a raised bog, and it contains plants, such as species of *Juncus* and *Carex*, which have no part in a typical raised bog. The stream running through a valley, however stagnant it may become in parts of its course and however acidic the rocks which it drains, brings a certain quantity of soluble salts from the upper reaches where erosion is taking place, and the water of a valley bog is typically less acid than that of a bog depending on precipitation alone. When a raised bog develops on the top of valley bog the stream is blocked or diverted by the growth of the bog vegetation and the peat which it forms. The original water course may thus be forced to one side, or split into two streams which find their way round the sides of the bog: part may run below the surface of the bog. The marginal watercourses form the *lagg* (a Swedish term) of raised bog, and the vegetation of the lagg is characteristically less extremely oxyphilous than that of the general surface of the raised bog.

Usually, however, raised bog is developed not on valley bog but on fen. As we saw in the last chapter, *Sphagnum* and associated plants, which require acid water in their habitat, locally colonise the surface of fen, especially where the large tussocks of certain fen plants have raised the level above the neutral or alkaline ground water, so that the natural acidity of the humus formed by the debris of the tussock is not neutralised. The small communities of oxyphilous plants arising in this way may remain very limited in extent, but there is abundant evidence in the British Isles, as well as on the continents of Europe and North America, that they have, in times past, spread over and superseded fen vegetation, replacing it by wide extents of the highly characteristic moss or bog formation. Where this formation has thus arisen on a fen localised in a basin it has itself remained restricted to the basin, though this may be of considerable extent, up to several miles in diameter. The surface of the moss is characteristically convex, sloping gently from the centre towards the periphery,[1] where it ends in a relatively steep bank bounded by a ditch or watercourse (lagg) repre-

[1] Pl. 115, phot. 283 *a*.

senting the original drainage channels of the fen basin and receiving the
water draining from the bog. Beyond the lagg fen vegetation, not yet
covered by the moss, may often be seen. Thus the moss as a whole is raised
above the immediately surrounding fenland, and this is the origin of the
German term *Hochmoor*. This is the type of bog so common in the great
central limestone plain of Ireland,[1] where they are based on the local fen
basins, and are known as "red bogs" from the red-brown colour of the
dominant vegetation. Similar raised bogs occur in Scotland, northern
England and Wales, and were formerly much commoner, many having
been destroyed by draining and peat cuttings.

Blanket bog. It seems probable that raised bogs are formed in climates
intermediate between that of the East Anglian fens, where the air seems too
dry for the bog-moss vegetation to extend vigorously in dependence upon
atmospheric moisture, and that of the west of Scotland and the west of
Ireland, where the rainfall is high and the air so constantly moist that bog is
the *climatic formation*, not necessarily arising in fen basins but covering the
land continuously except on steep slopes and outcrops of rock. This is the
third type of bog met with in the British Isles and may be called *blanket bog*,
because it covers the whole land surface like a blanket.[2] The contrast can be
well seen in the bogs developed east and west of Galway city in the west of
Ireland. To the east is the great plain of Carboniferous Limestone with
raised bogs developed in the numerous small fen basins. To the west is a
region of acidic rocks under an extreme oceanic climate, and bearing almost
continuous blanket bog. Blanket bog is independent of localised water
supplies, depending on high rainfall and very high average atmospheric
humidity. Raised bog, on the other hand, seems always to have an aquatic
or semi-aquatic origin, being built upon fen, marsh or valley bog, so that
the bog peat is often underlain by fen peat and lacustrine or estuarine silt.
The climate in which raised bog is developed must, however, provide
sufficient atmospheric humidity to make the upward growth of the bog
possible.

Bog or moss communities. The dominants and associated species of
blanket bog and raised bog are mostly, though not entirely, the same;
but, as we have seen, the habitats (in the wide sense) are different, and the
structure and development also differ in several respects. Blanket bog
has never been studied by modern methods, but we know more about the
development of raised bog, thanks to the labours of continental workers
and to some quite recent work in the British Isles. The main plant communities of bog or moss are six—Sphagnetum, Rhynchosporetum, Schoenetum, Eriophoretum, Scirpetum and Molinietum—and of these Sphagnetum
is the first and most fundamental.

(1) **Sphagnetum.** The species of the genus *Sphagnum*, as is well known,

---

[1] Pl. 115, phot. 283 and 283 a, p. 686.   [2] Pl. 121, p. 714.

have a highly specialised vegetative structure. The surface of the stem is
covered with a layer of large empty cells, whose walls are
strengthened by ribs and pierced by relatively wide holes; the
leaves consist of a single layer of cells, the framework composed
of cells similar to those covering the surface of the stem, and running between them are lines of narrow living cells containing chlorophyll. The network of fine capillary channels formed by the empty cells with their pierced walls results in the plants absorbing liquid water in contact with their shoots through the holes in the cell walls and holding it like a sponge, so that a considerable quantity can be easily squeezed out with the fingers from a living tuft of *Sphagnum*. Water is also held between the surfaces of the leaves and the stems. When this surface water has been lost by evaporation that contained in the cells is slowly evaporated into the air, the actual rate of evaporation depending of course primarily on the saturation deficit of water vapour in the air as well as on the structure of the plant.

Variation of habitat and structure. Some species of *Sphagnum* are aquatic mosses, growing immersed in water: others live on constantly wet soil or more commonly on the wet bases of the aerial shoots of other plants. The habitat of these must be pretty constantly wet, either from soil water or abundant precipitation, and the "terrestrial" Sphagna are most abundant and flourish most luxuriantly under conditions of very high average atmospheric humidity. The species which live in the less constantly wet habitats, as Watson (1918) points out, possess xeromorphic characters in greater or less degree—compactness of habit, close imbrication of leaves on the branches, infolding of the leaf edges—characters which check loss of water by evaporation to the air. The more aquatic species are least xeromorphic and least able to resist desiccation.

Acidity and mineral requirements. The Sphagna are well known to flourish only where they are in contact with more or less acid water and to be killed by exposure to alkaline solutions. The acid substances held in their cell walls absorb the bases of nutrient salts, setting free the acid ions and thus maintaining an acid medium in contact with the moss. While this is true of all the species they vary a good deal in sensitiveness to an alkaline medium, and this is correlated with the degree of their own acidity (Skene, 1915). Broadly speaking the least acid species grow in the less acidic habitats, where they can usually obtain a greater supply of mineral salts, while the extremely acid species, most readily killed by alkaline solutions, have a very low mineral requirement and grow in the most acidic habitats. Thus Skene found that *Sphagnum contortum*, a species which, judged from its habitats, is of relatively high mineral requirements, had an average "primary acidity" (measured in terms of grams of acid hydrogen per hundred grams of *Sphagnum* material after thorough washing out of the absorbed bases) of 0·0715, while *S. rubellum*, a hummock-forming species of highly acidic raised bog, had an average primary acidity of 0·1092. This correlation is not however exact. Skene found that the acidity of different samples of the same species varied a good deal and that their ranges of acidity overlapped considerably; and there is evidence that what appears to be the same species may differ markedly in acidity and habitat in different regions. Further research is required on the relation of acidity and mineral requirements to habitat, and on the possible existence of different ecotypes of the same "species".

Oxidation-reduction potential. Pearsall (1938) has very recently found that the pH values in the Sphagnetum of raised bog, where the conditions were optimal for the growth of the bog moss (species not stated), varied between 4·17 and 4·62 at a depth of 10 cm., most of the determinations giving values about 4·2 or 4·3. These soils are "reducing". As the surface of the bog dries, as a result of drainage or otherwise, *Eriophorum* and ultimately *Calluna* becoming abundant, oxidation occurs and the acidity rises, pH values well below 4·0 being recorded. Thus while mor and acid peat have on the whole a low oxidation-reduction potential, *Sphagnum* peat shows an increase, on drying, with increase of acidity (cf. pp. 82, 84).

Habitats of various Sphagna.[1] Aquatic species of relatively high mineral requirements are *S. inundatum* and *S. platyphyllum*, and these grow in water which is less acid, e.g. in some fens. Typical aquatic species of low mineral requirements are *S. plumosum* and *S. cuspidatum*,[2] the latter the common dominant in wet hollows of raised bog in Ireland, Wales and Scotland as well as on the continent.

Species inhabiting situations of intermediate moisture are the following:

Relatively high mineral requirements:

S. amblyphyllum	S. subsecundum
S. contortum	S. teres
S. squarrosum	S. warnstorfii

Lower requirements: Still lower:

S. angustifolium	S. apiculatum
S. cymbifolium (wide range of acidity)	S. balticum
S. papillosum[3]	S. tenellum (molluscum)

The following are species of the driest and most highly acid habitats, conspicuous in the formation of hummocks in raised bogs:

Less dry: Drier:

S. compactum	S. acutifolium
S. magellanicum (medium)	S. fuscum
S. plumulosum (subnitens)	S. imbricatum
	S. rubellum[3]

Role of Sphagna in vegetation. As has been said, certain species of the genus are the characteristic dominants and primary peat formers of the

---

[1] Kindly communicated by Prof. H. Osvald.   [2] Pl. 116, phot. 285, etc.

[3] *Sphagnum papillosum* and *S. rubellum* (Pl. 116, phot. 285) are the most important hummock formers in the British Isles: *S. tenellum* (*molluscum*) and *S. plumulosum* (*subnitens*) are also common species.

many mosses or bogs which cover wide areas of flat or gently sloping or undulating land in the cooler regions of the northern hemisphere, more especially the uneven morainic ground left by the Pleistocene ice-sheets. Starting with aquatic species in the innumerable pools or lakelets occupying the hollows, or with less hydrophytic species on valley bog, fen or forest vegetation in a sufficiently moist climate, the bog mosses may spread far and wide over surrounding vegetation, the more hydrophytic being succeeded by less hydrophytic species. In this way the moss bogs have destroyed and buried great areas of fen or forest more than once in post-glacial times, probably always as the result of change from a drier to a wetter climate. Owing to the peculiar structure of its tissues already described *Sphagnum* carries its own water with it as it grows upwards and outwards. Thus a pad or cushion of one of the less hydrophytic bog mosses forms an extending sheet saturated with water and with a convex upper surface, and the raised bog as a whole is an aggregate of such sheets, also with a slightly convex surface, since the centre of the bog represents its oldest and therefore its highest part.

The older, lower layers of the moss are cut off from light and air by the living surface layer, and progressively die. Compressed by the increasing superincumbent weight as the moss rises higher, and unable to decay completely owing to absence of free oxygen and of the normal action of soil bacteria, the lower layers of moss are converted into typical acid bog or moss peat, whose antiseptic properties are well known from the wonderfully complete preservation after many centuries of various objects, including the bodies of animals and men, that have been buried in the bog.

Such a raised bog is not however a simple mass of *Sphagnum*, but has a complex structure. In the first place it is composed of very numerous aggregated *Sphagnum* cushions or hummocks, in each of which progression from aquatic or subaquatic species at the base can be traced upwards to the more xeromorphic species at the summit of the cushion. And secondly it is inhabited by a number of species of oxyphilous vascular plants whose remains form part, sometimes the greater part, of the peat. The shape and growth of the individual cushions reflect *in petto* those of the bog as a whole. The structure of a raised bog will be described in more detail in connexion with the Irish raised bogs (pp. 686–96).

Sphagnetum in Great Britain. A great deal of *Sphagnum* peat has been formed in the British Isles during the wetter climatic epochs of the post-glacial period. But in the lowlands of England Sphagneta are now mainly met with in sandy heath areas where the drainage is impeded so that acid soil water accumulates. These "wet heath" Sphagneta with their associated species of flowering plants are really "valley bogs" of limited extent, and show a floristic composition quite similar to, though not identical with, those of raised and blanket bogs. Other species of Sphagna occur in certain wet woods and on certain fens under appropriate conditions. On the uplands Sphagnetum does not now cover any large areas of Great Britain. In many upland regions peat is now being actively eroded, though in the wettest climates, and locally where the necessary edaphic conditions are realised, peat is still being formed.

Sphagnum bogs of wet heath. The *Sphagnum* bog occupying the lowest portion of the "wet heath" described by Watson from Chard Common in Somerset (see Chapter xxxvi, pp. 740–1) is an example of the type of local lowland Sphagnetum (essentially a valley bog) referred to above, and similar bogs may be found in many parts of the country, though by far the greater number have been drained and destroyed. These of course are definitely local bogs determined by low lying ground with acid soil water overlying impermeable strata, or where the drainage is otherwise impeded. More extensive valley bogs have been described by Rankin (1911, pp. 259–64) from the New Forest in Hampshire. Here the low plateau of permeable Eocene beds has been dissected by small streams which have been cut down to the base level of erosion, so that their waters are extremely sluggish and in some places form chains of stagnant pools. Along the course of the stream reedswamps and alder thickets are developed, and outside these are Molinieta (see p. 523), but beyond the Molinietum, and abutting on the heath which covers the plateau, are Sphagneta. Besides these valley bogs or "valley moors", as Rankin calls them, he describes "spring moors" (better called "spring bogs"), also dominated by *Sphagnum*, which form round the springs of acid water that issue from the sides of the valleys along the lines of junction of permeable and impermeable strata. These valley and spring bogs tend to encroach on the adjacent heaths in the same way as typical raised bogs encroach on the fenland on which they are founded, and may in fact develop into raised bogs.

The following species are given by Rankin as composing the flora of the "spring" and "valley" bogs of the New Forest:

Sphagnum spp. (dominant, forming a matrix)

Eriophorum angustifolium	a	Juncus acutiflorus a
Eleocharis multicaulis	a	

The rhizomes and roots of these three species increase the firmness of the bog, and in the web so formed the following plants occur:

Drosera intermedia	Narthecium ossifragum
D. rotundifolia	Pinguicula lusitanica
Malaxis paludosa	

A narrow edge zone overlapping the adjacent heath includes:

Drosera intermedia	Rhynchospora alba
Lycopodium inundatum	R. fusca

Calluna is said to colonise the drier bogs abundantly.

Raised bogs. Apart from these examples no typical raised bog has been

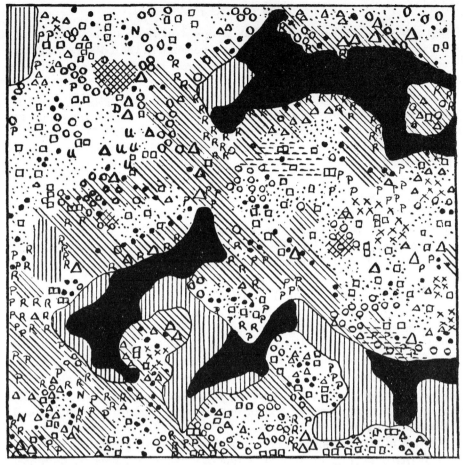

|||| *Sphagnum cuspidatum*

\\\\ *S. papillosum*

≡≡ *S. tenellum*

P *S. pulchrum*

※※ *Hypnum cupressiforme*

::::: *Cladonia sylvatica*

u *C. uncialis*

x *Rhacomitrium lanuginosum*

R *Rhynchospora alba*

N *Narthecium ossifragum*

D *Drosera rotundifolia*

• *Erica tetralix*

▵ *Eriophorum vaginatum*

△ *E. angustifolium*

□ *Scirpus caespitosus*

o *Calluna vulgaris.*

■ Open water or bare mud with or without *Zygogonium ericetorum*

Fig. 136. Quadrat (5 m. square) of "Regeneration Complex" (Active Development) in Tregaron Bog (Cardiganshire)

The hollows contain open water or are lined with *Sphagnum cuspidatum*. A little higher comes *S. pulchrum*. *S. papillosum* is the chief hummock-forming species, sometimes with *S. tenellum* capping the lower hummocks. *Hypnum cupressiforme, Rhacomitrium lanuginosum* and *Cladonia sylvatica* also occur on the hummocks. Of flowering plants *Rhynchospora alba* occupies the lowest and wettest positions, *Calluna* and *Scirpus caespitosus* the highest and driest. Godwin and Conway, unpublished.

fully described in England, and it is doubtful if an unspoiled example
now exists, though there may be some in the Lake District and
Tregaron bog neighbouring regions, but Tregaron bog in the Teifi valley in
central Wales has recently been studied by a party from the
Cambridge Botany School, and the following short description and figures
have been most kindly provided by Dr Godwin and Miss Verona Conway
from their as yet unpublished work on this bog (Figs. 136–7).

"The village of Tregaron, Cardiganshire, stands on extensive morainic
deposits which, after retreat of the ice, blocked the wide valley of the Teifi,
whose waters formed a shallow lake below the bar. In post-glacial times
this lake was filled by the growth of fen peat, above which developed three
large raised bogs, one to the west and two to the east of the river, which was
ultimately confined to a narrow central valley.

"The large western bog, which has been least affected by peat cutting,
shows a steeply sloping margin (*rand*) along the flood plain of the river.
This is drained by numerous channels from the bog itself and shows well-
marked zones of Scirpetum, Molinietum, and Callunetum. On the uncut
margin which abuts on the hillside is a small *lagg* (see p. 675), with characteristic species and a thin fen carr.

"The highest part of the bog surface shows the structure of a typical
active bog or 'regeneration complex' illustrated in Figs. 136 and 137 A.
The most abundant Sphagna in this area are *S. cuspidatum* in the pools,
S. pulchrum, an early colonist of the pools, and *S. papillosum*, the chief
hummock-forming species. The usual flowering plants, bryophytes and
lichens are present.

"Much of the bog surface is occupied, not by the regeneration complex,
but by Scirpeta (Fig. 137 B) and Molinieta in which Sphagna are unimpor-
tant and peat formation probably slow. The relation of these communities
to the regeneration complex is not yet certain, nor is it known how far they
reflect climatic or human influences."

Small bog at Loch Maree. The only available description of a Scottish
raised bog is that of a small but very perfect example completely enclosed
by native pinewood on the south-western side of Loch Maree in Ross-shire,
at about 57° 38′ N. lat. and 5° 24′ W. long., not far above the shore of the
lake. This is developed on a terrace of the hillside, and is only about 150 m.
long by 70 m. wide and oval-oblong in shape (Fig. 138). It appears to be
quite untouched and shows the typical features of such a bog very well
indeed. The surface is slightly convex. (1) The lagg or drainage channel,
initiated by a stream which descends the hill, is on one of the long sides of
the bog against the steep hill which drains into it, and just above the lagg
at the base of the steep slope are massive cushions of Sphagnum tailing out
above into the vegetation of the pinewood. Within the lagg there is (2) a
marginal zone only a few metres wide dominated by *Calluna*. This encloses
the main area of the bog which contains many pools and consists of a more

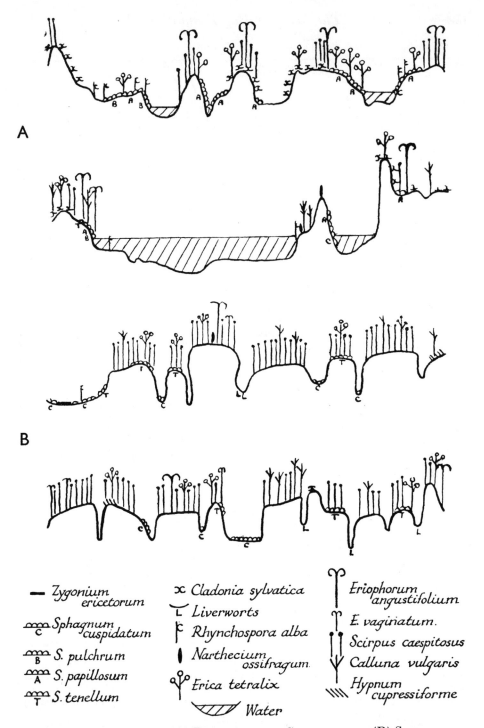

—	*Zygonium ericetorum*
⌒⌒⌒ C	*Sphagnum cuspidatum*
⌒⌒⌒ B	*S. pulchrum*
⌒⌒⌒ A	*S. papillosum*
⌒⌒⌒ T	*S. tenellum*
✕	*Cladonia sylvatica*
ᘁ	*Liverworts*
ꝑ	*Rhynchospora alba*
❙	*Narthecium ossifragum*
⚲	*Erica tetralix*
▱	*Water*
ꙮ	*Eriophorum angustifolium*
ꙷ	*E. vaginatum.*
❡	*Scirpus caespitosus*
⅄	*Calluna vulgaris*
⦚	*Hypnum cupressiforme*

FIG. 137. PROFILES OF (A) REGENERATION COMPLEX, AND (B) SCIRPETUM CAESPITOSI IN TREGARON BOG

The hummocks and hollows are shown, with *Cladonia* usually on the highest hummocks in A. In B there is no open water and the Sphagna are much less prominent. Each profile is 5 m. long. Vertical scale 5 times the horizontal. Symbols not drawn to scale. Godwin and Conway, unpublished.

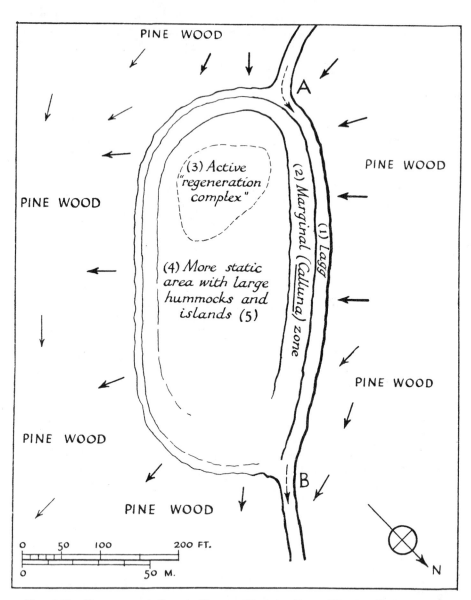

FIG. 138. DIAGRAM OF SMALL RAISED BOG AT LOCH MAREE (ROSS-SHIRE)

The bog is developed on a terrace of the steeply sloping hillside and is entirely surrounded by native pinewood. The thicker arrows show the steeper slopes. (This diagram is drawn from memory and the boundaries and scale are only approximate.) Cf. Pl. 115, phot. 282, which is taken from the hillside at the south-west end of the bog (top of the diagram), approximately from the position of the vertical arrow.

or less continuous carpet of *Sphagnum*. Part of this (3) appears to be actively growing (regeneration complex), the remainder (4) has reached a more static condition. It shows scattered raised "islands" (5), most of which are *Sphagnum* or *Rhacomitrium* hummocks, while the largest are rocky outcrops rising through the surface of the bog. Some of these bear dwarf pines.

The following species were recorded:

(1) *Lagg* (in and near the stream):

Molinia caerulea	d		
Sphagnum papillosum	cd	Calluna vulgaris*	f
S. recurvum	cd	Carex panicea	f
S. plumulosum	f	Potentilla erecta	f
Juncus effusus	o–la	Erica tetralix	o

* Not flowering freely

(2) *Marginal* Calluna *zone:*

Calluna vulgaris*	d	Potentilla erecta	f
Molinia caerulea	a	Drosera rotundifolia	o
		Sphagnum papillosum	f

* Flowering freely.

(3) *Regeneration complex:*

In pools and hollows		Forming hummocks	
Sphagnum cuspidatum var.		Sphagnum papillosum	a
submersum	a	S. plumulosum	f–a
S. inundatum	r	S. magellanicum (medium)	o
Rhynchospora alba	a	S. rubellum (summits)	f
Drosera anglica	a		
Eriophorum angustifolium	f–a		
Carex pulicaris	o		

Sphagnum tenellum (molluscum), Rhacomitrium lanuginosum and *Hypnum cupressiforme* were scattered on the hummocks.

(4) *More static area.* This contained the following additional species:

Eriophorum vaginatum	a–ld	Erica tetralix	f
Narthecium ossifragum	a	Pedicularis sylvatica	o–lf
Molinia caerulea	f–la	Juncus squarrosus	o
Potentilla erecta	f–la	Scirpus caespitosus	r–o
Drosera rotundifolia	f	Carex panicea	o
Calluna vulgaris	f	C. stellulata	r

Seedlings of *Pinus sylvestris* were abundant.

(5) *"Islands"* (rock outcrops):

Calluna vulgaris	a–d	Hypnum schreberi	f
Molinia caerulea	f	Dicranum scoparium	o
Pteridium aquilinum	f	Hylocomium loreum	o
Agrostis canina	o	H. splendens	o
Juncus effusus	o	Thuidium tamariscinum	o
Vaccinium myrtillus	o		
Pinus silvestris (dwarfed)	f	Cladonia sylvatica	o
Ilex aquifolium	o	C. uncialis	o

Raised mosses of Lonsdale. The mosses[1] of the Lonsdale lake and estuarine basins (north Lancashire, just south of the Lake District) were of raised bog type, but most of them have been partially or completely destroyed by draining and peat cutting and none is now dominated, according to Rankin's description (1911, pp. 247–59), by active Sphagnetum (regeneration complex). The best example extant in 1911 was Foulshaw Moss, which then occupied about two square miles (c. 5·2 sq. km.), though formerly, with adjacent mosses, it reached from the sea for seven miles inland. The surface sloped gently up from the edge where it abutted on the hillside, and was dominated by *Eriophorum vaginatum* mixed with *Scirpus caespitosus*. On the tussocks formed by these grew shrubby *Calluna*, *Erica tetralix* and *Andromeda*, while characteristic mosses, chiefly Sphagna, spread between them. "The occasional pools (writes Rankin) are filled with a matrix of *Sphagnum*, traversed by the rhizomes of *Eriophorum angustifolium* and *Rhynchospora alba*, and the lanky stems of the cranberry (*Oxycoccus quadripetalus*); here also the bog asphodel (*Narthecium ossifragum*) and the sundews (*Drosera rotundifolia*, *D. intermedia* and—most rarely—*D. anglica*) find a sufficiently firm substratum for their growth" (Rankin, 1911, p. 249). There was evidence however that Foulshaw was formerly (in the nineteenth century) much wetter, with a greater expanse of bog moss and it was then said to be quite impassable. The existence of several feet of *Sphagnum* peat below the surface *Eriophorum* peat (see Fig. 47, p. 168) shows that Foulshaw was at one time an actively growing raised moss.

Other raised mosses. Raised mosses are quite frequent in north-western England and south-western Scotland, and Pearsall (1938) has made observations on the acidity and oxidation-reduction potential of some of them, but they have not been systematically described.

Irish raised bogs. Raised bogs, in their actively growing condition dominated by species of *Sphagnum* and locally known as "red bogs", either because of the prevalence upon them of crimson-coloured bog mosses (mainly *Sphagnum rubellum*) or because of the warm reddish brown tint of the blended bog vegetation as seen from a distance, are a conspicuous feature of the Irish central limestone plain. In travelling through the plain one is constantly coming across these bogs as great brown uniform expanses of vegetation raised a few feet above the general level of the plain and with very slightly convex surfaces.[2] Thus between Athenry and Athlone, a distance of 42 miles, nine such bogs were seen to right or left of the road, and between Mullingar and Kinnegad (12 miles) at least three. South of Edenderry there is an area of about 20 square miles almost entirely covered with raised bog.

Distribution, climate and peat

The climate, though oceanic, is not so wet as in the far west, and the

---

[1] The distribution of some of these mosses is shown in Fig. 109 (p. 596).
[2] Pl. 115.

PLATE 115

Phot. 282. Small raised bog on terrace at Loch Maree, Ross-shire, entirely surrounded by native pinewood. The dark marginal zone of Callunetum is seen in the foreground: behind is the regeneration complex with hummocks. Cf. Fig. 138.

Phot. 283. Raised bog near Shannon Bridge, Co. Roscommon, in the Irish central plain, from the top of a neighbouring esker. Callunetum (dark) occupies the cut bog margin across the road. Other raised bogs in the distance. R. J. L.

Phot. 283a. Raised bog east-north-east of Athlone, West Meath, showing the convex surface in profile. The foreground, from which peat has been removed (turf stacks on the left) is occupied by Callunetum. The cut edge of the bog (vertical peat cliff) is seen as a dark band in the middle distance. R. J. L.

RAISED BOG

PLATE 116

Phot. 285. Close view of hollow and adjacent hummock. *Sphagnum cuspidatum* with a plant of *Eriophorum angustifolium* in the hollow. *S. papillosum, S. rubellum* and *Cladonia sylvatica* (white) on the hummock. *Calluna vulgaris* on each side. *R. J. L.*

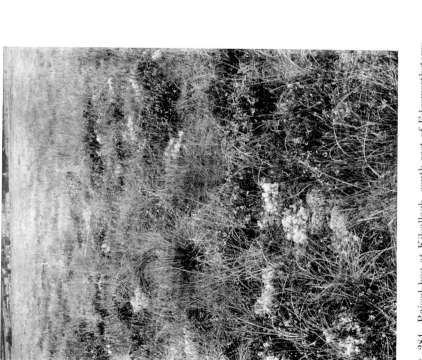

Phot. 284. Raised bog at Kilsallagh, south-east of Edgeworthstown, West Meath. Surface of "regeneration complex" showing hollows and hummocks. *Sphagnum cuspidatum* in hollow at centre of extreme front. Hummock with *S. papillosum* and *Cladonia sylvatica* (lightest) immediately above. The tufted plants are *Scirpus caespitosus* and *Eriophorum vaginatum. Calluna* is abundant. 2 ft. (60 cm.) rule in middle distance. *R. J. L.*

bogs are topogenous, either built upon old lake areas which have passed through the stage of fen, sometimes of fen wood ("carr"), or directly on the site of forest which has been flooded or which has succumbed to the invasion of marsh and bog plants favoured by increasing wetness of climate. This is proved by the nature of the underlying peat (see below).

Some idea of the enormous mass of peat contained in these bogs can be gained from the fact that the population have been cutting the edges for fuel for many centuries, so that it is very rare to find an untouched bog edge, and the result is that the margins are almost everywhere "nibbled", so to speak; but the great mass of the bog remains intact. From only a few small bogs has the peat been completely removed.[1]

Structure and development. The convexity of the surface is everywhere evident, but the angle of slope is very slight, in one case which was measured a gradient of about 1 in 100, or 3°. The few untouched or apparently untouched edges slope much more steeply, and are occupied by luxuriant and dominant *Calluna*, *Pteridium* or *Ulex europaeus*, occasionally sparsely colonised by seedling birch (*Betula pubescens*) or subspontaneous pine. Such bog margins have apparently stopped growing, but the general surface of many of the bogs seems still quite active ("regeneration complex"). Below the steep marginal slope there can generally be traced the remains of the "lagg" or marginal ditch which is a constant feature of raised bog and receives the drainage water from the peat, but no example of an entirely natural lagg (such as is described on pp. 682, 685) was met with. Even where there has apparently been no peat removed from the edge of the bog the adjacent ground (originally fenland) has been modified by pasturing and drain cutting.

Raised bog vegetation is mixed, like that of blanket bog, and contains most of the same species, but there is a definite genetic sequence, or rather series of sequences, in the layers of peat which have built up the bog. When the bog has been raised on a substratum of fen the lower layers of the peat record the structure of the fen, and above this come successive *cycles* of different kinds of bog peat. The existence of these cycles depends

Hollow-hummock cycle

upon the fact that the surface of a bog at any given time is not uniform, but consists of alternating *hummocks* and *hollows* inhabited by different species.[2] The peat at the bottom of each hollow is built up by the vegetation in the natural process of autogenic succession until it forms a new hummock, the surfaces of the adjacent pre-existing hummocks which have stopped growing thus coming to occupy a lower level than the new hummocks. The old hummocks thus become the sites of new hollows, the old hummock vegetation dies, and is replaced by species characteristic of hollows, these in their turn giving way to new

[1] The bogs are unlikely to remain much longer, since the systematic exploitation of the peat on a large scale is beginning.

[2] See Fig. 139; Pl. 116, phot. 284; and Pl. 117, phot. 286.

hummock formers. Thus the structure of the peat of a raised bog is lenti-cular,[1] each lenticle representing a complete cycle of "hollow-hummock" development, and all the phases of the cycle are represented at any given time on the surface of an actively growing bog ("regeneration complex"). The lenticular structure can be clearly seen in vertical section of the peat. The story of raised bog development was first worked out in detail by Osvald in his classical paper on the Swedish bog "Komosse",[2] and the Irish raised bogs are essentially similar, though some of the species are different.

Below are lists of the species occurring in different stages of the "hollow-hummock" development on a typical raised bog surface south of Athlone, in the middle of the central plain.

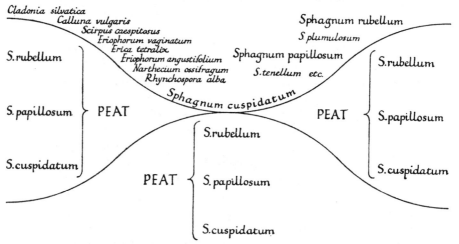

FIG. 139. DIAGRAM OF THE SUCCESSION OF SPECIES FORMING THE PEAT IN A TYPICAL "HOLLOW-HUMMOCK" CYCLE (REGENERATION COMPLEX)

Vegetation of seral stages. *Stage* 1. Semi-aquatic *Sphagnum* phase—wet hollow:

Sphagnum dusenii*	d	Rhynchospora alba	o
Eriophorum angustifolium	o		

    * The dominant species of the hollows is generally *S. cuspidatum.*

Sphagnum cymbifolium and one plant of *Narthecium* occurred along the margin.

This community is increasingly colonised by vascular plants, species of semi-aquatic bog moss remaining dominant in the lower layer; and thus develops into

[1] Fig. 139.

[2] H. Osvald, Die Vegetation des Hochmoores Komosse, *Akad. Abhandlungen*, Uppsala, 1923. In the summer of 1935 the author had the privilege of examining some Irish raised bogs in the company of Prof. Osvald, who interpreted the details of their vegetation and peat structure in the field: the present account was thus made possible.

PLATE 117

Phot. 286. Two pools, with hummock between, on raised bog near Edenderry. *Sphagnum cuspidatum* in pools, *S. papillosum*, etc. forming hummocks: *Scirpus caespitosus*, *Eriophorum vaginatum*, *Erica tetralix*, *Calluna*. *R. J. L.*

Phot. 287. Large hummock (height above water in pool 84 cm., diameter 137 cm.) on raised bog between Athlone and Ballinasloe. *Sphagnum cuspidatum* below; *S. magellanicum* and *S. rubellum* forming the mass of the hummock, which is capped by *Leucobryum glaucum*, *Cladonia sylvatica* and old *Calluna*. *Scirpus caespitosus* on the right. *R. J. L.*

IRISH RAISED BOGS—HUMMOCKS AND POOLS

PLATE 118

Phot. 289. Big tussock of *Scirpus caespitosus* on cut peat bank near Mullingar. *Calluna* to the right and below. *Molinia* behind. The peat is here drier than in untouched bog. R. J. L.

Phot. 288. Hummock and pool (in front) on raised bog between Athlone and Ballinasloe. *Sphagnum cuspidatum* and *Rhynchospora alba* in front. Most of the hummock is composed of *Sphagnum magellanicum*, darker (red) to right and *S. papillosum*, lighter (yellow) to left. *Cladonia sylvatica* and *Calluna* crown the hummock. R. J. L.

IRISH RAISED BOG

Stage 2. *Rhynchospora* phase—hollow:

Rhynchospora alba	d	Oxycoccus quadripetalus	
Narthecium ossifragum	f	var. microcarpus	o
Erica tetralix	f		
Andromeda polifolia	o	Sphagnum cuspidatum	d
Calluna vulgaris	o	S. tenellum	f
Drosera anglica	o	Lepidozia sp.	o
D. rotundifolia	o	Cladonia uncialis	o
Eriophorum vaginatum	o		

Stage 3. *Narthecium* phase—hollow. *Narthecium* tends to assume dominance at the expense of *Rhynchospora*, while the semi-aquatic Sphagna and other species retrogress. In this stage the following list was made:

Narthecium ossifragum	d	Rhynchospora alba	o
Erica tetralix	f	Scirpus caespitosus	o
Andromeda polifolia	o		
Drosera anglica	o	Sphagnum papillosum	f
D. rotundifolia	o	S. rubellum	o
Eriophorum vaginatum	o	S. tenellum	o
Calluna vulgaris	o	Lepidozia sp.	o
		Kantia sp.	o

Stage 4. *Calluna—Sphagnum rubellum* stage—active phase of hummock formation. In this *Sphagnum rubellum* rapidly assumes dominance in the lower, and *Calluna* in the upper layer:

Calluna vulgaris	d	Oxycoccus quadripetalus	
Erica tetralix	f	var. microcarpus	o
Eriophorum vaginatum	f	Rhynchospora alba	o
Andromeda polifolia	o		
Drosera rotundifolia	o	Sphagnum rubellum	d
Eriophorum angustifolium	o	Hypnum schreberi	o
Narthecium ossifragum	o	Kantia sp.	o
		Mylia sp.	o

Stage 5. *Calluna-Cladonia* stage—last stage of hummock, in which the Sphagna have lost dominance and are dying:

Calluna vulgaris	d	Cladonia sylvatica	a
Erica tetralix	f	C. uncialis	o
Eriophorum vaginatum	f	Sphagnum rubellum ⎫	
E. angustifolium	o	S. plumulosum ⎬ f, but dying	
Andromeda polifolia	o	S. tenellum ⎭	
Narthecium ossifragum	o	Kantia sp.	
Oxycoccus quadripetalus		Riccardia sp.	a
var. microcarpus	o	Lepidozia sp.	
Scirpus caespitosus	o		

Structure of hummocks. In the whole succession the Sphagna are of primary importance: first the species of high water requirement which dominate the wet hollows (*S. cuspidatum, S. dusenii, S. cymbifolium*), then the active hummock builders (*S. plumulosum, S. papillosum, S. rubellum*). *S. magellanicum*, better known in England as *S. medium*, is not so common in Ireland as the preceding species. *S. fuscum*, which is the characteristic species on the continent, appears to be rare in the Irish bogs. Some of the larger hummocks may be capped with *Leucobryum glaucum* (Pl. 117, phot. 287). Other mosses (e.g. *Hypnum schreberi, H. cuspidatum*) and liverworts may also settle among the Sphagna. *Calluna* may enter the less wet hollows, but it only becomes luxuriant and eventually dominant after active hummock formation has begun. As we shall see in the following chapters the ling has a very wide range of tolerance of soil water content, but it never becomes dominant unless there is good drainage for its roots and this it finds in the rising cushions of the second group of Sphagna. On the older hummocks, after the entrance of *Cladonia sylvatica*, the ling tends to become "leggy" and die. In view of the dryness of soil the ling can tolerate, it is difficult to attribute this to drought, and it may depend on insufficient nutrition or merely on the age of the plant. As we shall see later a similar phenomenon occurs with age on the English heaths and grouse moors.

The photographs on Pls. 116–118, taken by Dr Lythgoe on various raised bogs of the central Irish plain, show the positions in hollows and on hummocks of some of the more important species.

Peat profiles of raised bogs

Athlone bog. A boring was made by Prof. Osvald through this raised bog at a spot about 350 m. from its northern edge, in the middle of the typical "regeneration complex", as he calls the aggregation of active hummocks and hollows just described. With his permission a detailed record of this boring to a depth of 10 m., drawn to scale, is annexed (Table XXI, pp. 692–3).

It will be obvious, on consideration, that any given vertical section through a bog of the structure described will not always pass through the complete cycle of hollow-hummock development: in fact it will only do so if the bore strikes a hummock somewhere about its centre. Otherwise it will pass through the edges of the lenticular units, so that the series of phases represented in the bore will be incomplete. It will be noted that the basal layer of the cycle (the *S. cuspidatum* layer) and the layer immediately below it are generally the most "humified", i.e. decayed, and in these highly humified layers the *species* of *Sphagnum* present are rarely recognisable. The degree of humification is represented by numbers on a scale of 1 to 10. Thus 8 means highly humified, 2 very little.

The surface vegetation at the spot where the boring was made was that of a hollow partly grown up and dominated by *Narthecium* (see the list of Stage 3, p. 689).

Below the living vegetation cover came 20 cm. of moderately humified *Narthecium-Sphagnum* peat already formed in the hollow, and then 5 cm. of highly humified *Calluna-Sphagnum* peat marking the top of the preceding hummock, and underlain by alternating layers, much less humified, of peat consisting of *S. papillosum* and *S. rubellum*, with *Eriophorum vaginatum* and *Calluna*, to a total depth of 1·2 m. This represents the varying colonisation and growth of the typical hummock-forming bog-mosses, but the base of the hummock was not passed through.

Under this came a complete cycle, consisting, from above downwards, of 20 cm. of highly humified *Sphagnum-Calluna* peat, 45 cm. of *Sphagnum papillosum* peat, and finally 10 cm. of *S. cuspidatum* peat representing the vegetation of the hollow on which the hummock was built.

At 2 m. begins an incomplete cycle, followed by two which are fairly complete, each with typical *S. cuspidatum* peat at the base.

At 4 m., towards the bottom of the bog, we encountered half a metre of *Sphagnum* peat with *Eriophorum vaginatum* and *Calluna*, underlain again at 4·5 m. by a thin layer of *Sphagnum cuspidatum* and *Eriophorum angustifolium* peat, marking the wet conditions in which the raised bog began to develop.

Below this was the fen peat on which the bog was built, consisting of *Carex* and *Cladium* and passing down into reedswamp with *Phragmites*, *Menyanthes* and *Carex inflata*, the peat of the fen and reedswamp together having a thickness of more than 3 m. and becoming muddy towards the base.

At a depth of 7·77 m. from the present surface of the bog was a thin layer of yellow calcareous lake mud with remains of *Phragmites*, and below this 1·75 m. of cream-coloured lake marl with no remains of vegetation, representing the conditions of open water.

At 9·6 m. from the surface the bluish grey clay (glacial till) on which the lake was based was encountered, and at 10 m. the boring was stopped.

Edenderry bog. Another bore (Table XXII, pp. 694–5) was made in an extensive raised bog south of Edenderry. This was put down in a low hummock of *Calluna-Tetralix* (see Stage 4, p. 689) surrounded by *Sphagnum tenellum*. Immediately below came 25 cm. of *Sphagnum-Narthecium* peat, and then a fairly complete hollow-hummock cycle with *Sphagnum rubellum*, *S. papillosum* and *Eriophorum vaginatum* above and *Sphagnum cuspidatum* peat below. Then came about 2·5 m. of incomplete or doubtfully complete cycles, probably five in number, in which the basal *S. cuspidatum* peat could not be clearly identified, underlain by the first "regeneration cycle". Below this was peat composed of *Sphagnum, Calluna, Oxycoccus* and much *Eriophorum vaginatum*, and (at about 4·5 m. from the present surface of the bog) the topmost fen peat consisting of *Caricetum* colonised by *Sphagnum*. Under this was more than a meter of fen peat with alternating *Carex*, *Cladium* and *Equisetum*, and the moss *Amblystegium* towards the bottom. Below the fen peat we came to *Sphagnum* and *Calluna* again, then to more *Carex* peat with *Phragmites* and *Menyanthes*, and below also *Myrica*. Thus we have two fen horizons separated by a development of bog moss and heather. The upper fen was probably formed as the result of reflooding of

Table XXI. *Section of bog south-west of Athlone through moss (bog) peat and underlying fen and reedswamp peat to lake marl and basal glacial clay. H = degree of humification (scale of humification 1 to 10).*

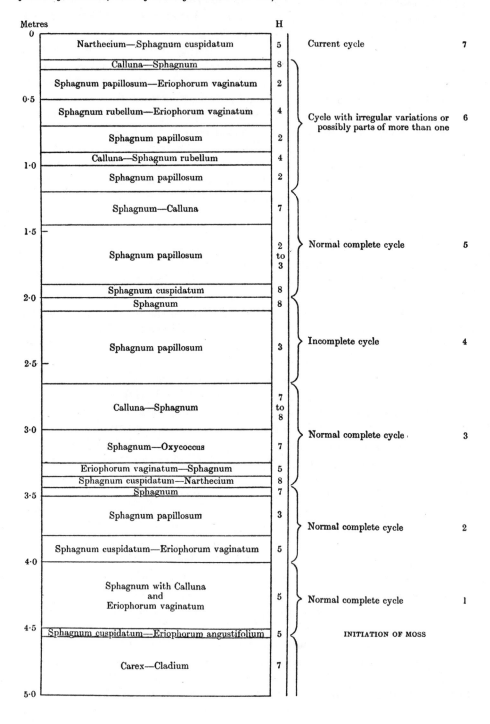

Table **XXI** *continued*

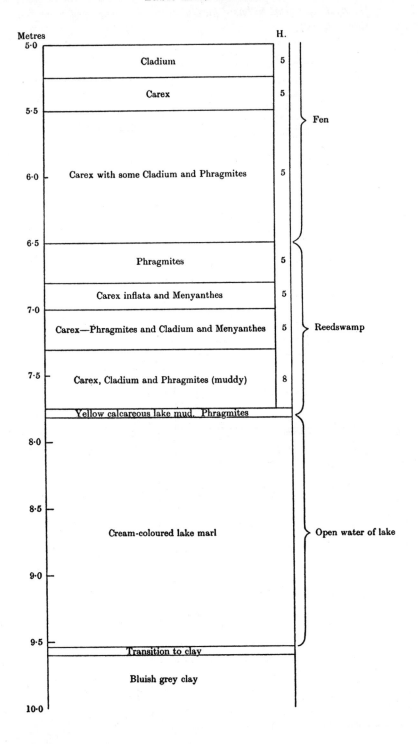

Metres		H.	
5·0	Cladium	5	
	Carex	5	Fen
5·5			
6·0	Carex with some Cladium and Phragmites	5	
6·5			
	Phragmites	5	
	Carex inflata and Menyanthes	5	
7·0	Carex—Phragmites and Cladium and Menyanthes	5	Reedswamp
7·5	Carex, Cladium and Phragmites (muddy)	8	
	Yellow calcareous lake mud. Phragmites		
8·0			
8·5	Cream-coloured lake marl		Open water of lake
9·0			
9·5	Transition to clay		
	Bluish grey clay		
10·0			

Table XXII. *Section of bog south of Edenderry through moss (bog) peat and underlying fen and carr peat to basal glacial clay. H = degree of humification (scale of humification 1 to 10).*

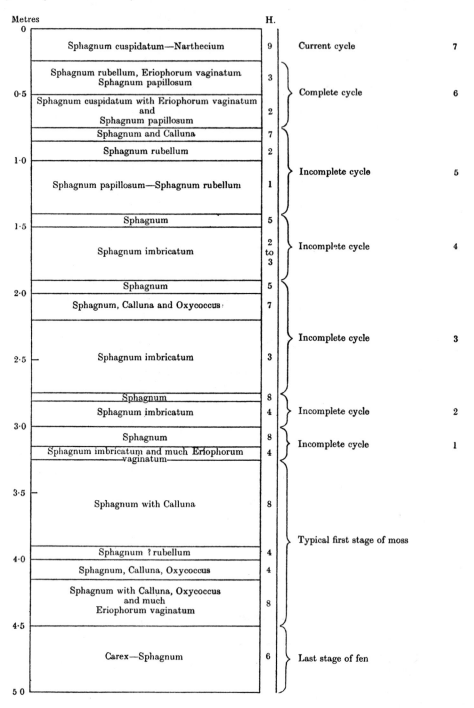

Table **XXII** *continued*

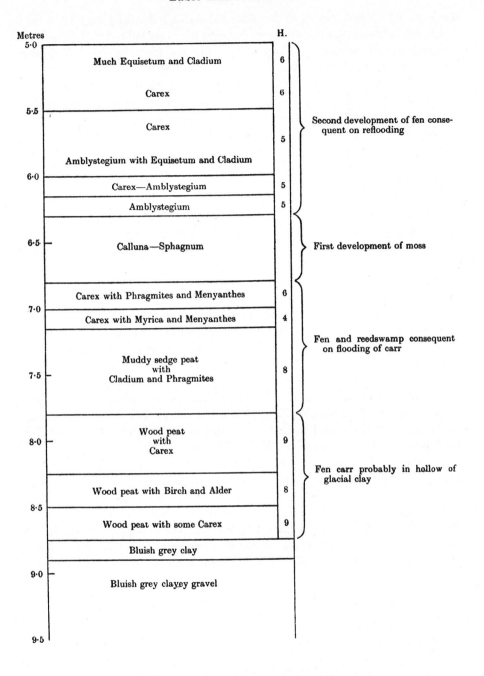

Metres		H.	
5·0	Much Equisetum and Cladium	6	
	Carex	6	
5·5	Carex	5	Second development of fen consequent on reflooding
	Amblystegium with Equisetum and Cladium		
6·0	Carex—Amblystegium	5	
	Amblystegium	5	
6·5	Calluna—Sphagnum		First development of moss
7·0	Carex with Phragmites and Menyanthes	6	
	Carex with Myrica and Menyanthes	4	Fen and reedswamp consequent on flooding of carr
7·5	Muddy sedge peat with Cladium and Phragmites	8	
8·0	Wood peat with Carex	9	Fen carr probably in hollow of glacial clay
8·5	Wood peat with Birch and Alder	8	
	Wood peat with some Carex	9	
	Bluish grey clay		
9·0	Bluish grey clayey gravel		
9·5			

the area, which had begun to develop moss: this reflooding may have been due to some topographical cause or perhaps to increased rainfall. The lower fen horizon passes down into a considerable thickness (85 cm.) of muddy sedge peat with *Cladium* and *Phragmites*, and this into wood peat with *Carex*; lower down was a denser wood peat of birch and alder, with *Carex* again below that. This basal sedge marsh and wood (fen carr) was based on the same bluish grey clay (here at 8·75 m. from the surface) that underlay the lake marl in the Athlone bog. Here, then, the fen underlying the bog started with carr on a wet clay, with no preceding lake as at Athlone, the sedges and the trees probably colonising a wet depression in the ground moraine left on the disappearance of the last ice-sheets.

Physiognomic dominants. The lists given on p. 689 do not give at all a just idea of the part played by *Scirpus caespitosus* in the physiognomy of the raised bog. The great tussocks of this plant (which is everywhere present except in the wet hollows) together with those of *Eriophorum vaginatum*, from which at first glance they are not readily distinguishable, are a conspicuous feature of the surface, and the bogs are *physiognomically* dominated by a mixture of *Calluna* with *Scirpus* and *Eriophorum*. In the neighbourhood of drainage channels *Scirpus* is the main dominant on wide belts of peat. Towards the edge of the bog the Sphagna are less prominent, hummock formation is less active ("Stillstandkomplex"), and the vegetation of the now flatter general peat surface tends to approximate to a rather sparse Callunetum in which *Scirpus* and *Eriophorum vaginatum* are abundant. The comparative water relations of *Scirpus caespitosus* and *Calluna* are discussed in the next chapter (p. 710). The Scirpetum seems to represent a drier phase of the raised bog, in which active growth has ceased, than that dominated by the Sphagna (cf. Fig. 137*b*, p. 683).

Vegetation of drainage channels. Besides the "lagg" which serves to drain its edges there are usually present, in a large raised bog, underground drainage systems, and the existence of such lines of drainage is often made evident by a series of patches or strips of vegetation quite different from that of the general bog surface. These depend on the movement of water into the drainage channel below and consist of marsh, fen or wet meadow species. The following were met with along the course of such a drainage system in the Athlone bog:

Cardamine pratensis	Hydrocotyle vulgaris
Carex inflata	Menyanthes trifoliata
Comarum palustre	Molinia caerulea
Holcus lanatus	Myrica gale

It is noteworthy that *Molinia*, which occurs here, is absent from the regeneration complex of the raised bog. At intervals along the course of the underground water channel there were deep holes, with steeply sloping or precipitous sides, at the bottom of which flowing water could sometimes be seen. On the well-drained peat of the sides the following species occurred:

Crataegus monogyna	Pteridium aquilinum
Orchis sp.	Rubus fruticosus (agg.)
Osmunda regalis	Ulex europaeus
Potentilla erecta	Vaccinium myrtillus

Sphagnetum of blanket bog. In the blanket bogs of western Scotland and Ireland Sphagneta form quite local patches and hummocks, and in the elevated blanket bogs of the plateaux of western and northern England, Scotland and Ireland also they rarely dominate any considerable continuous area. The structure and development of these Sphagneta need detailed study by modern methods, and the following scattered data are all that can be furnished at present. Blanket bog is dealt with in the next chapter.

Sphagneta of the northern Pennines. Lewis (1904, p. 325) described extensive upland *Sphagnum* bogs in the northern Pennines on the borders of Westmorland and Yorkshire, on either side of the Darlington and Tebay railway, about 6 to 10 miles south-west and south-south-west of Middleton-in-Teesdale. These mosses (e.g. Shacklesborough Moss, Red Gill Moss, Beldoo Moss) have an aggregate area of several square miles and are developed at an elevation of about 1300–1500 ft. ($=c.$ 400–450 m.) on the upland plateau in which arise the parent streams of the Balder, a tributary of the Tees, towards the head of its drainage basin on the eastern side of the main Pennine watershed. Elsewhere on the Pennines *Sphagnum* bogs are rare, and those which still exist are quite small and local, nor is *Sphagnum* peat at all common in the upland Pennine peat moors.[1]

The dominant species of *Sphagnum* in a similar north Lancashire Sphagnetum was said by Wheldon and Wilson (*Flora of Lancashire*, 1907) to be *S. recurvum* (one of the *cuspidatum* group), though the following species also occur (*Naturalist*, 1910):

S. rubellum	S. cuspidatum
S. acutifolium	S. undulatum
S. plumulosum	S. tenellum (molluscum)

Apart from the bog mosses the flora was reported by Lewis to be almost entirely composed of the following plants:

Erica tetralix	f	Rubus chamaemorus	f
Eriophorum angustifolium	f	Empetrum nigrum	o
Oxycoccus quadripetalus	f	Calluna vulgaris	r

Wheldon and Wilson add the following as "marginal or very subordinate constituents" in similar upland Sphagneta in north Lancashire:

Carex canescens	Rhynchospora alba
C. stellulata	Viola palustris
Juncus effusus	

[1] These northern Pennine Sphagneta and those mentioned below require re-investigation to determine their relation to the types of bog described in this chapter and the next.

Sphagnetum in Cornwall. On Bodmin Moor in Cornwall, according to Magor (unpublished), Sphagnetum is a very local community and almost always restricted in area. Very wet spots form the so-called "piskie pits", usually of a vivid emerald green, in which *Sphagnum* peat may reach a depth of 12 ft. (3·5 m.) or more. Small quantities of the following plants may occur in the Sphagnetum:

Calluna vulgaris	Juncus effusus
Carex stellulata	Rhynchospora alba
Erica tetralix	Viola palustris
Eriophorum angustifolium	

Flora of Sphagnetum. According to Watson (1932, p. 293) Sphagnetum is well represented in Perthshire, North Wales, and on the higher parts of Exmoor, at 1100–1600 ft. (*c.* 335–480 m.), and the commonest species of bog moss "in the drier Sphagneta" are usually Cymbifolia and Acutifolia, especially *Sphagnum papillosum* and *S. quinquefarium*. According to Skene (1915, p. 79) these are species of medium to high acidity. The other species listed by Watson as abundant are the mosses *Polytrichum commune* and *Aulacomnium palustre*, while *Calluna* is occasional and the following are said to be frequent:

Anagallis tenella	Menyanthes trifoliata
Carex stellulata	Molinia caerulea
C. pulicaris	Myrica gale
Drosera rotundifolia	Narthecium ossifragum
Eriophorum angustifolium	Oxycoccus quadripetalus
E. latifolium	Pinguicula lusitanica (local)
Hydrocotyle vulgaris	Rhynchospora alba
Hypericum elodes	Viola palustris
Juncus acutiflorus	

Calypogeia fissa	Hypnum stramineum
Cephalozia connivens	H. cuspidatum
Gymnocolea inflata	Odontoschisma sphagni

Watson remarks that it is usual to find one particular sub-genus of *Sphagnum* dominating the whole or part of a bog, and one species may become the dominant or even the sole species over a given area. Thus over half an acre of bog on the Blackdown Hills in south Somerset *S. girgensohnii* was almost the only species present. *S. papillosum*, *S. cymbifolium* and *S. quinquefarium* often occur in almost pure patches. These are among the more acid species (Skene, 1915, p. 79), while *S. rubellum*, another common member of the Acutifolia, is the most acid of all. *S. cuspidatum* (one of the less acid) is the "most hydrophilous species and is frequently associated with *Drepanocladus fluitans*, *D. exannuletus*,... *Cephalozia fluitans*... and *Gymnocolea inflata*".

Many of the bryophytes accompanying *Sphagnum*, particularly those which normally live in drier communities, are characterised, in this habitat, by the increased length of their shoots (Watson, 1932).

REFERENCES

LEWIS, F. J. Geographical distribution of vegetation of the basins of the rivers Eden, Tees, Wear and Tyne. Part I. Southern portion. *Geogr. J.* 1904.

LEWIS, F. J. and MOSS, C. E. In *Types of British Vegetation*, Chapter XII, "The Upland Moors of the Pennine Chain". Pp. 266–82. 1911.

MAGOR, E. W. Geographical distribution of vegetation in Cornwall: Camelford and Wadebridge district (unpublished).

The Naturalist, 1910, pp. 265, 313.

OSVALD, H. Die Vegetation des Hochmoores Komosse. *Svenska Växtsoc. Sällsk. Handl.* I. *Akad. Abhandl.*, Uppsala. 1923.

PEARSALL, W. H. The soil complex in relation to plant communities. III. Moorlands and bogs. *J. Ecol.* **26**, 298–315. 1938.

RANKIN, W. MUNN. In *Types of British Vegetation*, Chapter XI, pp. 249, 255, 259–64. 1911.

SKENE, MACGREGOR. The acidity of *Sphagnum* and its relation to chalk and mineral salts. *Ann. Bot.* **29**, 65–87. 1915.

WATSON, W. Sphagna, their habitats, adaptations and associates. *Ann. Bot.* **32**, 535–51. 1918.

WATSON, W. The bryophytes and lichens of moorland. *J. Ecol.* **20**, 284–313. 1932.

WHELDON, J. A. and WILSON, A. *Flora of West Lancashire*. Eastbourne, 1907.

Chapter XXXV

THE MOSS OR BOG FORMATION (*continued*)

(2) RHYNCHOSPORETUM. (3) SCHOENETUM. (4) ERIOPHO-RETUM. (5) SCIRPETUM. (6) MOLINIETUM. IRISH BLANKET BOG. SUMMARY

(2) **Rhynchosporetum albae.** *Rhynchospora alba*, as we have seen in the last chapter, tends to dominate the hollows of raised bog in the phase immediately succeeding *Sphagnum cuspidatum*. It also dominates the wetter parts of the (blanket) bogs of Connemara (Co. Galway)[1] which lie but a few feet above sea level and are interspersed with rocky outcrops[2] and innumerable lakelets.[3] It seems to be a phase of succession preceded by a wetter stage in which the peat is colonised by *Zygogonium ericetorum*, often with a little *Carex limosa*, and followed by the dominance of other species, such as *Schoenus nigricans* or *Molinia*. The following species have been recorded·

	Rhynchospora alba		d
Calluna vulgaris	l	E. latifolium	f
Cladium mariscus	l	Menyanthes trifoliata	l
Drosera anglica	f	Myrica gale	f
D. intermedia	f	Narthecium ossifragum	f
Erica tetralix	f	Phragmites vulgaris	
Carex panicea	f	Rhynchospora fusca	r
Potamogeton polygonifolius	l	Schoenus nigricans	va
Eriophorum vaginatum	f	Scirpus caespitosus	f
Campylopus atrovirens	la	Sphagnum magellanicum	l
Leucobryum glaucum	l	S. plumulosum	l
Rhacomitrium lanuginosum	l	S. rubellum	l
Eriophorum angustifolium	f	S. subsecundum	l

The *Phragmites* and *Cladium* form narrow reed swamps round parts of the lakelets, and where they occur in local patches elsewhere in the bog they probably represent, like *Menyanthes* and *Potamogeton*, relicts of this reed swamp community occurring on the sites of former pools, though in some cases they may indicate the sites of underground drainage water from neighbouring knolls.

(3) **Schoenetum nigricantis.** Considerable tracts of the west Irish blanket bogs are dominated by *Schoenus nigricans*,[4] which in England (except the extreme south-west) is mainly characteristic of calcareous swamps or fens. Such a Schoenetum may follow Rhynchosporetum in

---

[1] Pl. 122, phot. 298.   [2] Pl. 121, phot. 296.
[3] Pl. 42, phot. 89.   [4] Pl. 122, phot. 298.

succession, but the list of species differs scarcely, if at all, from the list already given for Rhynchosporetum, or from that for mixed bog on p. 715. It has recently been suggested that the abundance or dominance of *Schoenus* in blanket bog near the western coasts may be favoured by the falling of sea spray, driven by inshore gales, on the surface of the bog, thus changing the soil reaction in the direction of its more normal habitat.

(4) **Eriophoretum.** Cotton-grass bogs, or mosses as they are always called, are characteristic of the whole Pennine chain, occupying very extensive areas on the flat or gently sloping plateaux at elevations between 1000, or more commonly 1200, and 2200 ft. (*c.* 370–670 m.) (Fig. 140). The actual altitude depends on the heights of the plateaux, the former representing the average lower limit in the southern, the latter the average upper limit in the northern Pennines.

Moss (1911) emphasises the dreary and monotonous character of the *Eriophorum* mosses. The cotton-grass is not only the dominant, but frequently the only vascular plant over wide areas. The cotton-grasses are active in the formation of peat, which may reach an extreme depth of about 30 ft. (9 m.) and is said to consist almost wholly of the remains of the dominant species.[1] It is usually saturated and frequently supersaturated with water, though the surface layer may become quite dry during a drought. Like the *Sphagnum* bogs these far more extensive Eriophoreta are always called "mosses" locally. "Moss" is much the commonest place name on the Pennine uplands. Of the two commoner species, *Eriophorum angustifolium* and *E. vaginatum*, the latter, which occupies drier peat than the former, is very much more commonly dominant, especially in the southern Pennines. Both species are relatively deep-rooted, and abundantly supplied with intercellular channels, but the former has creeping rhizomes and more rapidly colonises very wet peat (Pl. 120, phot. 293).

(*a*) **Eriophoretum vaginati.** Moss (1911, 1913), Adamson (1918) and Watson (1932) have studied this community, which is the characteristic bog or moss of the southern Pennines, almost everywhere covering the higher plateaux above 1000–1200 ft. (300–360 m.), though it is often in various stages of retrogression (Fig. 140 and Pls. 119, 120).

Soil. The community normally occurs on deep peat which it has itself formed. There is no evidence as to its beginnings: at present it is only extending locally, as in hollows where pools have existed. The peat, which varies in depth from 2 to 3 ft. up to 30 ft. (0·6–9 m.), is always relatively pure and free from admixture of sand or other mineral matter. The roots of all the plants of the community are entirely confined to the peat. Below this the mineral soil is podsolised: from 3 in. to 2 ft. (7–60 cm.) below the base of the peat there is nearly always a thin layer of "pan", which may be

---

[1] Doubt has recently been cast on this statement. The peat underlying the existing cotton-grass mosses requires re-investigation.

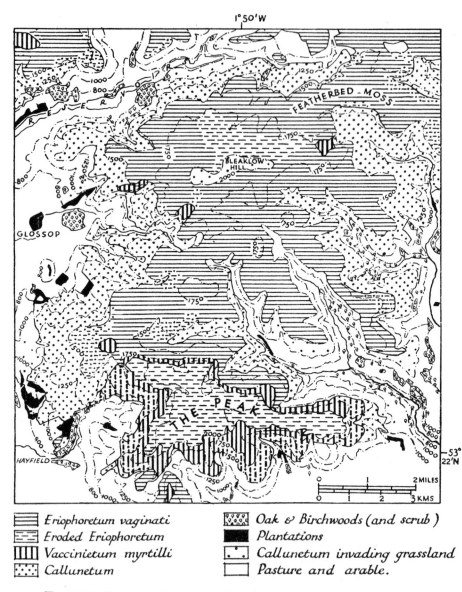

Eriophoretum vaginati
Eroded Eriophoretum
Vaccinietum myrtilli
Callunetum

Oak & Birchwoods (and scrub)
Plantations
Callunetum invading grassland
Pasture and arable.

FIG. 140. PLATEAU (1500–2000 ft.) OF THE PEAK DISTRICT (NORTH
DERBYSHIRE) OCCUPIED BY ERIOPHORETUM VAGINATI

Large areas of Eriophoretum are eroded and invaded by Vaccinietum myrtilli and Callunetum
from the neighbouring "edges" and slopes. Contours in feet. After Moss, 1913.

quite hard and stone-like (Adamson, 1918). This "*B* horizon" is coloured red, presumably by ferric salts. Pearsall (1938) found that the *p*H values of pure undisturbed Eriophoretum vaginati varied from 2·98 to 3·40, and is thus markedly more acid than Sphagnetum. The oxidation-reduction potential varies round the critical point and in summer, when the surface dries out, the peat is feebly oxidising.

Floristic composition. Moss (1911) and Watson (1932) each gives fifteen species of vascular plants as occurring in this community: nine of these (marked with an asterisk *), including the dominant, are common to the two lists. Below is the composite list:

*Eriophorum vaginatum	d		
Agrostis stolonifera	o	Juncus squarrosus	o
A. tenuis	o	*Molinia caerulea	r
Andromeda polifolia	r	Narthecium ossifragum	r
Calluna vulgaris	lsd	Oxycoccus quadripetalus	r
*Carex canescens	o, la	Pinguicula vulgaris	r
*Empetrum nigrum	lsd	*Rubus chamaemorus	lsd
*Erica tetralix	r–la	*Scirpus caespitosus	r–la
*Eriophorum angustifolium	l	*Vaccinium myrtillus	r–la
Juncus effusus	o	V. vitis-idaea	o

Watson records thirty-five species of bryophytes and nine lichens from the community. The following bryophytes he marks as abundant or frequent:

Calypogeia trichomanis	f	Gymnocolea inflata	f
Campylopus flexuosus	a	Lepidozia repens	f
Cephaloziella starkii	a	Lophozia floerkii	f
Dicranella heteromalla	a	L. ventricosa	f
Diplophyllum albicans	f	Tetraphis pellucida	a
Drepanocladus fluitans var. atlanticus	f	Webera nutans	sd

Three species of *Sphagnum* (*S. recurvum, cymbifolium* and *papillosum*) which are of medium to high acidity, are recorded as occasional, and the following three lichens are frequent or abundant:

Biatora granulosa	f	Cladonia coccifera	a
Cetraria aculeata var. hispida	f		

Other areas. Lewis (1904) briefly describes Eriophoretum vaginati in the northern Pennines with a very small quantity of stunted *Calluna* only 2 or 3 in. high.

On the Cleveland Hills, on the eastern side of Yorkshire, pure Erio-phoretum of any extent is unknown, but on the highest parts of the watershed (1450 ft.) on "peat of great thickness", the prevalent Callunetum may be mixed with an equal amount of *Eriophorum vaginatum*, and the latter species is locally dominant. Interspersed are "numerous pools full of floating *Sphagnum*, probably *S. cuspidatum*, which when drier also contain *Eriophorum angustifolium*" (Elgee, 1914, p. 10).

The Eriophoretum vaginati occupying the surface of Foulshaw Moss (a raised bog) and described by Rankin (1911) has already been mentioned (p. 686).

(*b*) **Eriophoretum angustifolii.** This community is very local in the southern Pennines, but more extensive on the northern part of the chain. According to Watson (1932) it is much more frequent on the Somerset moorlands than E. vaginati, and limited areas are described by Pethy-bridge and Praeger (1905) at high altitudes (about 500–700 m.) on the Wicklow mountains in eastern Ireland. The Wicklow Eriophoreta are small in extent and "below the uniform waving foliage of the cotton-grass is a continuous dense stunted growth of *Calluna* with several of the plants of the *Calluna* and *Scirpus* associations". The peat is "sopping wet" and spongy, unlike that of the Scirpetum caespitosi (see below), which is the characteristic community of most of the Wicklow mountain plateau. The species recorded from three sample areas were as follows:

Eriophorum angustifolium	d
Calluna vulgaris	sd

All three areas:

Eriophorum vaginatum	Cladonia sp. (two areas)
Empetrum nigrum	Erica tetralix (one area)
Scirpus caespitosus	Vaccinium vitis-idaea (one area)
Vaccinium myrtillus	Sphagnum spp. (one area)

The Eriophoretum angustifolii described by Lewis (1904) from the northern Pennines has *Eriophorum vaginatum* mixed in smaller quantity, together with a few individuals of *Rubus chamaemorus*, *Erica tetralix* and *Empetrum nigrum*.

Eriophorum angustifolium often occupies wet channels and depressions in the blanket bogs of *E. vaginatum*, and colonises cracks in the bare peat of such hollows by means of its creeping rhizomes (Pl. 120, phot. 293).

Bryophytes. Watson (1932) remarks that many Bryophytes present in the Eriophoretum vaginati are absent or rarer in the wetter E. angustifolii. He records the following species from the last-named community (of the Somerset moors) not listed from the E. vaginati:

Aulacomnium palustre	Lepidozia setacea
Campylopus brevipilus	Leptoscyphus anomalus
Cephalozia fluitans	Odontoschisma sphagni
Drepanocladus aduncus	Sphagnum acutifolium
D. fluitans (type)	S. plumulosum
D. revolvens	

Succession. There is no contemporary evidence as to any seral stages leading up to the formation of the upland Eriophoreta, but Rankin (1911, pp. 249, 254–5 and Fig. 16) showed that beneath 3 in. (7·5 cm.) or so of *Eriophorum* peat, on which the existing cotton-grass of Foulshaw Moss (see p. 686) is growing, there are several feet of very well preserved *Sphagnum* peat, the Sphagna of which it is composed belonging to the Cymbifolium

PLATE 119

Phot. 290. Eriophoretum at Wessenden Head, near Huddersfield, West Yorkshire (August). *Mrs Cowles.*

Phot. 291. Eriophoretum near Huddersfield. *E. vaginatum* in fruit (June). *W. B. Crump.*

ERIOPHORETUM VAGINATI ON THE SOUTHERN PENNINE PLATEAU

PLATE 120

Phot. 292. The peat of Eriophoretum vaginati dissected by stream channels. *Vaccinium myrtillus* capping some of the peat haggs. Plateau of Fairsnape Fell, Lancashire. *A. Wilson.*

Phot. 293. South Pennine plateau. *Eriophorum angustifolium* in foreground, colonising peat cracks. *E. vaginatum* (tufted) fringing bare peat, *Empetrum nigrum* beyond. Eriophoretum vaginati on the plateau behind.

J. Massart

Phot. 294. Large tussock of *Eriophorum vaginatum* and carpet of *Empetrum nigrum* on side of peat channel. Eriophoretum on plateau beyond. *J. Massart.*

Phot. 295. *Rubus chamaemorus* on side of peat channel. *J. Massart.*

RETROGRESSIVE ERIOPHORETUM VAGINATI

group. Thus there can be no doubt that a Sphagnetum immediately preceded the existing Eriophoretum vaginati. And there is independent evidence that in the middle of the nineteenth century Foulshaw Moss was much wetter, quite impassable, and bore a greater expanse of bog moss than it did at the beginning of the twentieth century when Rankin described it. It must however remain an open question how far the change from Sphagnetum to Eriophoretum was a natural developmental change caused by the growing up of the bog moss above the level to which it could raise water. The drying of the surface was certainly greatly accelerated and may have been entirely determined by the trenching and removal of peat which had long been carried on.

The *Sphagnum* peat at Foulshaw is again underlain by other types of peat, showing that the existing and recently existing raised bogs were preceded by phases of forest and fen (see p. 168, Fig. 47).

Retrogression of Eriophoretum. Crowberry, ling and bilberry (*Empetrum nigrum, Calluna vulgaris* and *Vaccinium myrtillus*) occur repeatedly in the floristic lists of the Eriophoreta, and the first two also in some Sphagneta, though they do not grow with the luxuriance they show in drier habitats. These species are indicative of drier conditions than those which are optimal for the plants active in the formation of deep peat, and their presence in increasing quantity and vigour marks the beginning of conditions which when carried to their term will lead to the supersession of the typical "mosses" and their replacement by drier moorland types. Mixed communities of cotton-grass and ling, or at the higher levels above 2000 ft. (c. 600 m.), of cotton-grass, ling and bilberry, cover very great areas in the northern Pennines, and all of these may be interpreted as transitional from Eriophoretum to Callunetum or Vaccinio-Callunetum. The ling itself, as we shall see (Chapter xxxvi), has a very wide range of tolerance in respect of the total water content of the soil; and it seems that the purer the peat the more water the ling will tolerate, for Crump (1913) found that the W/H coefficient (i.e. water content divided by humus content) of *Calluna* soils varied within quite narrow limits.

The process of replacement of Eriophoretum by Callunetum or Vaccinietum is however often associated with active drainage and destruction of the cotton-grass peat. One way in which this can occur is by the cutting back towards their source of moorland streams rising in a peat-covered plateau. This process was carefully followed by Moss (1911, 1913) and Adamson (1918) on the plateau of the Derbyshire "Peak" (Fig. 140) and similar physiographic habitats in the southern Pennines, and it is a common phenomenon in other regions also. As the stream cuts back into the peat its banks become drier and the first sign of the change in vegetation is the appearance of a lining of *Empetrum* round its head. Lower down its course, i.e. nearer the edge of the peat plateau, where the drainage is freer and the peat of the stream banks becomes drier, *Empetrum* is followed by *Vaccinium myrtillus.* Meanwhile the bed of the stream is widened by erosion, and tributary streamlets cut back into its banks, increasing the general drainage of the surface peat, which then gradually becomes colonised by the bilberry.[1] When erosion is continuously active some of the tributary stream channels are cut back until they impinge upon others, so that a network of water courses is formed. The mounds (peat "haggs") left in the meshes of this network of channels are drained on all sides and the surface is occupied by dominant bilberry with crowberry (*Empetrum nigrum*),[2] cloudberry (*Rubus chamaemorus*),[3] and sometimes *Calluna*. The evidence that this succession has occurred on a large scale is quite convincing: the peat on which the bilberry grows is entirely composed of the remains of *Eriophorum vaginatum*, and the Ordnance Survey maps published between 1870 and 1880 show the stream channels about 1·2 km. longer than in the maps of 1830, while in the early years of this century they were about 0·4 km. longer still.

On the lower and more isolated hills of the southern Pennines Adamson (1918, pp. 98–9) describes a process of gradual drainage and drying of the peat without actual erosion of stream channels. Here *Eriophorum vaginatum* is gradually replaced by *Calluna*. "Passing from the pure Eriophoretum the first stage is a lack of luxuriance of the dominant plant, an almost complete absence of *Eriophorum angustifolium* and an admixture in increasing quantity of *Scirpus caespitosus* and *Calluna vulgaris*. With increasing dryness *Vaccinium myrtillus* and *Juncus squarrosus* appear, with some *Nardus stricta* and *Deschampsia flexuosa*, though the last seems seldom more than an occasional species. In certain cases this succession reaches the relatively stable condition of co-dominance of *Nardus* and *Juncus squarrosus*, with *Eriophorum* existing only on the wettest parts." The factor which leads to the establishment of this *Nardus* grassland instead of a *Calluna* climax is perhaps the binding of the old peat surface by the close tufts of the grass, which makes the invasion of *Calluna* difficult.

The action of wind may also result in the destruction of the Eriophoretum on isolated and projecting summits and on the crests of ridges. This may be so violent as to lead to the removal of the peat and the survival only of isolated tufts of *Eriophorum angustifolium*, which though primarily a more hydrophytic species, is much more resistant to such conditions than *E. vaginatum*. More commonly however the removal of the peat is less rapid and the crowberry and bilberry come in just as they do on the desiccating peat on the edges of stream courses.

All these forms of retrogression of Eriophoretum are assisted by burning, which is occasional on the cotton-grass moors, though not a deliberate and regular practice as on the grouse (heather and bilberry) moors. After destruction of the Eriophoretum by fire the peat is removed faster by the streams and there is more rapid immigration of bilberry, *Nardus* and *Juncus squarrosus*. The two last penetrate the moors along the lines of footpaths,

[1] Pl. 120, phot. 292.   [2] Pl. 120, phots. 293, 294.   [3] Pl. 120, phot. 295.

which can often be traced from a distance by the difference in colour of the vegetation.

(5) **Scirpetum caespitosi.** Scirpetum and Eriophoretum are in a sense alternative communities, for they occupy quite similar physiographical habitats and they are rarely both developed extensively in the same region. But the *Scirpus* peat ("pseudo-fibrous") is somewhat drier than *Eriophorum* peat, and is said to be markedly less fertile under cultivation. Like the Eriophoreta Scirpetum occurs mainly on the western side of

Distribution   the British Isles. According to Lewis (1908, p. 257) *Scirpus caespitosus* dominates many hundred square miles in the northwest Highlands (Sutherland, Ross and Inverness), the flat basaltic plateau of northern Skye, and the valley floors and gently sloping hillsides of North Uist and Lewis (Hebrides) and of the Shetlands. This Scirpetum also extends southwards through western Perthshire and Argyllshire.

Scirpetum of Argyllshire. G. K. Fraser (1933) has recently given some account of the Scottish Scirpetum, particularly from the point of view of the peat which it forms and of its silvicultural possibilities; and he describes it as "the ultimate (climax) moorland vegetation in the west and north-west of Scotland". It is in fact typical "blanket bog". As the climatic climax it is the form of vegetation to which all other types tend to give place in this climatic region, and the process has been hastened and extended, according to Fraser, by the widespread grazing and burning which have taken place in the past. In the development of the Scirpetum the growth of *Sphagnum* cushions plays an important part.

Inverliver Forest. The "forest" of Inverliver on Loch Awe in Argyllshire, where Fraser's researches were largely conducted, has an annual

Climate   rainfall varying from 74 to 88 in. (1850–2200 mm.), the monthly precipitation ranging from 7 to 12 in. from October to January, when rain falls on four days out of five, to not less than 4 or 5 in. in the drier months from March to July, when rain falls on two days out of three. Periods of ten days without rain are infrequent. Snow is of little account, as the small amount which falls usually melts within a few hours. Data as to atmospheric humidity are not available, but even after an exceptional 20 days of continuous bright weather with no rain and in the warmest season the soil is still thoroughly wet.

Days of bright sunshine average nine per month in winter and eleven in summer. The maximum summer temperature is about 21° C. but the mean monthly minima for July and August are less than 9° C., in January and February just under 0° C. The annual and daily ranges of temperature are slight. Frost may occur in nearly every week of the year, and it does occur frequently during the growing season except for about 7 weeks in July and August. The temperatures in the peat itself are considerably higher than in the air above it, the range is much less, and at a depth of 1 ft. (30 cm.) the thermometer remains stationary at about 7° C. throughout the year. But during bright sunshine very marked differences occur between the *Sphagnum* cover, the air above, and the peat below. Thus when the temperature 30 cm. above the bog moss was 9° C. and that at the surface of the peat below the *Sphagnum* 4° C., in the moss itself the thermometer recorded 22°C.

Vegetation. Under these climatic conditions the country is covered with a continuous sheet of blanket bog, except where special physiographic or edaphic factors are present. Since the topography varies widely there is considerable variation in vegetational detail, but the general character remains the same. The dominant *Scirpus caespitosus* forms an open, even or diffuse, tufted cover. *Calluna* is present in abundance as scattered open patches of low growth with short flowering shoots. *Narthecium ossifragum* is said to be always abundant, but of small size. *Potentilla erecta*, *Erica tetralix* and *Eriophorum angustifolium* constantly occur. Other species of vascular plants are considered by Fraser to be relict plants from previous vegetation, though this interpretation may be open to question. The moss vegetation, shows a fairly equal distribution of species of *Sphagnum* such as *S. magellanicum* (= *S. medium*), *S. papillosum* and *S. rubellum*, and of common heath mosses such as *Hypnum schreberi*, *H. cuspidatum* var. *ericetorum*, *Hylocomium splendens*, *Brachythecium purum*, with *Rhacomitrium lanuginosum* forming hoary pads on the tufts of heather. The liverwort *Pleurozia purpurea* forms compact masses in slight hollows.

Peat "hagging" by the formation of deep cracks and channels is due, according to Fraser, to the cutting away of the peat along drainage channels and subsequent erosion of their sides. Though this process leads to local and temporary change in the vegetation, such as increase of *Calluna* and *Rhacomitrium* along the edges of the haggs, it has no permanent effect on the bog as a whole.

Scirpetum in the northern Highlands. Superficial examination of several areas of blanket bog in the northern and north-western Highlands (September 1937) resulted in the following lists (see Table XXIII). The different areas, spread over a fairly wide region, are briefly characterised below. The peat in all except (8) appeared to be deep.

(1) Above Glen Garry (Inverness-shire), lat. 57° 4' N., long. 4° 56' W.:[1] undulating ground, alt. 600 ft.

(2) Above Guisichan Forest (Inverness-shire), lat. 57° 15' N., long. 4° 50' W.: nearly flat, alt. 1000 ft. Unusual species such as *Carex flava*, *C. pulicaris*, *Equisetum palustre* and *Selaginella selaginoides* were present, but the area had been somewhat disturbed by partial drainage in places.

(3) Dundonnell Forest (Sutherlandshire), lat. 57° 45' N., long. 5° 7' W.: undulating, alt. 1000 ft.

(4) "The Crask", 7 miles south of Altnaharra (Sutherlandshire): nearly flat, big pools and eroded haggs, alt. 800 ft.

[1] Latitudes and longitudes approximate only.

Table XXIII. *Composition of Scirpetum in different Highland areas*

	1	2	3	4	5	6	7	8	Rannoch Moor 9	10	11
Calluna vulgaris	f	cd	a	f, ld	f, la	lcd-ld	cd	va	f-a	f	f
Carex panicea	·	o-f	·	·	r	·	·	·	la	o	·
C. stellulata	·	o	·	·	·	·	·	·	·	·	·
Drosera anglica	o	o	·	f	r	·	o	·	·	·	·
D. rotundifolia	·	·	·	·	·	·	·	·	·	·	·
Erica cinerea	a	f	a	f	f	f	f-la	·	f	d	f
E. tetralix	lf	o	o	f (ld)	f	f-la	o	·	a-ld	·	f
Eriophorum angustifolium	·	f	o	o	o-f	f	f-la	·	a	·	a
E. vaginatum	·	o	lf	·	o	f	o	o, l	·	·	·
Juncus squarrosus	·	·	·	·	·	·	·	·	ld	l	·
Menyanthes trifoliata	f	·	f-d	·	f	lcd	·	edge, f	f-a	f	f
Molinia caerulea	la	a	·	a	f	·	·	a-va	f	o	a
Myrica gale	la	o-lf	f-la	·	a	·	·	o, l	f	a	·
Narthecium ossifragum	f	la	·	d	lf	f-la	o	f-a	o-la	o	f-a
Potentilla erecta	·	·	·	·	·	·	·	·	·	·	·
Scirpus caespitosus	d	cd	a-d	·	d	lod	cd	d	·	l	·
Aulacomnium palustre	·	·		·	·	·	o	·	·	·	·
Campylopus flexuosus	·	+		·	·	·	r	·	·	·	·
Hylocomium loreum	·	·		·	·	·	f	·	·	·	·
H. splendens	o	+		·	f-a	f	a	·	·	·	·
Hypnum cupressiforme	·	·		·	·	·	·	·	·	·	·
H. schreberi	o	·		·	·	·	·	·	·	·	·
Rhacomitrium lanuginosum	·	+		a	o-f	·	o	l	o	·	a
Sphagnum compactum	a	+		o-f	o-f	·	o	·	·	·	·
S. cuspidatum var. submersum	·	·		f	·	·	o	·	+	·	·
S. cymbifolium	·	·		·	·	·	·	·	·	·	·
S. fallax	o	·		·	·	·	r	·	+	·	·
S. molluscum (tenellum)	a	·		f	f	·	f-a	·	·	·	·
S. papillosum	·	·		a	f	·	·	·	·	·	·
S. plumulosum	a	·		f	o	·	·	·	·	·	·
S. rubellum	·	·		a	f	·	·	l	·	·	·
S. subsecundum	·	·		·	·	·	·	·	·	·	·
Cladonia coccifera	·	·		·	·	·	·	·	·	·	·
C. floerkiana	·	·		·	·	·	·	o	·	·	·
C. sylvatica	f	f		a	f	+	f	o	f	·	f
C. uncialis	·	·		a	f	+	f	f	+	·	f

(In column 3, for the bryophytes and lichens: "Bryophytes and lichens not noted".)

(5) Allt-a-Chraisg, 6 miles south of Altnaharra: slope of 10°. Wet, alt. 500 ft.

(6) Three miles south of Altnaharra; lat. 58° 16' N., long. 4° 30' W.; slope of 20° (more than 18 in. of peat); rather dry. Note the absence of *Sphagnum, Eriophorum, Narthecium, Myrica* and *Drosera*.

[(4)–(6) form a series of increasing dryness with increasingly steep slope.]

(7) Ord of Caithness (border of Sutherland and Caithness on the east coast); lat. 58° 8' N., long. 3° 35' W.; flat, not far from cliff edge, rather dry (more than 44 in. of peat), alt. 750 ft.

(8)–(11) Closely adjacent small areas on Rannoch Moor, head of Glencoe, Argyllshire, lat. 56° 39' N., long. 4° 58' W.; flat, slight differences of level, alt. 1000 ft.

> (8) Higher ground, peat shallow over rock (Eriophora absent, *Sphagnum* local).
>
> (9) Lower ground, flat peat (Eriophora abundant, Sphagna frequent).
>
> (10) Depression (*Eriophorum angustifolium* dominant, *Scirpus* absent).
>
> (11) Low haggs and pools with bare peat.

The characteristics of these lists are (a) the very restricted number of species of flowering plants, (b) the high constancy of nearly half the phanerogamic species present, (c) the presence of *Calluna* in every list. *Scirpus caespitosus* is present and usually dominant in every list but (10), which is from a local wet depression dominated by *Eriophorum angustifolium*. In the three driest areas, (2), (6) and (7), *Scirpus* and *Calluna* are co-dominant. This seems to indicate that the requirements of deer-sedge and heather are very nearly the same in this type of habitat, but nevertheless *Scirpus* is a true blanket bog *dominant*, while *Calluna* is not, in spite of its absolute constancy. *Scirpus* is typically dominant in the less wet areas of Highland blanket bog. When the dryness increases a very little further *Calluna* becomes co-dominant, but only attains its greatest luxuriance and overwhelming dominance on still drier peats which cannot be reckoned as bog.

Exmoor and Bodmin Moor. In England Scirpetum caespitosi is rare, but restricted areas occur in the south-west—Exmoor (Watson, 1932) and on Bodmin Moor (Magor). In north Devon *Scirpus caespitosus* is often more frequent on upland moors but is rarely dominant (Watson, 1932): in Cornwall it is rare (Magor). Occasionally, however, Scirpetum is formed in both counties and the following lists are given by Watson and by Magor from Exmoor and Bodmin Moor respectively. In the Bodmin Moor area the *Scirpus* is tufted, with bare intervals of black peat, as on the Wicklow Mountains, and one may infer from the lists that this area was decidedly wetter than that on Exmoor, which contains such species, unusual for Scirpetum caespitosi, as *Anthoxanthum odoratum* and *Deschampsia flexuosa*. Including the margin of the Exmoor area nine phanerogams are common to the two areas. Of these, four (*Calluna, Erica tetralix, Eriophorum angustifolium* and *Narthecium*) are the most constant in the eleven Highland areas listed above, quite constant in the five examples of Wicklow Scirpetum (p. 712), and are four out of the five given by Fraser as constant at Inverliver. *Potentilla erecta* comes next in constancy, and then *Drosera rotundifolia*.

Scirpus caespitosus   d

	Exmoor	Bodmin Moor		Exmoor	Bodmin Moor
Anthoxanthum odoratum	o	.	Polygala serpyllifolia	o	r
Calluna vulgaris	o	f	Potentilla erecta	o	o
Carex binervis	.	o	Rhynchospora alba	.	a
C. caryophyllea	.	o	Scirpus fluitans	.	o
C. diversicolor (flacca)	.	r	Sieglingia decumbens	.	o
C. inflata	.	la	Vaccinium myrtillus	.	r
C. oederi	.	o			
Deschampsia flexuosa	a	.	Calypogeia fissa		
Drosera rotundifolia	m	va	f. aquatica	o	.
Empetrum nigrum	m	.	Campylopus flexuosus	a	.
Erica cinerea	r	.	Cephalozia bicuspidata	a	.
E. tetralix	r	a	Cephaloziella starkii	f	.
Eriophorum angusti-			Cladonia sylvatica	m	o
folium	m	va	Hylocomium squarrosum	o	.
E. vaginatum	.	f	Hypnum cupressiforme		
Juncus bulbosus	.	o	var. ericetorum	a	.
J. squarrosus	.	f	Polytrichum commune	.	a
Luzula multiflora			Rhacomitrium lanugin-		
f. congesta	o	o	osum	.	o
Molinia caerulea	.	va	Sphagnum cymbifolium		
Nardus stricta	a	o	f. congestum	o	.
Narthecium ossifragum	m	a	S. quinquefarium	o	.
			Sphagnum spp.	.	o

m = marginal.

Cleveland moors. On the eastern side of England Scirpetum is practically absent. Elgee (1914) describes it as "very rare" on badly drained clayey shale with thin peat on the Clevelands in north-eastern Yorkshire. *Erica tetralix* is said to be usually in about equal proportion to the *Scirpus*, and either may be dominant. "Other frequent though quite subordinate species are *Eriophorum vaginatum*, *Molinia caerulea* var. *depauperata*, *Juncus squarrosus* and *J. conglomeratus*."

Wicklow Mountains. In Ireland Scirpetum caespitosi is well developed on the Wicklow Mountains south of Dublin, where it was first described by Pethybridge and Praeger (1905). Here it occupies the flatter, less well-drained slopes between about 1250 and 2000 ft. (c. 380–610 m.).[1] It is "as a rule mixed with a considerable amount of stunted and apparently poorly thriving *Calluna*. The soil is thick peat and except in the hottest part of summer, when the soil may be comparatively dry, it is thoroughly saturated with water". In its "purest form" this Scirpetum "resembles a lawn",

[1] See Fig. 89, p. 505.

having a uniform height of about 6 in. (15 cm.) and contains the following species (five samples):

Andromeda polifolia	(2)	Eriophorum vaginatum	(2)
Calluna vulgaris	(5)	Narthecium ossifragum	(5)
Drosera rotundifolia	(2)		
Empetrum nigrum	(1)	Cladonia sp.	(2)
Erica cinerea	(1)	Rhacomitrium lanuginosum	(1)
E. tetralix	(5)	Sphagnum spp.	(4)
Eriophorum angustifolium	(5)		

More commonly, however, the *Scirpus* forms large separate tufts, the interspaces of "soppy peat" containing considerable amounts of *Calluna*,

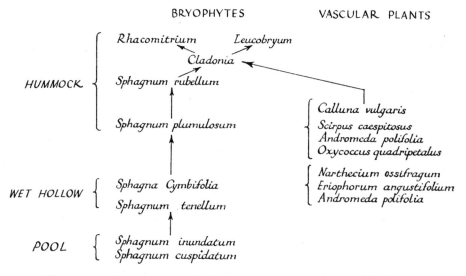

FIG. 141. SERAL STRUCTURE OF A HUMMOCK IN THE SCIRPETUM
(PLATEAU BLANKET BOG) OF THE WICKLOW MOUNTAINS

The flowering plants associated with the Sphagna of succeeding developmental phases are shown in the right-hand column in approximate positions of greatest frequency.

Erica tetralix and *Eriophorum angustifolium*. *Narthecium ossifragum* is characteristic, and much more abundant than in other wet peat communities of the region. On the Wicklow Mountains *Andromeda polifolia* is confined to this community, and on its leaves the ascomycete *Rhytisma andromedae* Pers. occurs (Pethybridge and Praeger, 1905).

There is a good deal of hummock formation in the Wicklow Scirpetum, as in parts of the blanket bog of the west. The hummock succession as we[1] saw it in 1935 is quite similar, and the structure may be represented as in Fig. 141. The bryophytes are the actual builders of the hummocks, the vascular plants colonising at the levels indicated.

[1] Godwin, Osvald and Tansley.

Besides the species noted above we recorded in 1935 *Hylocomium splendens*, *Hypnum schreberi* and *Antitrichia curtipendula* below the *Calluna*. On flat ground at an altitude of about 1800 ft. (550 m.) at Sally Gap the Scirpetum had much less *Calluna*, with *Sphagnum subsecundum*, *S. papillosum* and *S. tenellum* in fairly extensive patches: also a little hummock building by *S. plumulosum* and *S. rubellum*.

This Scirpetum is still actively forming peat, and where it abuts on Callunetum the surface of the soil is seen to be several feet higher than in the latter. It is from the Scirpetum alone, according to Pethybridge and Praeger, that peat is cut for fuel in the Wicklow Mountains, and the increased drainage caused by the turf-cutting leads to invasion by the Callunetum.

In this area, as elsewhere in the British Isles, the peat is being extensively eroded in places and many of the Wicklow plateau bog communities are in process of degeneration. Remains of *Pinus sylvestris* and *Betula* (Sub-Boreal) have been found in the *Scirpus* peat at 1250 and 1700 ft. respectively, indicating drier climatic conditions anterior to the formation of the sub-Atlantic peat.

(6) **Molinietum caeruleae.** This community has already been described (Chapter XXVI, pp. 518–23) as a type of acid grassland. *Molinia* has a wide range of habitats, particularly in regard to pH value. We may recall that it resembles *Nardus* in rooting depth and in its ability to colonise mineral soil *or* peat, but requires both more water and a greater degree of root aeration, probably also a larger supply of mineral salts. On mineral soils it often forms, like *Nardus*, a peaty sod about 9 in. thick and easily separable from the mineral soil below. But like *Nardus* also it colonises peat formed by other species; and Molinietum, like Nardetum, as W. G. Smith remarked, may be described as "marginal", in a wide sense, to the great peat areas.

In the wet acid peat succession it is one of the "drier" communities, colonising in abundance, as it does, the drier parts of the mixed blanket bog of western Ireland (pp. 715 ff.), and also the sides of the older *Sphagnum* hummocks. It apparently plays a similar part in the Scottish Highlands. But the plant does not always become dominant in such situations, and Molinietum is probably always to be regarded as a transitional or marginal community. It is also said to cause the breakdown of peat already formed by other plants.

Callunetum vulgaris. When the surface of any bog or moss becomes dry the ling, which is nearly always already present, often in abundance, but is not luxuriant, tends to establish dominance; but this community is not to be regarded as part of the bog or moss formation. [See p. 710 and Chapter XXXVI, p. 724.]

IRISH BLANKET BOG

The plateau bog on the Wicklow Mountains in the east of Ireland has already been described.

Parts of the extreme west of Ireland are very largely covered by a thick layer of more or less wet acid peat, wherever the ground is flat or not too steeply sloping, supporting a carpet of moss or bog vegetation.[1] This is the same kind of bog which covers much of western and north-western Scotland and the Hebrides (there dominated mainly by *Scirpus caespitosus*), and belongs to the same general type as the Scirpetum caespitosi of the Wicklow Mountains and the cotton-grass bogs of the Pennine plateaux. The name "blanket bog" was suggested (in 1935) for this kind of vegetation, since it covers the country like a blanket in regions of high precipitation and very high atmospheric moisture; and there is no reason to suppose that it is built up (except locally) like the raised bogs of the central plain of Ireland, on fen or on the sites of old lakes through a stage of fen. Nevertheless the vegetation it supports is mainly composed of the same species as those of the raised bog.

Connemara and western Mayo. Western Galway and Mayo are typical regions in which by far the greater part of the surface is covered by blanket bog.[1] This is a country of acidic metamorphic rocks rising in bold rounded mountains between 2000 and 3000 ft. (600–900 m.) above the great stretches of flat or gently undulating land which lie but little above sea level. The bog covers the flat country and the greater portion of the lower mountain slopes up to an angle of at least 15°. Only where the rock stands out above the peat blanket[2] is the vegetation heathy, mainly dominated by *Calluna* and *Ulex gallii*, and in the more sheltered places by dwarf wood of *Quercus sessiliflora* with *Betula pubescens*, or by scrub composed of *Corylus*, *Crataegus*, *Salix aurita*, *Sorbus aucuparia* and *Ilex*. The blanket bog is the real climatic climax, the scrub and dwarf woodland are local communities belonging to the xerosere and conditioned by better drainage. At one time, however, probably in the Sub-Boreal period, this country was covered with pine forest, as is shown by the layer of pine stools below the bog peat which belongs to the more recent period (Sub-Atlantic), and has been formed during the last two and a half millennia.

The Connemara bog in western Galway is studded with innumerable lakelets of all sizes, from small pools in the peat to considerable areas of water, enclosed, or partly enclosed, by rocky rims supporting heath or a little scrub. In places there are low-lying basins, evidently the sites of old lakes which have become covered with bog vegetation, and some of these are marked by relict *Phragmites* and *Cladium*, the vestiges of old marginal reedswamps. Poorly developed reedswamp of the same species occurs on the margins of some of the existing lakes, with characteristic aquatic

[1] Pl. 121.   [2] Pl. 121, phot. 296.

PLATE 121

Phot. 296. The blanket bog of Connemara, near Roundstone. The expanse of bog is broken here and there by rocky outcrops. The "Twelve Beinns" behind. *Mrs Cowles.*

Phot. 297. The blanket bog of north-western Mayo with bog pools. Near Sheskin Lodge. *R. J. L.*

WEST IRISH BLANKET BOG

PLATE 122

Phot. 298. Rhynchosporetum albae (white heads) with *Myrica gale* (right and left centre) in foreground. Schoenetum nigricantis (black heads) behind. Large hummock capped with *Leucobryum glaucum* in background. *R. J. L.*

Phot. 299. Closer view of the same large hummock, eroded on the right by prevailing wind and covered with *Leucobryum glaucum* and *Molinia caerulea*. *Schoenus nigricans* and *Myrica gale* in foreground. *R. J. L.*

DETAIL OF CONNEMARA BLANKET BOG

plants such as *Lobelia dortmanna* and the American species *Eriocaulon septangulare*.

Peat. The pH values obtained from Connemara mixed blanket bog peat with *Schoenus* were 4·4–4·6. In bogs developed round small lakes with *Molinia* and *Myrica* present the pH was as high as 5·36. These permanently wet blanket bog peats appear to be all reducing (Pearsall, 1938).

Flora and vegetation. Over much of the blanket bog there is no general dominant. The vegetation consists mainly of a varying mixture of the following abundant species:

Eriophorum vaginatum	Schoenus nigricans
Molinia caerulea	Scirpus caespitosus
Rhynchospora alba	

Other more or less frequent species are:

Calluna vulgaris	Eriophorum angustifolium
Carex lasiocarpa	Menyanthes trifoliata†
C. panicea	Myrica gale
Drosera anglica	Narthecium ossifragum
D. intermedia	Pedicularis sylvatica
D. rotundifolia	Potamogeton polygonifolius†
Erica tetralix*	

* *Erica mackaiana*, which is found only here and in the Pyrenean region, occurs mainly on shallow peaty soil round rocks or by the wayside, where it hybridises freely with the closely allied *E. tetralix* to form the variable *E. praegeri*.

† These two species, which are of fairly frequent local occurrence in the general bog community, are probably, like *Cladium* and *Phragmites*, relicts from more open water conditions.

Of the five more abundant species mentioned, *Rhynchospora*, *Schoenus* and *Molinia*[1] are dominant, as we have seen, over considerable areas. The wetter parts of the bog are often dominated by *Rhynchospora alba*, and the drier by *Molinia caerulea* var. *depauperata*. In the shallow hollows bare peat is often covered by the purple-black conjugate alga *Zygogonium ericetorum*, and sparsely colonised by *Carex limosa*. *Rhynchospora alba* follows, and there may be a succession Zygogonietum →Rhynchosporetum → Molinietum as the general surface of the bog rises. *Schoenus nigricans* is a local dominant of parts of the bog, but not *Eriophorum*, and rarely *Scirpus caespitosus*.

The bryophytes, and especially species of *Sphagnum*, certainly play an important part in the formation of local hummocks,[2] though hummock formation does not constitute a universal genetic feature of the blanket bog, as it does of the raised bog. Pockets of *Sphagnum* are frequent on the surface of the peat among the higher plants, *S. subsecundum* on the lower ground, while *S. plumulosum*, *S. papillosum* and *S. rubellum* are the three most important hummock-building species, as on the raised bog. As the hummock increases in height, establishing a substratum better drained than

[1] Pl. 122.   [2] Pl. 122.

that of the general bog peat, it is colonised by *Molinia*, and by *Calluna*, which here grows and flowers much more luxuriantly than in the general bog. Finally *Cladonia sylvatica*, sometimes associated with *C. uncialis*, often joins *Calluna*, or the hummock is capped with a cushion of *Rhacomitrium*, more rarely of *Leucobryum*. Other bryophytes of the bog are *Campylopus atrovirens*, which forms conspicuous black patches, *Aulacomnium palustre* and *Dicranum* spp.

Blanket bog in north-western Mayo. The vegetation of the blanket bog on the hills near Glen Cullin, in the north-west of Mayo, was analysed. Parts of this were flat with a fairly even surface, and from a comparison of several neighbouring areas the following list may be taken as typical:

Flat bog between Carrefull and Glen Cullin (Co. Mayo)
(altitude 250 ft. (c. 75 m.))

Vascular plants:		Bryophytes and lichens:	
Schoenus nigricans	d	Campylopus atrovirens	f–a
Rhynchospora alba	ld	Dicranum sp.	o
Molinia caerulea	va	Leucobryum glaucum	o
Eriophorum vaginatum	a	Rhacomitrium lanuginosum	f
Scirpus caespitosus	a	Sphagnum cuspidatum	f
Narthecium ossifragum	f	S. cymbifolium	o
Calluna vulgaris	f	S. platyphyllum	o
Erica tetralix	f	S. plumulosum	f
E. cinerea	l	S. rubellum	f
Anagallis tenella	o	S. subsecundum	o
Drosera rotundifolia	o	S. tenellum	o
D. anglica	o	Blepharosia sp.	o
Eriophorum angustifolium	o	Cladonia sylvatica	f
Pinguicula vulgaris	o	C. uncialis	o
Potentilla erecta	o		
Polygala vulgaris	r		

Hummock formation, based mainly on the activity of *Sphagnum plumulosum* and *S. rubellum*, is local in these bogs. There is very often a capping of *Rhacomitrium lanuginosum*, which may form the bulk of the hummock. *Sphagnum plumulosum* usually starts the hummock by settling in a tussock of *Schoenus, Scirpus* or *Eriophorum vaginatum*, and still occurs round the base of the hummock when growth has been completed. It is followed by *Sphagnum rubellum* or directly by *Rhacomitrium*, and the sides of the mound so formed are colonised by *Molinia, Scirpus caespitosus* and *Eriophorum vaginatum*, the summit by *Calluna* and *Cladonia sylvatica*, sometimes by *Erica cinerea*.

Hummock structure and flora. The hummocks are usually quite small in size, rarely as much as a metre in diameter and 30–40 cm. in height; but occasionally a very large one is met with. One of these (Fig. 142) was 6 m. in diameter, the summit 75 cm. above the higher and 1·3 m. above the lower general surface of the sloping bog. Its base was formed as usual of *Sphagnum plumulosum*, and higher up its sides were *S. rubellum* and some

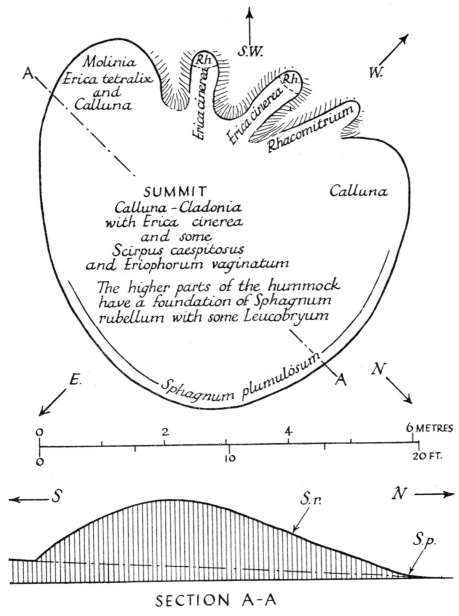

FIG. 142. DIAGRAM OF EXCEPTIONALLY LARGE HUMMOCK IN THE BLANKET BOG OF WESTERN MAYO. Drawn to true scale

Leucobryum glaucum. The slope of the exposed western side reached 25°, while the gentler eastern slope did not exceed 20°. The slopes were occupied by *Molinia*, *Eriophorum vaginatum* and *Scirpus*, with some *Erica tetralix*, the summit by *Calluna* and *Erica cinerea* with *Cladonia sylvatica*. The exposed western face was strongly eroded, cut into miniature valleys and headlands with bare peat sides. The headlands, i.e. the situations which were best drained and most exposed to the wind, were occupied by *Erica cinerea* with *Rhacomitrium* on the extreme ends. This wind erosion of the exposed western sides of hummocks is a very common phenomenon throughout the region. A somewhat smaller hummock in the Connemara bog is shown in Pl. 122.

The enormous tract of flat bog a little over 250 ft. in altitude, which stretches uninterruptedly for 12 or 14 miles north and south of Bellacorick,[1] has almost exactly the same list of species. *Schoenus nigricans* tends to dominate the wetter, *Scirpus caespitosus* the drier, parts. Hummock formation is local,[2] and the summits are generally occupied by *Rhacomitrium*. *Empetrum nigrum* occurs on some of the hummocks and *Lycopodium selago* on bare patches of peat—two mountain plants at quite a low altitude.

SUMMARY OF THE BOG OR MOSS FORMATION

From the constant recurrence of the same species in Sphagnetum, Rhynchosporetum, Schoenetum, Eriophoretum, Scirpetum, Molinietum and the mixed communities of the blanket bog, it is obvious that these communities are ecologically very closely related and should be placed together in one formation. Detailed and accurate seral studies are largely lacking, but from the data recorded in this chapter the main outlines of classification and succession are clear.

We may summarise the chief *forms* of the moss or bog formation as follows:

(1) **Blanket moss or bog**—ombrogenous, the climatic climax (except where drainage is quite free) in regions of cool summers, high rainfall and very high atmospheric humidity, i.e. extremely oceanic cool temperate climate. Surface flat or with a slight slope (under 15°): hummock formation local.

(*a*) Western Scotland, western Ireland and outlying islands, on almost every terrain lacking free drainage, up to considerable (undetermined) altitudes. Dominants usually mixed, with the following plants prominent: *Sphagnum* spp., *Rhynchospora alba*, *Eriophorum angustifolium*, *E. vaginatum*, *Scirpus caespitosus*, *Schoenus nigricans* (west Ireland), *Molinia caerulea*. Rhynchosporetum, Schoenetum, Scirpetum or Molinietum often developed over considerable areas.

(*b*) Plateaux and gentle slopes of mountain masses (Pennines, Wicklow

[1] Pl. 121, phot. 297.　　　　　　　　　　[2] Pl. 123, phot. 300.

PLATE 123

Phot. 300. A local group of hummocks, each capped with *Rhacomitrium lanuginosum*. Pool with *Menyanthes trifoliata* in foreground. Near Sheskin Lodge, Co. Mayo. *R. J. L.*

Phot. 301. Hummock with *Sphagnum* spp., *Cladonia sylvatica*, *Molinia caerulea*, *Eriophorum angustifolium* and *Rhynchospora alba* (at base of hummock in front). *Potentilla erecta* on the right. Between Moylaw and Eskeragh, Co. Donegal. *R. J. L.*

HUMMOCKS IN BLANKET BOG

Mountains, Dartmoor, etc.) at elevations of 1500–2200 ft. (*c.* 450–650 m.) in less extremely maritime regional climates, where the elevation considerably lowers temperature and increases precipitation and atmospheric humidity compared with surrounding lowlands. Scirpetum, Eriophoretum, Calluno-Eriophoretum, Vaccinio-Eriophoretum.

(2) **Valley bog**—topogenous, formed in valleys and depressions where water, draining from acidic rocks, stagnates, and bog plants establish themselves. Surface flat or concave. In addition to the bog species proper other species of more exacting requirements are generally present. Insufficiently known.

(3) **Raised moss or bog**—topogenous, often based on the site of a former lake and built up above fen peat or valley bog peat, in less extremely humid climates. Surface more or less convex as a whole, with a marginal *lagg* or drainage channel which receives the drainage water. Surface composed in detail of hummocks and hollows determined by the constituent units of Sphagnetum, each built up from hollow to hummock by a succession of different species of *Sphagnum*, and colonised by various vascular plants, of which *Rhynchospora alba* and *Narthecium ossifragum* dominate the hollows, *Scirpus caespitosus, Eriophorum vaginatum, Calluna vulgaris* with *Cladonia sylvatica* the hummocks. *Scirpus, Eriophorum* and *Calluna* give the general tone to the bog.

Characteristic of the central plain of Ireland: a few in England and Wales and a good many in Scotland, especially the south-west, the Midland Valley and parts of the eastern coastal plain.

The blanket moss formation of the Pennine plateaux, the Wicklow Mountains, and certain other areas, is in one sense a relict of conditions introduced by the Pleistocene ice-age, for it occupies plateaux and gentle slopes which have been planed by ice and are often covered by a layer of impermeable boulder clay or "till", the ground moraine of the ancient ice-sheets, which holds up the drainage water. The peat made by the bog plants, and on which they still grow, began to be formed in wet post-glacial climates and often contains the remains of birch and pine forests developed during drier climatic phases. The latest (Sub-Atlantic) climate was cool and moist on these sub-oceanic mountains and has favoured moss development, but to-day many of the peat bogs are subject to severe erosion, probably due to the combined action of wind and water, so that the peat is extensively "hagged", large areas laid bare and sometimes removed down to the mineral soil below. Other mosses are progressively desiccated and occupied by other communities such as Vaccinietum, Callunetum and Nardetum, a process hastened of course by draining and burning.

The "lowland mosses" at quite slight elevations, are of the raised moss type, except in western Scotland and western Ireland, and were originally formed on the sites of old estuaries and lakes. In England most of them have been destroyed by drainage.

The accompanying scheme (Fig. 143) attempts to relate the main bog types to the hydroseres, to Callunetum, and to forest.

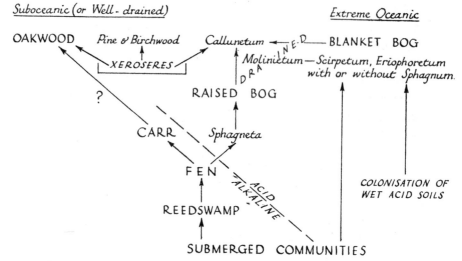

FIG. 143. SCHEME OF HYDROSERES IN RELATION TO FEN, BOG OR MOSS, CALLUNETUM AND FOREST

REFERENCES

ADAMSON, R. S. On the relationships of some associations of the Southern Pennines. *J. Ecol.* **6**, 97–109. 1918.

CRUMP, W. B. The coefficient of humidity: a new method of expressing the soil moisture. *New Phytol.* **12**, 125–47. 1913.

ELGEE, FRANK. The vegetation of the eastern moorlands of Yorkshire. *J. Ecol.* **2**, 1–18. 1914.

FRASER, G. K. Studies of Scottish moorlands in relation to tree growth. *Bull. Forest. Comm.* No. 15. 1933.

LEWIS, F. J. Geographical distribution of vegetation in the basins of the rivers Eden, Tees, Wear and Tyne. *Geogr. J.* 1904.

LEWIS, F. J. In "The British Vegetation Committee's Excursion to the West of Ireland". *New Phytol.* **7**, 253–60. 1908.

LEWIS, F. J. and Moss, C. E. Chapter XII, "The Upland moors of the Pennine Chain", in *Types of British Vegetation*, pp. 266–82. Cambridge, 1911.

MAGOR, E. W. Geographical distribution of vegetation in Cornwall: Camelford and Wadebridge district (unpublished).

Moss, C. E. *Vegetation of the Peak District.* Cambridge, 1913.

Moss, C. E. See LEWIS and Moss. 1911.

PEARSALL, W. H. The soil complex in relation to plant communities. III. Moorland bogs. *J. Ecol.* **26**, 298–315. 1938.

PETHYBRIDGE, G. H. and PRAEGER, R. LLOYD. The vegetation of the district lying south of Dublin. *Proc. Roy. Irish Acad.* 1905.

RANKIN, W. M. "The Lowland Moors of Lonsdale" in *Types of British Vegetation*, pp. 247–59. 1911.

WATSON, W. The bryophytes and lichens of moorland. *J. Ecol.* **20**, 284–313. 1932.

Part VII

HEATH AND MOOR

Chapter XXXVI

THE HEATH FORMATION

The heath formation of western Europe, of which the chief consociation is characteristically dominated by the common heather or "ling", *Calluna vulgaris* Salisb., occurs throughout the British Isles, and its presence is determined by a complex of factors. Of these the relatively moist air associated with an oceanic or suboceanic climate is one, while at least moderately free soil drainage and soil conditions (apparently mainly a low base status) correlated with high acidity, which permit the existence of the symbiotic fungus *Phoma* on which the ling seems to depend in nature (Rayner, 1915), are others.

For the dominance of heath some factor which prevents the successful gregarious establishment of trees must also be present, e.g. violent winds, recurrent fires, or a certain intensity of grazing, because in the absence of such factors trees will grow and may become dominant on most of the soils which are covered by heath when the factors hostile to trees are effective (see Chapter XVII). On the western coasts of Europe (apart from vegetation determined by salt spray) and on the exposed slopes of Irish and British oceanic and suboceanic hills where more or less violent winds are frequent, up to a level of about 2000 ft. (*c.* 600 m.), heath is commonly dominant. This is in the absence of *intensive* grazing, for while heath will support a certain number of animals, anything like heavy and continuous pasturing, either by sheep or rabbits, will quickly convert it into grassland.

On the western coasts and on more inland hills these conditions are satisfied on very various types of rock other than limestone, provided the slope is steep enough or the soil is sufficiently permeable to secure fair drainage, and heavy grazing is absent. Even on flat surfaces of limestone the strong leaching of the surface soil resulting from high rainfall, and the concurrent accumulation of acidic humus by colonising lichens and mosses, may produce acid conditions enabling *Calluna* and its symbiotic fungus to establish themselves, and once established the ling can even send its roots down into calcareous soil. Thus where the climate is exceptionally favourable to it we may have heath developed over limestone rock.

The lowland heaths so characteristic of acid sandy soils in Great Britain and north-western Europe generally depend for their maintenance on the anthropogenic factors of fire or felling.

Distribution. The general European distribution of heath follows the cool temperate oceanic and suboceanic climatic regions pretty closely. Thus it occurs throughout the British Isles, in southern Scandinavia, Denmark, north-west Germany, Holland, Belgium, and through most of

France. Northwards it is replaced by the arctic dwarf scrub vegetation, southwards by the Mediterranean *máquis*. Eastwards heath becomes rarer and rarer till it disappears altogether as the climate becomes thoroughly continental in type, though *Calluna* itself occurs as far as the Ural Mountains and even beyond, but there it is said to be confined to woods.

Though forming luxuriant Callunetum only on soil with free drainage, the ling itself can nevertheless tolerate a considerable quantity of soil water and occurs mixed in various proportions with *Nardus, Molinia, Scirpus caespitosus*, and *Eriophorum*, over practically the whole of our northern and western peat moors and also, as we have seen, habitually colonises the drier parts of *Sphagnum* bog. If the surface dries out the ling commonly becomes dominant and forms genuine Callunetum. These "heather moors" lack certain species found on the southern English heaths with their drier climate and possess others which are characteristic of the wetter and colder climate and soil (see Chapter XXXVII).

LOWLAND HEATH (CALLUNETUM ARENICOLUM)

Southern heaths. On the Neogenic rocks of various age in the suboceanic climate of the midlands, eastern and south-eastern England heath is almost always developed on sandy or gravelly soil poor in bases and therefore acid in reaction and freely permeable to water—much more rarely on clays and loams; and the same is true of its occurrence on the continent of Europe. As we have already seen (Chapter XVII, p. 354) this lowland heath is in general a biotic or a fire climax and gives way to woodland in the absence of arresting factors, though there are apparently some heath soils which will not support tree growth.

Soil—the podsol profile. The usual subsoil of a southern English heath is a sandstone, sand or gravel, poor in basic salts and acid in reaction. Where a typical podsol profile is developed the horizons are as follows.

The soil surface is covered by a layer of raw humus (compacted dry peat or *mor*) varying from a centimetre or two to 4 or 5 in. (c. 10 or 12 cm.) in thickness (A0), usually derived largely from the mosses and especially the lichens—mainly species of *Cladonia*—which constitute the ground layer of the Callunetum. In extreme cases, as on the top of Hindhead (800–900 ft. = c. 250–270 m.) in south-west Surrey the peat may reach a depth of 12 in. (30 cm.) (Haines, 1926). Below the dry peat comes a layer of humous sand (A1) commonly stained a grey-chocolate colour by the humous substances carried down from the surface peat, and containing white ("bleached") sand grains. This may pass into a pale whitish or ash-coloured layer of sand (A2) practically devoid of humus. The A2 horizon is however often absent, the "bleached" podsol layer being represented only by A1. Below this sand come the "pan" layers (B): first the "humus pan" (B1) which may be quite black but is usually dark brown, soft and friable in consistency, though often sharply bounded; and then the "iron pan" (B2) containing ferric salts and typically a dark red-brown. *B*1 and *B*2 are, however, not always well separated. The pan is often underlain by some-what light reddish sand (*B*3) into which some of the iron salts have been carried. This passes down into the unaltered sand (*C*) below. Fig. 144 is

reproduced from Farrow's diagram of a typical soil profile from the Cal-lunetum of Cavenham Heath in Breckland (north-west Suffolk).

Here we have all the features of a typical podsol, developed in a climate which is not a typical podsol climate, and probably owing its characters to the combination of the characteristic vegetation and the highly permeable soil initially very poor in basic salts. Great variation however occurs in the extent and definiteness of the podsol profile under Callunetum. Very often, as has been said, there is no distinct "bleached" layer, the whole of the *A* horizon being coloured by humous substances. Often also the "pan" is not sharply differenti-ated, though at least an ill-defined illuvial layer (*B*) can generally be made out. In gravelly soils, par-ticularly, the differentiation of the horizons may be badly defined. In other cases there may be several *B* horizons with bleached layers be-tween. The causes of these variations are still largely obscure. It has been suggested that many of the existing podsol profiles are "fossil", i.e. owing their origin to climatic conditions no longer present. Many have certainly been "truncated", i.e. the top layers eroded with subsequent fresh pod-solisation.

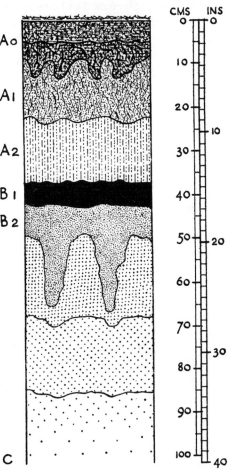

FIG. 144. A HEATH PODSOL PROFILE

For description see text. (Reprint of Fig. 28.)
After Farrow, 1915.

Hindhead Common. The soil of the Callunetum on Hindhead Com-mon in Surrey, developed over the Hythe Beds of the Lower Greensand and very thoroughly studied by Haines (1926), consists of a surface peat very dense and dark, especially in the lower part, slightly reddish brown, and 6–12 in. (15–30 cm.) thick, or even more—extremely deep for a southern English heath. This peat has a very low base status, is definitely poor in dissociable mineral matter, probably possessing no clay-forming material at all, and has a mean pH value of 3·42 at 2 in. (5 cm.) and of 3·9 at 9 in. (22·5 cm.), very few samples falling outside the range 3·0 to 3·6 at the former, and 3·5 to 4·3 at the latter depth. An extreme pH value of 2·3 was found in one place at 2 in.[1] Below the peat there was usually a thin layer of mixed humus and disintegrated sandstone not more than about a centimetre deep. Then came a layer of sand increasingly consolidated below, till the hard compact parent sandstone was reached. No trace of "pan", nor any clear illuvial layer, was found anywhere on the heath. There is probably little or no leaching, owing to the thickness and denseness of the peat. A rise in salt percentages found in some cases at 9 in., especially in the valleys, is probably due to disintegration of parent material and to lateral percolation from the higher ground in the layer of unconsolidated sand. No trace of calcium could be found in any of the samples, nor could nitrates be detected in the ordinary way; but from the appearance of a trace in an extract of unusual strength it was estimated that the nitrate content (as KNO_3) may have been of the order of 0·00027 per cent.

Heath fires occur at Hindhead every few years and consume most of the surface peat. The data given above refer mainly to an area in which 15 years had elapsed since the last fire and a considerable depth of peat had accumulated, but even here Haines regards the profile as still immature.

Structure and Flora. The structure of the Callunetum at Hindhead is typically two- or three-layered. The heather, which is generally overwhelmingly dominant, varies in height from a few inches to about 2 ft. 6 in. or 75 cm. (even more in old "leggy" growth) according to the soil and the age of the community, corresponding with the time that has elapsed since the last heath fire. A second ill-defined and interrupted layer developed beneath the dominant consists of flowering plants which can endure shade, such as the bilberry (*Vaccinium myrtillus*) and the dwarf gorse (*Ulex minor*). Both of these are local on the south English heaths but may be very abundant where they occur. Neither flowers under these deeply shaded conditions. The third or ground layer consists of mosses and lichens, *Polytrichum juniperinum*, *Hypnum cupressiforme* var. *ericetorum* and *H. schreberi* being the commonest mosses, while species of *Cladonia* are dominant among the lichens. It is largely this ground layer and especially *Cladonia*, typically forming by far its greater portion, that produces the raw material of the surface peat.

Erica cinerea[2] fairly often shares dominance with *Calluna*, and is sometimes dominant alone on rather dry sunny slopes. According to Fritsch and his co-workers (1913, 1915) the purple heath plays a prominent part in the development of the Callunetum after fire at Hindhead, and is therefore presumably somewhat more xerophytic than *Calluna*, at least when the

---

[1] This seems to be the highest acidity recorded for a British soil.
[2] Pl. 124, phot. 303; Pl. 125, phots. 307–8.

PLATE 124

Phot. 302. Young vigorous Callunetum with scattered Pteridium fronds of low growth. Toys Hill, Kent, on Hythe Beds. *S. Mangham.*

Phot. 303. Sandy heath on the raised beach of the Ayreland of Bride, Isle of Man, with *Erica cinerea* and *Ulex gallii*. From *E. J. Moore* (1931).

Phot. 304. Dense Pteridietum of low growth fringed with *Calluna* and *Vaccinium myrtillus* on edge of a chert quarry. Hythe Beds, Crockham Hill Common, Kent.

FACIES OF LOWLAND HEATH

PLATE 125

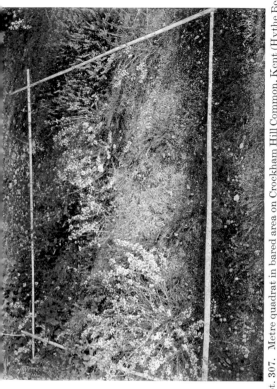

Phot. 305. Dry area of Esher Common (Bagshot Sand) 1922. Stumps of pinewood felled in 1916–17. Invasion of *Calluna*. From Summerhayes and Williams (1926).

Phot. 307. Metre quadrat in bared area on Crockham Hill Common, Kent (Hythe Beds). *Erica cinerea*, *Calluna* and *Teucrium scorodonia* displacing *Polytrichum piliferum* and *Agrostis canina* turf (centre and right). Rabbit grazed; note abundant dung.

Phot. 306. The same area in 1925 with nearly closed Callunetum. Marked growth in height of birches. From Summerhayes and Williams (1926).

Phot. 308. Metre quadrat, later seral stage. *Erica cinerea* dominant, with scattered fronds of *Pteridium* here but 1½ m plants are young. Fritsch in fact is of opinion that the purple heath would always give way eventually in competition with the ling. *Vaccinium myrtillus*[1] may also replace *Calluna* as the dominant, but what determines such a replacement is not clear. *Erica tetralix* often dominates areas of wetter soil, but on some heaths whose soil is not wet, as on the ridge at Hindhead, it is scattered through the general Callunetum (Fritsch and Parker, 1913). This may be due to high precipitation and atmospheric moisture at the relatively high altitude (270 m.) of this summit.

Ulex minor is very abundant on many heaths of the south-eastern and south central English counties. It occurs in two growth forms, a prostrate or semi-prostrate plant, which may be subdominant to *Calluna* or occupy gaps in the Callunetum, and an erect form 70–120 cm. high forming small clumps of dwarf scrub (Fritsch and Parker, 1913; Skipper, 1922). The former, typically prostrate, but sometimes with erect branches reaching a height of 45–75 cm., is the common form, and is often an important member of the Callunetum: the tips of the shoots dry off if they rise above the general level of the *Calluna*. At Hindhead the latter form is confined to the valleys and enters into the common gorse community, mainly dominated by *Ulex europaeus*. This last is not a member of the Callunetum except as a sporadic invader, mainly along paths where its seeds are probably carried by ants (Weiss, 1908; 1909). It is doubtful if it ever colonises undisturbed heathland. Juniper is not uncommon on some south-eastern heaths.

The bracken fern (*Pteridium aquilinum*), like the common gorse, is frequently found in and about heaths, but is hardly a true member of the Callunetum. This aggressive plant spreads actively by rhizome growth from centres of infection, and heath soils suit it very well. But it does not as a rule appear to invade undisturbed heath and its growth is severely restricted by the dominance of the heath undershrubs, the fronds in thick young heath being small, widely spaced and not more than a foot or so high.[2] With increasing growth of the heaths it seems to succumb altogether, perhaps because the highly compacted surface peat interferes with soil aeration. Furthermore *Pteridium* cannot stand wind nearly so well as the heaths, so that it is absent from exposed plateaux and slopes (cf. Chapter XXVI, p. 502).

On the other hand Farrow (1917) found *Pteridium* advancing along a broad front and invading Callunetum on Cavenham Heath in Breckland, apparently killing the *Calluna*.[3] As soon as the *Calluna* was thickly interspersed with bracken fronds the plants became unhealthy and soon died, seemingly as the result of the dead fronds falling on and smothering the *Calluna* bushes (Fig. 145). The surface peat is here very thin and the sand loose, conditions favouring the horizontal growth of bracken rhizomes.

The following list is taken from the adult Callunetum of Hindhead Common on the Hythe Beds of the Lower Greensand (Fritsch and Salisbury,

---

[1] Pl. 124, phot. 304. [2] Pl. 124, phot. 302; Pl. 125, phot. 308.
[3] Pl. 126, phot. 311.

1915) in which the *Calluna* was 75–100 cm. high (Fritsch and Parker, 1913).
The shrubs (other than *Ulex* and *Sarothamnus*) and trees are omitted (see
below under *Succession*):

Calluna vulgaris	d	Ulex minor	sd
Vaccinium myrtillus	la	Erica tetralix	l
Ulex europaeus	l	Sarothamnus scoparius	r
Agrostis tenuis	r	Molinia caerulea	o
Blechnum spicant	lr	Polygala vulgaris	r
Carex pilulifera	f	Potentilla erecta	r
Deschampsia flexuosa	o	Pteridium aquilinum	l
Galium saxatile	r	Sieglingia decumbens	r

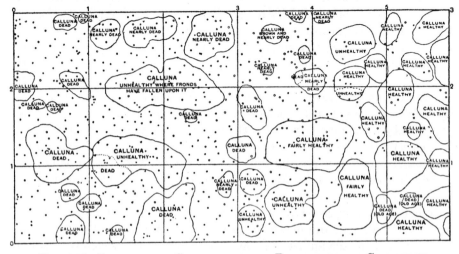

FIG. 145. INVASION OF CALLUNETUM BY *PTERIDIUM* ON CAVENHAM
HEATH (BRECKLAND)

The invasion is from left to right. The bracken is just reaching the right hand edge of the
chart, where its fronds are still sparse and the heather is thick and healthy. On the left
where the bracken fronds are thick the heather is all dead. Each square represents a square
metre. Cf. Pl. 126, phot. 311. From Farrow, 1917.

Mosses and lichens

Dicranum scoparium	f	Cladonia floerkiana	
Hypnum cupressiforme		f. trachypoda	l
var. ericetorum	f	C. pyxidata	lf
Leucobryum glaucum	f	C. sylvatica	f
Cladonia coccifera	r	Parmelia physodes	f

Algae and fungi

Gloeocystis vesiculosa	f	Mesotaenium violascens	f
Hormidium flaccidum	lr	Clitocybe sp.	f

Esher Common. An even poorer flora is recorded from the Callunetum
of Esher Common and Oxshott Heath (Summerhayes, Cole and Williams,
1924) on Plateau Gravel and Bagshot Sand. This area is also in Surrey,

PLATE 126

Phot. 311. Callunetum (on the left) invaded and destroyed by Pteridietum from the right. Cavenham Heath, Breckland. From *E. P. Farrow* (1917).

Phot. 309. Displacement of Ammophiletum by Callunetum on the coast of South Haven Peninsula, Dorset. From *R. d'O. Good* (1935).

Phot. 310. Mature Callunetum invaded by young pines. Wych Cross, Ashdown Forest, Sussex, on Ashdown Sand. *S. Mangham.*

DEVELOPMENT AND DESTRUCTION OF CALLUNETUM

about 23 miles (*c.* 37 km.) north-east of Hindhead Common and at a much lower altitude. It is evidently drier than Hindhead Common. The dominant *Calluna* was about 45 cm. high.

Calluna vulgaris	d	Deschampsia flexuosa	
Erica cinerea	f	Potentilla erecta	
Ulex minor	lf	Cuscuta epithymum	
Dicranum scoparium	a	Polytrichum juniperinum	f
Hypnum cupressiforme		Leucobryum glaucum	o
var. ericetorum	a		
H. schreberi	f	Cladonia pyxidata	a

Cavenham Heath. From Cavenham Heath in Breckland—drier still—Farrow (1915) records *no* other angiosperms, except occasional seedlings of shrubs and trees, in *thick* Callunetum (though a good many grass heath species are found between the *Calluna* bushes where these are less crowded), but the following mosses and lichens occur in the ground layer:

Ceratodon purpureus	Cladonia alcicornis
Dicranum scoparium	C. cervicornis
Hypnum schreberi	C. coccifera
Leucobryum glaucum	C. furcata
Polytrichum piliferum	C. sylvatica
	C. uncialis

Sherwood Forest. From Sherwood Forest in Nottinghamshire, on the Bunter Sandstone, Hopkinson (1927) records a similar paucity of species in mature Callunetum. *Erica cinerea* accompanies the dominant where the latter is not completely closed. In other places the purple heath is dominant and the ling subdominant. This combination may perhaps represent a seral stage (see below under *Succession*). *Ulex minor* and *Genista anglica* are both also recorded, while *Ulex europaeus* is said to invade and displace the ling in some spots, "growing up from the centre of many well-grown shrubs of *Calluna*". Several of the mosses and lichens mentioned by Fritsch and by Farrow are also recorded by Hopkinson from the Sherwood Callunetum.

Thus it is clear that the mature closed Callunetum is exceedingly poor floristically. The lists of fairly numerous species given for the "heath association" in *Types of British Vegetation* and elsewhere include many which only occur in immature or modified Callunetum and properly belong to some seral stage or to the much richer "grass heath" flora described in Chapter XXVI (pp. 506 ff.).

SUCCESSION

(1) *The Burn Subsere*

This subsere is very commonly met with on heaths owing to the ease with which the vegetation is set alight and burns after prolonged dry weather. Fritsch and Salisbury (1915) give an excellent account of it as seen on Hindhead Common in Surrey. Hindhead Common, in fact, as they clearly show, is nothing but a mosaic of different seral stages, corresponding with the different periods which have elapsed since the last fire. Started sometimes from negligence or mischief, sometimes with deliberate intent to destroy the old "leggy" *Calluna* and make way for new and more vigorous growth, these fires constantly sweep across areas which have already been recently burned. Thus it is possible to recognise the effect of burning on different stages of the subseral development. When tall adult Callunetum is burned, there is a great mass of combustible material, the heat developed is greater, and more of the plants are killed outright by the fire. The proportion of ling and purple heath killed on a fire-swept area varies a good deal from this cause. A majority of the plants of *Ulex minor* always manage to survive. The sprouting of the surviving bases of the dwarf gorse after the fire aids materially in the rapid redevelopment of the formation, which, for the rest, is carried out by the establishment of sporelings of algae, lichens and mosses, and of seedlings of the undershrubs.

The pioneers—algae and lichens. According to Fritsch and Salisbury (1915, p. 129) the first colonists of the charred peat of burned Callunetum are "*Cystococcus humicola*, forming dark green granules in most of the countless small depressions of the soil", and the gelatinous unicellular algae, *Gloeocystis vesiculosa*, *Trochiscia aspera* and *Dactylococcus infusionum*, forming a thin, macroscopically invisible layer over the greater part of the surface. At a very early stage *Ascobolus atrofuscus* grows over the whole gelatinous layer in which its hyphae ramify profusely, enveloping groups of the *Cystococcus* in a dark pseudoparenchymatous investment and thus forming minute, rounded, olive-brown protuberances which give a velvety texture to the surface. Apothecia arise in quantity from the hyphae and are not specially developed from those associated with the *Cystococcus*. The *Ascobolus* persists on burned areas of all ages and also on parts of the mature heath. *Humaria melaloma* also occurs, but is rare.

The dark green *Cystococcus* granules also become pale yellowish green owing to investment by the colourless hyphae of another (unnamed) fungus, and thus lichens are formed. Of these "*Cladonia delicata* is at first the only one to be found in fruit, but other sterile forms are probably associated. The small scale-like lobes of their pale green thalli form numerous patches on the soil, and these are dominant for some years on the barer parts, until in fact the phanerogamic vegetation begins to close, when they become subordinated to other forms or completely disappear." Meanwhile the *Gloeocystis* and its associates, with *Mesotaenium violascens*, multiply over the general surface, making increasingly conspicuous dirty green jelly-like lumps which occupy practically the whole surface left bare by other forms. After about five years these lumps become visible to the naked eye on close observation. Finally, in the tall Callunetum the gelatinous algae form conspicuous masses a quarter to half an inch in diameter. Thus the algae and lichens form a closed layer from the first. The only other algae met with were *Hormidium flaccidum* and *Zygogonium ericetorum*, both of which were rare and local. Diatoms appear to be entirely absent.

Three or four years after a fire the following Cladoniae become associated with *Cladonia delicata*:

Cladonia furcata (first to appear)   C. squamosa
C. floerkiana f. trachypoda   C. sylvatica (rare at first)
C. pyxidata

It is suggested that the Cladoniae accumulate peat and thus allow of increasing luxuriance of *Calluna*. Measurements indicate that the average rate of growth of the ling is more rapid where the peat is deeper.

As the phanerogamic vegetation becomes more and more closed *Cladonia sylvatica* increases in amount, and with it are associated *C. pyxidata* and *C. coccifera*. The lichens reach their greatest development, both in number of species and individuals, in a phase of relatively poor development of *Calluna*, which on some heaths seems permanent, and in which *Cladonia sylvatica* is dominant in a richly developed and varied lichen flora (cf. Chapter XXVI, p. 511). In the tall Callunetum of Hindhead Common *C. sylvatica* is the only common lichen, apart from *Parmelia physodes* on the older stems and branches of the ling (see list on p. 728).

Mosses. Mosses occur in the earliest stages of the subsere, including *Ceratodon purpureus*, which is very abundant, and small patches of *Tortula subulata* and *Funaria hygrometrica*. In slightly later stages these are associated with *Campylopus brevipilus* and *Polytrichum piliferum*. The mosses form considerable sheets, but are not uniformly distributed over burnt areas like the algae and lichens. About two or three years after a fire *Ceratodon* tends to disappear and for a time the mosses are quite insignificant compared with the lichens. In the final stages a few species of moss are frequent (see list on .p. 728), *Hypnum cupressiforme* var. *ericetorum* "attaining great development where light penetrates the *Calluna* canopy".

Vascular plants. Certain species which do not belong to the Callunetum (though they are often reckoned as "heath plants"), namely *Pteridium aquilinum*, *Molinia caerulea* and *Deschampsia flexuosa*, spread considerably after a fire, mainly from the valleys in which they are abundant, on to the peat bared of the ling, but they do not in the long run appear to extend their area at the expense of the redevelopment of the Callunetum (Fritsch, 1927), except perhaps when fires succeed one another at very short intervals.

Early herbaceous colonists are seedlings of *Carex pilulifera*, and *Polygala vulgaris*, *Rumex acetosella*, *Galium saxatile*, *Potentilla erecta*, *Epilobium angustifolium*. *Rumex acetosella*, which on many heaths is often locally dominant after burning, is never abundant over extensive areas on Hindhead. The only grasses whose seedlings occur in any quantity are *Agrostis tenuis*, *Deschampsia flexuosa*, *Festuca ovina* and *Poa annua*.

Cuscuta epithymum is also abundant in the early phases of the subsere, attacking a great variety of hosts, including:

Agrostis tenuis	Galium saxatile
Betula (sprouting)	Molinia caerulea
Calluna vulgaris	Potentilla erecta
Deschampsia flexuosa	Pteridium aquilinum
Erica cinerea	Sieglingia decumbens
E. tetralix	Ulex minor

In the adult Callunetum it appears to be absent.

Undershrubs. Of the undershrubs which enter into the structure of the climax *Ulex minor* is seldom killed by the fires, and sprouting at once from the base it secures a start in the recolonisation by vascular plants. Seedlings appear in great numbers but mostly die. A large number of the original *Calluna* plants are killed by the intense heat of the fires consuming the tall Callunetum, and here the ling regenerates mainly by seed. When the fire sweeps over the younger growth much of the *Calluna* survives and regeneration is largely by sprouting from the stocks. *Erica cinerea* more rarely survives burning but it produces very numerous seedlings. It seems that this species, at any rate in the young stages, can resist dry conditions better than *Calluna*, and hence may become dominant alone on slopes facing south and south-west. It is also normally prominent in the subfinal stage of the developing Callunetum ("CUE" phase of Fritsch), eventually yielding, except locally, to the dominance of the ling, with the dwarf gorse subdominant ("CU" phase of Fritsch).

Erica tetralix and *Vaccinium myrtillus* are much less abundant, but they both commonly survive the fires, the former because its tufts accumulate a protective covering of litter and humus, the latter because its rhizomes are 2 or 3 in. below the soil surface. These plants, then, regenerate *in situ* from the old stocks.

Summary of burn subsere. The following is a summary of the phases of succession in the burn subsere at Hindhead:

(1) Algal phase (*Cystococcus, Gloeocystis, Trochiscia*), with the fungi *Pyronema confluens* and *Ascobolus atrofuscus*, and seedlings and sprouting stools of *Ulex minor* which have survived the fire.

(2) Algae as in Phase 1, with soredial groups formed by *Ascobolus* and *Cystococcus*: mosses (*Ceratodon, Funaria*); sprouting stools as in Phase 1; seedlings of *Calluna* and *Erica cinerea*; a number of herbaceous species, especially *Deschampsia flexuosa, Carex pilulifera, Polygala vulgaris, Rumex acetosella* and *Cuscuta epithymum*.

(3) *Lichen phase* with *Cladonia* spp.: algae as before, with *Mesotaenium violascens*: increase in mass of phanerogamic vegetation, especially *Calluna* and *Erica cinerea*.

(4) Closing *Calluna, Ulex minor* and *Erica cinerea*, with numerous lichens: disappearance of herbaceous species ("CUE" phase).

(5) Adult Callunetum: *Calluna* dominant, *Ulex minor* subdominant, *Erica cinerea* suppressed; mosses, lichens and algae forming a ground layer ("CU" phase); considerable peat development. *Vaccinium myrtillus* co-dominant with *Calluna* in places with thickest peat.

(2) *Gravel or sand subsere*

The pioneer on this substratum, when it is bared at Hindhead by any means, is the terrestrial form of the conjugate alga *Zygogonium*

At Hindhead *ericetorum*, which forms extensive dark purple sheets binding the surface layers of sand. These may be overgrown by the bright green threads of *Hormidium flaccidum*. The lichen *Baeomyces rufus* and the moss *Polytrichum piliferum* also play important parts, while scattered individuals of the fungus *Clavaria argillacea* occur in the autumn (Fritsch and Salisbury, 1915).

No information is available as to later stages of the gravel slide sere at Hindhead, but on horizontal gravel surfaces on the same rock (Hythe Beds) laid bare by chert diggings on Crockham Hill Common (700 ft.)

At Crockham Hill in west Kent, the protonema of *Polytrichum piliferum* seemed to be the pioneer in recolonisation. It is possible however that the moss may have been preceded by algae. The protonema covered the bare soil in places with a close network and became studded with the dwarf leafy shoots of the moss. The thin sheet of incipient peat formed by the *Polytrichum* was then colonised by *Cladonia* spp., *Agrostis canina* (considerable patches) and numerous seedlings of *Calluna* and *Erica cinerea*, together with *Teucrium scorodonia*, and (in the square metre recorded) a few plants of *Galium saxatile*, *Veronica officinalis*, and one seedling each of *Rumex acetosella* and *Ulex europaeus*. Two years later the plants of *Erica* and *Calluna*, especially the former, had greatly increased in size and formed rapidly extending tufts, some of which had come into lateral contact, greatly restricting the extent of *Polytrichum* and *Agrostis*.[1] More heath seedlings had appeared and *Teucrium scorodonia* had also increased greatly in numbers, especially in the shelter of the heath tufts. Though the history of this quadrat could not be followed further, as it was destroyed by fresh chert digging, there is no doubt that the heaths would have closed up and eliminated practically all the *Polytrichum* and the other flowering plants. By following three separate quadrats, laid out in different places and representing different phases of succession, for 3 years, it was very roughly estimated that the whole sere, from bare ground to adult Callunetum, would have taken approximately 15–20 years to complete its development. Judging from the much faster progress during a wet than during a dry summer it is probable that the period occupied would be greatly affected by a run of wet or of dry years.

On an area in Breckland bared of vegetation by wind-blown sand and on

[1] Pl. 125, phot. 307. A later stage of the sere is seen in Pl. 125, phot. 308.

which the sandblast had ceased, Farrow (1919) records *Cladonia coccifera,
C. cervicornis, C. uncialis, C. aculeata, Polytrichum piliferum* and *Campylopus flexuosus* as the agents of colonisation. Nothing is said of later stages in the succession (see Watt's account of the development of Agrostido-Festucetum, Chapter XXVI, pp. 512–14).

Thus the stages of succession appear to follow the same general course in the bare gravel and in the burn subseres, though there are minor differences between them.

(3) *Felling subsere*

Summerhayes and Williams (1926) briefly describe the colonisation by
At Oxshott Callunetum of an area of pinewood on Oxshott Heath felled in 1917. They make no mention of pioneer algae, lichens and mosses, such as were found at Hindhead in the early stages of heath regeneration. *Calluna* itself comes slowly at first, since the seed is mainly derived from plants surviving in openings of the wood or under the original pines. It is not until these surviving plants of *Calluna* have recovered sufficiently to flower freely and produce abundant seed that colonisation is at all rapid. Also the undecayed pine needle litter is unfavourable to ecesis of seedlings of the ling. Thus after 5 years from felling (Pl. 125, phot. 305) the *Calluna* is seen to be quite sparse, but 3 years later (Pl. 125, phot. 306) the Callunetum is nearly closed. The frequency of *Erica cinerea* in the early development of the Callunetum is noted, in agreement with Fritsch. Of mosses, *Campylopus flexuosus, Webera nutans, Polytrichum juniperinum* and *P. formosum* occurred in very small quantity under the young heather canopy, while *Dicranum scoparium* and *Hypnum cupressiforme* var. *ericetorum*, abundant in the mature Callunetum, seemed to be spreading slowly in from what had been open heather areas in intervals of the original pinewood.

WET HEATH COMMUNITIES

In constantly wet places within an area of heath, for example where the ground water reaches the surface in a local hollow of the ground and forms a pond or bog, or where water oozes from the surface of a slope and trickles downhill, or in a low lying water-logged area where the nature of the subsoil determines acid conditions, characteristic wet-heath communities occur. These have a great deal in common with the wet peat moor, "bog" or "moss" communities developed in the extremely humid climates of western Scotland and Ireland, as well as on the mountain plateaux (see Chapter XXXIV).[1]

If the water level remains approximately constant in relation to the soil surface, the wet heath community may be permanent. It is not clear how

[1] On strictly vegetational lines these "wet-heath" communities are indeed to be considered as "valley bogs" belonging to the formation described in Chapters XXXIV and XXXV, but their occurrence in the lowlands as parts of the physiographic heath complex makes it convenient to describe them in this chapter.

far an actual autogenic hydroseral succession occurs through the gradual raising of the soil surface as a result of humus accumulation, though erosion and deepening of water channels (as for example between the tussocks of Molinietum on Chard Common described below) and the consequent increased drainage of the intervening areas undoubtedly leads to colonisation by a drier heath community. Where a dry heath slopes down to a lower-lying wet area there is of course a zonation of transitional vegetation between the two, but this does not necessarily represent an actual hydrosere.

Molinietum. Perhaps the most characteristic plant of damp or wet heath is *Molinia caerulea*, which is nearly as widespread in this habitat in southern England as it is on certain types of fen and on the margins of blanket bog in western Scotland and Ireland. Summerhayes and Williams describe the Molinietum of the low-lying damp parts of Esher Common: "...the community was strikingly uniform in appearance, consisting of large tussocks growing close together and producing inflorescences up to 3 ft. 6 in. (1 m.) high....The chief associates of the *Molinia* were *Calluna* and *Erica tetralix*, the latter being especially abundant locally, while *Calluna* was commoner nearer the upper edge of the community....It is probable...that permanent Molinieta do not exist on heaths in the low rainfall districts of England. Where the drainage is good the *Molinia* is replaced by other herbaceous or dwarf shrub species, while on damper soils woodland is soon established" (Summerhayes and Williams, 1926, p. 220). There were numerous associated species, but most of them were "far from common; indeed they rarely affect the uniform appearance of the *Molinia*" (p. 221).

Succession. During the long drought of the summer of 1921, there were numerous severe fires on Esher Common, some of which burned themselves out through having consumed all the combustible material. On the damper bare areas thus produced the early stages of recolonisation were traced from 1922 to 1925 and are represented in Fig. 146*a–d*. The leading features of the subsere were the colonisation of the burned soil first by *Funaria hygometrica* and then by *Epilobium angustifolium*, which became completely dominant in the second year but was quite transient, and followed by the invasion of *Molinia*, *Betula* and *Calluna*, with *Polytrichum*, as soon as the alkalinity and intense nitrification of the soil resulting from the fire gave place to increasing acidity. The Molinietum would presumably give way to birch-heath and perhaps ultimately to pinewood.

Chard Common. The best existing account of wet-heath zonation is given by Watson (1915) from an area of Chard Common, Somerset. This wet heath occupies about 3 acres (1·2 hectares) in the middle of drained pasture land on non-calcareous Lias. The area drains to a stream on its edge by means of a complicated system of narrow water channels separating small polygonal areas or islands (Fig. 147), which vary from a few deci-

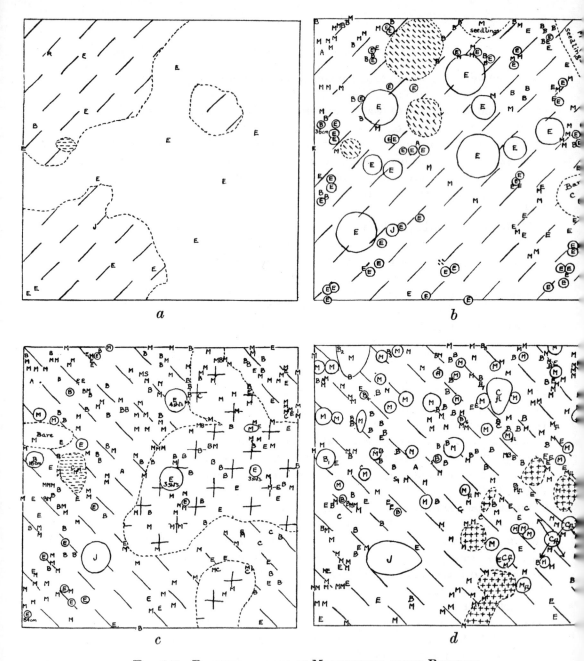

FIG. 146. REDEVELOPMENT OF MOLINIETUM AFTER BURNING

a, b, c, d. The same quadrat in four successive years, 1922–25. Esher Common, Surrey. From Summerhayes and Willi 1926.

BRYOPHYTES. //*Funaria hygrometrica*. = *Marchantia polymorpha*. *Polytrichum* spp. (mostly *P. juniperinum formosum*). + *Ceratodon purpureus*.

PHANEROGAMS. E, *Epilobium angustifolium*. J, *Juncus conglomeratus*. A, *Sorbus aucuparia*. B, *Betula* spp. M, *Mo caerulea*. C, *Calluna vulgaris*. Sq., *Juncus squarrosus*.

(*a*) Bare area colonised by *Funaria hygrometrica* and *Epilobium angustifolium*.

(*b*) *Epilobium angustifolium* dominant. Invasion of *Molinia*, with *Betula* seedlings and a little *Calluna*: invasio *Polytrichum* spp. *Funaria* still dominant in the ground layer.

(*c*) Recession of *Epilobium*: *Molinia* and *Betula* increasing: *Polytrichum* and *Ceratodon* dominant in the ground l

(*d*) *Molinia* becoming dominant: increase of *Betula* and *Calluna*: *Polytrichum* dominant in the ground layer.

metres to nearly two metres in diameter, and are dominated by *Molinia caerulea*. In rainy weather the whole area is very wet and the channels all contain free water, but in dry summers the polygonal *Molinia* tussocks become dry and are separated by damp furrows. A certain amount of erosion of the sides of the tussocks occurs during wet weather when water is

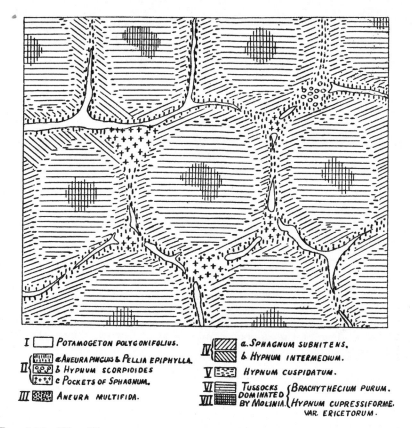

I ▭ *Potamogeton polygonifolius*.

II ⟨ *a. Aneura pinguis & Pellia epiphylla.*
   b Hypnum scorpioides
   c Pockets of Sphagnum. ⟩

III ▦ *Aneura multifida.*

IV ⟨ *a. Sphagnum subnitens.*
    b Hypnum intermedium. ⟩

V ▤ *Hypnum cuspidatum.*

VI ▤ *Tussocks dominated by Molinia* ⟨ *Brachythecium purum.*
VII ▦ *Hypnum cupressiforme. var. ericetorum.* ⟩

FIG. 147. WET HEATH COMPOSED OF TUSSOCKS SEPARATED BY CHANNELS

Chard Common, Somerset. Note the bryophyte zonation round each tussock. Scale 1 : 50. "*Sphagnum subnitens*"=*S. plumulosum*. From W. Watson, 1915.

actively draining away down the channels, and this tends to deepen the furrows and increase the height of the tussocks.

Zonation. The following zones were distinguished by Watson in passing from a permanent water channel to the summit of the adjacent tussock (Figs. 147, 148). Owing to the small size of the tussocks the whole of the zonation described only occupies about seven or eight decimetres, and some of the zones cannot in any case be considered as representing phases of succession, since they owe their origin to special conditions arising on the slightly eroded edges of the water channels.

(1) In the channels where water is constantly present *Potamogeton polygonifolius* occurs—alone where the depth of water exceeds 10 cm. This is the characteristic hydrophyte of the acid waters of wet heath. Desmids are very abundant and characteristic.

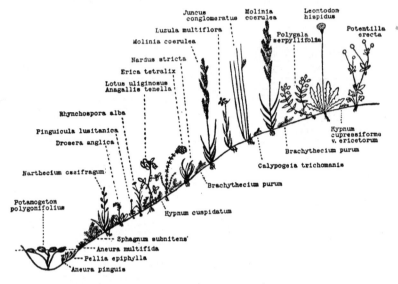

FIG. 148. PROFILE TRANSECT THROUGH A TUSSOCK

"*Sphagnum subnitens*" = *S. plumulosum.* Scale 1: 10. From W. Watson, 1915.

(2) On the edges of the deeper channels or in furrows where the water is always very shallow:

(*a*) Where the water is constantly being replaced, i.e. along definite drainage channels:

Aneura pinguis	a	Hydrocotyle vulgaris	o
Pellia epiphylla	a		
Potamogeton polygoni-folius	o		

The two liverworts typically have long narrow thalli.

(*b*) In shallow stagnant peaty pools, where the surface water may disappear:

Hypnum scorpioides (reddish)	d	Hypnum revolvens	o
		H. stellatum	o

(*c*) In wet pockets larger than those of (*b*):

Sphagnum cymbifolium (and varieties)	d	Sphagnum rufescens	o
S. recurvum	la	Aulacomnium palustre	a

(3) At the water level a line of *Aneura multifida* practically pure.

(4) Just above the water level:

(a) By the smaller (more stagnant) furrows: *Sphagnum plumulosum* is dominant, especially var. *violascens*, with *S. rufescens* (f) and other species of *Sphagnum*. Associated with the *Sphagnum* are *Cephalozia connivens* (a), *Calypogeia fissa* (a) and *Hypnum intermedium* (o), and the following flowering plants:

Drosera rotundifolia	a	Narthecium ossifragum	f
D. anglica	f	Pinguicula lusitanica	o
D. intermedia	f	Rhynchospora alba	o
Hydrocotyle vulgaris	f	Scutellaria minor	o
Mentha aquatica	o	Viola palustris	f

(b) By the sides of channels where there is a slow but constant current *Hypnum intermedium* is dominant, associated with *H. giganteum* (f), *H. revolvens* (f), *H. cuspidatum* (o) and *H. scorpioides* forma (o). Other mosses are *Sphagnum plumulosum* and *S. rufescens*, *Bryum pseudotriquetrum* and *Mnium affine* var. *elatum* (all o); and the following vascular plants:

Carex oederi	a	Equisetum fluviatile	
C. hornschuchiana	o	(limosum)	l
C. pulicaris	o	Galium uliginosum	a
C. inflata	l	G. palustre var. witheringii	l
C. paniculata	l	Glyceria plicata	l
C. stellulata	o	Hydrocotyle vulgaris	a
Eleocharis multicaulis	o	Juncus acutiflorus	a
Equisetum palustre	a	Mentha aquatica	a
var. polystachyum	o	Narthecium ossifragum	f
E. telmateia (maximum)	o	Pedicularis palustris	f
(in and near alder thickets	a)	Viola palustris	f

(5) At a slightly higher level *Hypnum cuspidatum* is sometimes dominant in a fairly definite zone, associated with *H. stellatum*, and occasionally with *H. intermedium* and *H. giganteum* belonging to 4 (b), from which this zone is not always separable in respect of its bryophytes. *Sphagnum subsecundum* is locally dominant in this zone, and *Calypogeia trichomanis* frequent.

The vascular plants are however characteristic; the cross-leaved heath being dominant, and the bog pimpernel very abundant:

Anagallis tenella	va	Erica tetralix	d
Carex helodes	l	Lotus uliginosus	f
Cirsium palustre	f	Lychnis flos-cuculi	f
C. anglicum	f	Lythrum salicaria	o
Epilobium palustre	a	Orchis ericetorum	o
Equisetum palustre	f	O. latifolia	o
E. telmateia (shaded			
places)	l		

(6) The next zone is dominated by *Molinia caerulea*, and the bryophytes are much less abundant owing to the competition of the grass. The characteristic moss is *Brachythecium purum*, one of the commonest grassland mosses in Britain. This is abundant, while *Sphagnum papillosum* and its

variety *confertum* are locally frequent. *S. plumulosum* and its variety *purpurascens*, *Hypnum cupressiforme* var. *ericetorum*, and the liverworts *Aneura multifida*, *Cephalozia connivens*, *Lophocolea bidentata* and *Calypogeia trichomanis* are occasional.

In addition to the dominant *Molinia* the following vascular plants occur in this zone:

Agrostis canina	f	Juncus effusus	a
Angelica sylvestris	f	Lotus uliginosus	o
Cirsium palustre	o	Luzula multiflora	a
C. anglicum	o	Lychnis flos-cuculi	o
Erica tetralix	o	Nardus stricta	f
Eupatorium cannabinum	o	Platanthera bifolia	l
Genista anglica	o	Rubus idaeus (thicket)	l
Juncus conglomeratus	a		

(7) In the driest parts of the *Molinia* tussocks, we have the following community:

Molinia caerulea   d

Agrostis tenuis	f	Leontodon hispidus	f
Blechnum spicant	o	Linum catharticum	o
Dryopteris dilatata	o	Luzula multiflora var.	
D. spinulosa	o	congesta	f
Galeopsis tetrahit	f	Myrica gale	ld
Galium saxatile	o	Polygala serpyllifolia	f
Holcus lanatus	f	Potentilla erecta	a
Hypericum humifusum	o	Rubus spp. (fruticose)	f
H. pulchrum	o	Succisa pratensis	f
Juncus squarrosus	o	Senecio sylvaticus	o
Leontodon nudicaulis	o	Ulex gallii	a

The only mosses are *Hypnum cupressiforme* (o) and its variety *ericetorum* (f), and *Thuidium tamariscinum* (o). The common agaric *Laccaria laccata* also occurs.

Here we have a definitely "dry land" community, though the specific heath element is only partly represented, and *Calluna* has not succeeded in obtaining a footing in the *Molinia* tussocks.

A community with a drier soil than that of the tussocks occurs on the drier edges of the area and includes several species of fruticose Rubi, but the habitat of this has undoubtedly been much altered by human activity. An adjoining "dry heath" shows a more natural late stage of the heath succession, including such characteristic heath species as *Calluna vulgaris* and *Polytrichum juniperinum*, besides *Pteridium aquilinum*, *Ulex gallii* and *U. europaeus* and many of the heath species occurring on the driest parts of the *Molinia* tussocks, in addition to species of "grass heath" and others commonly found on relatively dry siliceous soil.

Sphagnetum. A small low-lying portion of the area bordering the stream into which the channels drain is occupied by a swamp which is not definitely marked out into *Molinia* tussocks and intervening channels and furrows, as is the sloping portion described above. This swamp is regarded by Watson as the most primitive stage of the succession (apart of course from the submerged community of *Potamogeton polygonifolius* occupying the deeper channels). But as has already been remarked, it is not clear how far actual succession has occurred except through the agency of erosion between the *Molinia* tussocks. The wetter pockets of the swamp are occupied by a community of *Sphagnum recurvum* (which is not dominant between the *Molinia* tussocks). *S. recurvum* (with other species of *Sphagnum, Hypnum cuspidatum, H. stellatum, H. intermedium* and *H. giganteum*) forms a matrix in which local societies of *Hypericum elodes, Menyanthes trifoliata, Mnium affine* and *Bryum pseudotriquetrum* occur, with an abundance of *Hydrocotyle vulgaris, Galium uliginosum* and *Anagallis tenella*.

The general zonation in this area may therefore be summarised as follows, omitting alternative phases and the liverwort communities confined to the edges of the water channels between the tussocks:

Potamogetonetum polygonifolii

|

Sphagnetum recurvi

|

Sphagnetum plumulosi

|

Molinietum caeruleae

|

Molinietum with abundant invasion of *Ulex gallii, Potentilla erecta*, etc.

|

? Callunetum

Other wet heath species. Besides the species occurring in the area described above, the following are recorded from various English wet heath communities:

Andromeda polifolia	vr	Malaxis paludosa	r
Carex leporina	f	Osmunda regalis	l
C. hudsonii	l	Oxycoccus quadripetalus	vr
Erica ciliaris (Dorset and		Peplis portula	o
Cornwall)	r	Pinguicula vulgaris	o–f
Eriophorum angustifolium	l	Ranunculus flammula	f
Gentiana pneumonanthe	l	Salix repens	la
Juncus acutiflorus	f	Schoenus nigricans	l
J. bulbosus (supinus)	la	Scirpus setaceus	r
Lycopodium inundatum	r	Wahlenbergia hederacea	o
Hypnum fluitans	f	Polytrichum commune	f

REFERENCES

FARROW, E. P. On the ecology of the Vegetation of Breckland. I. General description of Breckland and its vegetation. *J. Ecol.* **3**, 211–28. 1915. V. Observations relating to competition between plants. *J. Ecol.* **5**, 155–72. 1917. VII. General effect of blowing sand upon the vegetation. *J. Ecol.* **7**, 55–64. 1919.

FRITSCH, F. E. The heath association on Hindhead Common, 1910–26. *J. Ecol.* **15**, 344–72. 1927.

FRITSCH, F. E. and PARKER, WINIFRED M. The heath association on Hindhead Common. *New Phytol.* **12,** 148–63. 1913.

FRITSCH, F. E. and SALISBURY, E. J. Further observations on the heath association on Hindhead Common. *New Phytol.* **14,** 116–38. 1915.

HAINES, F. M. A soil survey of Hindhead Common. *J. Ecol.* **14,** 33–71. 1926.

HOPKINSON, J. W. Studies on the vegetation of Nottinghamshire. I. The ecology of the Bunter Sandstone. *J. Ecol.* **15,** 130–71. 1927.

RAYNER, M. C. Obligate symbiosis in *Calluna vulgaris. Ann. Bot., Lond.,* **29.** 1915.

SKIPPER, E. G. The ecology of the gorse (*Ulex*) with special reference to the growth-forms on Hindhead Common. *J. Ecol.* **10,** 24–52. 1922.

SUMMERHAYES, V. S., COLE, L. W. and WILLIAMS, P. H. Studies on the ecology of English heaths. I. *J. Ecol.* **12,** 287–306. 1924.

SUMMERHAYES, V. S. and WILLIAMS, P. H. Studies on the ecology of English heaths. II. *J. Ecol.* **14,** 203–43. 1926.

WATSON, W. A Somerset heath and its bryophytic zonation. *New Phytol.* **14,** 80–93. 1915.

WEISS, F. E. The dispersal of fruits and seeds by ants. *New Phytol.* **7,** 27. 1908.

WEISS, F. E. The dispersal of the seeds of the Gorse and the Broom by ants. *New Phytol.* **8,** 81–9. 1909.

Chapter XXXVII

THE HEATH FORMATION (*continued*)

UPLAND HEATHS AND "HEATHER MOORS"
BILBERRY MOORS
DISTRIBUTION OF BRITISH HEATHS

SUMMARY OF PEAT COMMUNITIES

"Heath" and "Heather Moor". It has been usual to separate "heaths" from "heather moors", corresponding with the German distinction of *Heide* and *Heidemoor*, the former described as developed on sandy or gravelly soil with a minimum of peaty humus, the latter on deep acid peat, and most commonly in Britain only at higher altitudes. This distinction was drawn by Robert Smith (1900), by W. G. Smith in *Types of British Vegetation* (1911), and by Elgee (1914); and it is maintained by Watson (1932) in his account of the bryophytes and lichens of moorland. W. G. Smith separated the "upland heaths" of Cleveland and of Scotland from the "heather moors" of the Pennines, and Watson similarly separates "upland heaths" in Somerset and north Devon from "heather moor" on Dartmoor and in other more northern regions.

Adamson, however (1918), contended that the so-called heather moors of the southern Pennines are really for the most part upland heaths, "very closely allied to the heaths of the lowlands and the south". The peat is generally quite thin and the roots penetrate to the sandy layer described by Crump (1913) as the "sub-peat" (the *A* horizon of modern pedology). Adamson held that the Calluneta (and Vaccinieta) of these moors (or upland heaths) are most closely related, not to the wet peat communities (Eriophoreta, etc.), but to the acid grasslands dominated by *Nardus* and *Deschampsia flexuosa*, reversible changes between heath and grassland occurring in one direction or the other in correspondence with the increase of grazing or of peat accumulation. *Vaccinium myrtillus*, however, as Moss showed, and under some circumstances *Calluna* also, may colonise drying *Eriophorum* peat and thus produce what some would call a "true" heather moor on deep peat.

Floristic differences. With regard to the floristic differences between "heath" and "heather moor", these are mainly functions of geographical position or marked difference of altitude, and therefore ultimately of climate, and they really affect a very small proportion of the common species, as can be readily seen by consulting the extensive comparative lists given by Watson (1932, pp. 288–92). The differences shown are mainly rather slight differences in frequency of the less common species. The "heather moors" however do possess, according to Watson, a number of rarer bryophytes and lichens not occurring on "upland heaths" (see p. 746).

Of the species confined to southern lowland heaths we have *Ulex minor* (south-eastern and south-central England), *E. ciliaris*[1] (wet heaths in Dorset and Cornwall) and *Erica vagans*[2] (Lizard peninsula). All three are "southern Atlantic" species (Matthews), i.e. species centred in the southern portion of the Atlantic coastline of Europe. *Pinguicula lusitanica*, another species of the same European distribution, which may be abundant on wet heaths in south-western England, is also widely distributed in Ireland (where it ascends to 1500 ft.—*c.* 450 m.—in the west) and reaches western Scotland. *Agrostis setacea* is another south-western plant, which ranges rather farther east than the western species previously mentioned, both in England and on the continent. The peculiar Breckland species are plants of Central European affinity, and have no connexion with the Callunetum as such, though this is one of the leading plant communities of Breckland. They are indeed almost confined to the "better" types of grassland and cannot properly be considered heath plants at all (see Chapter XXVI, p. 512).

The species associated with raw humus which are confined to the north or much commoner there are more numerous, and include the following:

Undershrubs

Arctostaphylos uva-ursi, south to Derbyshire only. *Arctous alpina*, Highlands only.
Betula nana, mainly in the Scottish Highlands, not south of Northumberland.
Empetrum nigrum, abundant in the north and west, rare in the south-west.
Rubus chamaemorus, not south of Derbyshire and Wales.
Vaccinium vitis-idaea, much commoner in the north, but extending to the south-west. *V. myrtillus* is much more abundant in the north and west but by no means rare (rather local) in the south-east. *V. uliginosum* not south of Durham.

Herbs

Pyrola media, rare in the midlands of England (woods), more frequent in Scotland.
P. minor, much commoner in the north: confined to woods in the south of England. Other species of *Pyrola* (p. 451) are also less rare in Scotland.
Trientalis europaea, not south of Yorkshire, frequent on Scottish heaths as well as in woods.
Listera cordata, very rare in the south.
Saxifraga hirculus, wet moorland, rare and local in northern England and central Scotland.
Cornus suecica, not south of Yorkshire, where it is very rare: practically Arctic-alpine in Scotland where it is said to be almost always associated with *Vaccinium myrtillus*.

Of these only the undershrubs (especially *Vaccinium myrtillus*) play any considerable part in the Callunetum.

When we come to consider the actual lists of species published from particular examples of different types of heath community, we find that the most useful comparative data are those of Watson (1932). He gives a list of sixty-four angiosperms, six pteridophytes and no less than 232 bryophytes and lichens, from five "upland heaths" in Somerset, seven "heather

[1] Pl. 127, phots. 312, 313.   [2] Pl. 127, phot. 314.

moors" (mainly western) ranging from Perthshire to Dartmoor, and four lowland "wet heaths". Table XXIV shows the distribution of these 302 species in the three groups of communities.

Table XXIV

	"Upland heaths"	"Heather moors"	"Wet heaths"	Totals
Angiosperms	59	58	63	64
Pteridophytes	4	5	6	6
Mosses	70	83	72	83
Liverworts	37	46	37	48
Lichens	76	99	50	101
Totals	245	291	228	302

These figures make it clear that all three groups of communities contain the great majority of species of vascular plants recorded from any of them: of the mosses about seven-eighths occur on the "upland heaths" and the same proportion on the "wet heaths", all on the "heather moors": of the liverworts three-quarters of the whole number are recorded from "upland heaths", three-quarters from the "wet heaths" and all but two from the "heather moors": of the lichens three-quarters occur on the "upland heaths", only half on the "wet heaths", and all but two on the "heather moors".

Thus there is a very substantial community between the three types of vegetation, but the "heather moors" show a marked preponderance of species of non-vascular plants, which depend closely on damp air; and this is most conspicuous among the lichens.

If we consider the particular species which Watson found on one or more "upland heaths" but not on any "heather moor" and vice versa, we find that these are but few among the vascular plants, while the list of bryophytes and lichens occurring on "heather moors" but not on "upland heaths" is considerable. These however are almost all species occurring in one of the lower degrees of frequency ("occasional" or "rare"): the "abundant" and nearly all the "frequent" species are common to the two types.

The first of the following lists shows the very few species recorded from one or more of the upland heaths and from none of the heather moors, the second the numerous species of non-vascular plants recorded from heather moors and not from upland heaths.

Species recorded from "upland heaths", but not from "heather moors"

Vascular plants:

*Agrostis setacea	a	* Frangula alnus	r–f
*Pinguicula lusitanica	o	*Ulex minor	r

Bryophytes: none

Lichens:

Biatorina littorella	o	Usnea florida	o

* Also found on wet heaths.

Species recorded from "heather moors", but not from "upland heaths"

Vascular plants:

*Osmunda regalis	(r)†	*Comarum palustre	o
*Pinguicula vulgaris	(r–lf)		

Mosses:

*Campylopus atrovirens	o–f	Mnium serratum	r–o
*C. brevipilus	r	Oligotrichum hercynicum	f
*Dicranum spurium	r–o	Orthodontium heterocarpum	o
*Drepanocladus aduncus	(a)	Polytrichum alpinum	o
*D. sendtneri	o	*P. strictum	o
*D. uncinatus	o	Swartzia montana	o
*Hypnum imponens	r–o		

Liverworts:

*Cephalozia fluitans	r	Lophozia atlantica	o
*C. francisci	o	Marsupella funckii	r–o
C. leucantha	o	M. ustulata	o
*Leptoscyphus taylori	f	Odontoschisma denudatum	–
*L. anomalus	o	Sphenolobus minutus	o

Lichens:

Alectoria bicolor	o–f	Icmadophila ericetorum	o
A. chalyberiformis	o–f	Lecidea kochiana	r
*Biatora gelatinosa	r	Microglaena bredalbanensis	r
*Cetraria islandica	f	Nephromium parile	o
*C. var. tenuifolia	–	Peltidea apthosa	r–o
Cladonia bellidiflora	o	Polychidium muscicolum	o
C. degenerans	o	Psora demissa	o
Ephebia hispidula	o	*Pycnothelia papillaria	–
Gyrophora cylindrica	o–f	Sphaerophorus fragilis	r–o
G. polyphylla	o–f	S. melanocarpus	r–o
G. polyrhiza	o–f	Stenocybe bryophila	o
Haematomma coccineum	o	Stereocaulon condensatum	o
H. ventosum	o–f		

* Also found on wet heaths.

† The frequency letters enclosed in parentheses belong to species recorded from "non-typical habitats".

Furthermore, when we compare the frequencies with which species occurring both on "upland heaths" and "heather moors" are recorded for examples of the two groups we find that comparatively few show any markedly greater frequency on one or the other. Of the more widely distributed species *Polytrichum juniperinum* and *P. formosum* show greater abundance on the "upland heaths" while *P. commune* is more abundant on "moors". This is in accord with the preference of the last for wetter habitats.

Taking all these facts into consideration, we cannot but conclude that the *vegetational* distinction between upland heath and heather moor is very slight. What we have to deal with is a series of communities developed under a considerable range of conditions of climate and soil and thus showing minor differences in floristic composition of which the most conspicuous is the much greater number of species of bryophytes and lichens in the moister habitats, especially in the moister climates; but with all the communities essentially agreeing in the great majority of their species and in the frequencies with which they occur. We have no lists comparable with Watson's for the dry lowland south-eastern and eastern heaths, though there is no doubt that these are very much poorer in bryophytes and lichens, not only than the "heather moors" but also than the "upland heaths".

Depth of peat. There remains the criterion of depth of peat. Adamson would distinguish "upland heaths" on shallow peat in which the roots of the vascular plants penetrate into the "sub-peat" layer, from "heather moors" on deep peat in which they are confined to the peat itself. Heather moors so defined are based on peat not formed by *Calluna* but colonised by the ling as its surface layers dry out. In other words their successional development is different. We have no adequate study of this type of Callunetum, but there is no evidence that its flora and vegetation differ from those of Callunetum on shallow peat.

In what follows, therefore, no technical distinction will be drawn between heather moor and upland heath.

UPLAND HEATHS OR MOORS

(*Callunetum*)

"Upland Heaths" of Somerset. The following account is taken from Watson's paper (1932) already quoted. The chief vascular plants of the Callunetum of this south-western county are the following:

	Calluna vulgaris	d	
Agrostis setacea	a	Potentilla erecta	f
Deschampsia flexuosa	a	Pteridium aquilinum	la
Erica cinerea	f	Rumex acetosella	lf
E. tetralix	f*	Ulex gallii	a
Galium saxatile	f	Vaccinium myrtillus	f
Melampyrum pratense	f	Veronica serpyllifolia	f

* In slightly damper places, mixed with *E. cinerea*.

Cuscuta epithymum often attacks *Ulex gallii*, forming brown patches several yards square and visible from a considerable distance.

Where the *Calluna* is dense very little else grows except *Cladina* (*Cladonia*) *sylvatica*, its forma *implexa*, and *Hypogymnia* (*Parmelia*) *physodes*. The last-named is usually epiphytic, together with *Lecanora varia*, on the stems of the ling. Where the Callunetum is a little more open any of the following mosses may become abundant:

Brachythecium purum	Hypnum cupressiforme var.
Campylopus flexuosus	ericetorum
Dicranum scoparium	Webera nutans
Hylocomium splendens	

General flora. The following is a composite list (Watson, 1932) of species of the upland heaths (taken in the widest sense), from five regions of hill heathland in Somerset (Brendon, Blackdown, Mendip, Quantock and Exmoor) of an average altitude of 800–1000 ft. (*c.* 250–300 m.), the highest parts reaching 1300–1500 ft. (400–450 m.) and more (Watson, 1932). The underlying rocks are siliceous and belong to various geological systems. Near the summits of the higher plateaux the heath may give place to the moss (blanket bog) formation (see Chapter xxxv).

Agrostis canina	o–f	L. pilosa	f–a
A. setacea	a	Melampyrum pratense	r–f
A. tenuis	a	Molinia caerulea	la
Aira caryophyllea	o	Myrica gale	o
A. praecox	r–f	Nardus stricta	lf–la
Anthoxanthum odoratum	o–f	Pedicularis sylvatica	f
Betula pendula (alba)	o	Pinguicula lusitanica	o
B. pubescens	o–f	Polygala serpyllifolia	f
Calluna vulgaris	a–ld	Potentilla erecta	f
Carex binervis	o–f	Radiola linoides	r
Deschampsia flexuosa	f–a	Rumex acetosella	o–f
Empetrum nigrum	r–o	Sarothamnus scoparius	r
Erica cinerea	f	Scirpus caespitosus	o–la
E. tetralix	r–o	Sieglingia decumbens	o
Festuca ovina	a	Ulex europaeus	f
Frangula alnus	r–f	U. gallii	f–a
Galium saxatile	f	U. minor	r
Genista anglica	r	Vaccinium myrtillus	o–lf
Juncus squarrosus	o–f	V. vitis-idaea	r
Linum catharticum	o–f	Veronica serpyllifolia	f
Luzula multiflora	o–f		

Pteridophytes

Blechnum spicant	o	L. selago	r
Lycopodium clavatum	r–o	Pteridium aquilinum	lf–la

Mosses

Aulacomnium androgynum	o	Climacium dendroides	o
Bartramia pomiformis	o–f	Dicranella heteromalla	f–a
Brachythecium purum	f–a	D. cerviculata	o
B. rutabulum	f	Dicranum bonjeani	o
B. velutinum	o–f	D. scoparium	f–a
Bryum atropurpureum	o	Ditrichum homomallum	o
B. capillare	o	Eurhynchium myosuroides	f
B. erythrocarpum	o	E. praelongum	o–f
B. inclinatum	o	Funaria hygrometrica	o
B. pallens	o	Hylocomium loreum	o
B. pendulum	–	H. squarrosum	f–la
B. roseum	r–o	H. splendens	a
Campylopus flexuosus	a–f	H. triquetrum	o–f
C. fragilis	r–o	Hypnum cupressiforme	o
C. pyriformis	f–a	var. ericetorum	a–d
C. subulatus	o	H. schreberi	o–a
Catharinea undulata	f	Leptodontium flexifolium	r–o
Ceratodon purpureus	a	Mnium affine	o

M. cuspidatum	o	P. juniperinum	a
M. hornum	f	P. nanum	r
M. undulatum	f–a	P. piliferum	f–a
Plagiothecium denticu-		P. urnigerum	o
latum	o	Rhacomitrium canescens	o–f
P. elegans	o–f	R. fasciculare	r–o
P. sylvaticum	r–o	R. heterostichum	o
P. undulatum	o	R. lanuginosum	o
Pleuridium subulatum	o	Tetraphis pellucida	o
Polytrichum aloides	o–f	Thuidium tamariscinum	o–f
P. formosum	a	Webera nutans	f–a
P. gracile	o		

Liverworts

Alicularia geoscypha	r	L. setacea	r
Alicularia scalaris	o–a	Lophocolea bidentata	o
Aplozia crenulata	o	Lophozia bicrenata	r
Calypogeia arguta	o	L. attenuata	r
C. fissa	o	L. excisa	o
C. trichomanis	o–f	L. floerkii	r
Cephalozia bicuspidata	o–f	L. ventricosa	o
C. connivens	r	Plagiochila asplenioides	o
C. media	r–o	Ptilidium ciliare	o
Cephaloziella bifida	o	Scapania compacta	o
C. starkii	o–f	S. curta	o
Diplophyllum albicans	f	S. umbrosa	o
Eucalyx hyalinus	r–o	Sphenolobus exsectiformis	r
Frullania tamarisci	o–f		
Lepidozia reptans	o–f	Marsupella emarginata	r–f

Lichens

Acarospora fuscata	o	C. furcata	f
Bacidia umbrina	o	C. gracilis	r–o
Baeomyces roseus	o–f	C. foliacea	o
B. rufus	o–f	C. macilenta	o
Biatora coarctata	o–f	C. ochrochlora	o
B. granulosa	f–a	C. pityrea	o
B. uliginosa	f–a	C. pyxidata	f–a
Biatorina littorella	o	var. chlorophaea	f–a
Bilimbia lignaria	o	C. rangiformis	o–f
B. melaena	o	C. squamosa	o–f
B. sabuletorum	o	C. subcervicornis	f
Buellia myriocarpa	o	C. subsquamosa	r–o
Candelariella vitellina	o	Cetraria aculeata	o–f
Catillaria chalybeia	o	Coniocybe furfuracea	r
Cladina rangiferina	r	Coriscium viride	r
C. sylvatica	a	Crocynia lanuginosa	r–o
C. uncialis	f	Ephebe lanata	o
Cladonia bacillaris	r	Hypogymnia (Parmelia)	o–f
C. caespiticia	r	physodes	
C. coccifera	f	Lecanora polytropa	o–f
C. crispata	o	L. varia	o
C. fimbriata	f	Lecidea contigua	a
C. flabelliformis	o	L. crustulata	o
C. floerkeana	f–a	L. dicksonii	o

Lecidea expansa	o	P. polydactyla	o
L. lithophila	r	P. rufescens	f
L. lygea	r	Pertusaria dealbata	o–f
L. rivulosa	o–a	Physcia hispida	o
L. sorediza	f–a	Porina chlorotica	o
L. sylvicola	o	Rhizocarpon confervoides	o–f
Leptogium lacerum	o–f	R. geographicum	o
L. sinuatum	o–f	R. petraeum	o
L. microscopicum	o	Sphaerophorus globosus	o
Microglaena nuda	o	Stereocaulon coralloides	r
Parmelia fuliginosa	o	S. denudatum	o
P. omphalodes	o	Usnea florida	o
P. saxatilis	f	Verrucaria maculiformis	o
Peltigera canina	o	V. mutabilis	o

Alga

Botrydina vulgaris   f

The Cleveland heaths. Elgee (1914) has described the extensive Calluneta, always called "moors", of the elevated Cleveland district of northeast Yorkshire. These are developed on sandstones, coarse grits and sandy shales of Inferior Oolite age, rising to over 1400 ft. (*c.* 430 m.) on the highest parts of the plateaux, and almost completely destitute of overlying glacial drift except in the valleys. Callunetum "covers by far the widest areas on the high ridges and plateaux between the dales" but passes into moss with a mixture of *Calluna* and *Eriophorum* on the highest tracts of the watershed.

The soil of the Callunetum may be sandy and stony with a minimum of peaty humus ("thin moor"), or the peat may be as much as 2–4 ft., i.e. up to a metre or more in thickness ("fat moor"). The Callunetum attains its greatest luxuriance on peat at least 6 in. (15 cm.) deep, and frequently more, towards, but not on, the central watershed. This would be included as "fat moor", and here very few other species occur among the closely growing vigorous ling.

On the "thin moor", where *Calluna*, though dominant, is not so luxuriant, the following are the most characteristic associates:

Empetrum nigrum	Potentilla erecta
Erica cinerea	Vaccinium myrtillus
Hypnum cupressiforme var. ericetorum	Cladonia spp.

As the ling becomes old and straggly, leaving a space in the centre of the clump, this is colonised by *Hypnum* and *Cladonia*, by *Juncus squarrosus*, or by other species. On the wetter edges of "thin moor" *Nardus, Erica tetralix, Juncus squarrosus*, and *Cladonia* are interspersed with the *Calluna*, and on the slopes the following grasses become frequent:

Agrostis canina	Festuca ovina
Aira praecox	Sieglingia decumbens
Deschampsia flexuosa	

Molinia also occurs freely among the heather where the soil is damp, both on deep peat and where there is a thin layer of glacial drift, usually sand and gravel.

Burn subsere. The Cleveland moors, like all grouse moors, are regularly fired at intervals of several years. The first plants to appear on the burned areas ("swiddens") are lichens, liverworts and mosses: *Cladonia* spp., *Lophozia inflata*, *Sphagnum papillosum* var. *confertum*, *Webera nutans*, and at a later stage *Ceratodon purpureus* and *Polytrichum commune*. *Calluna* regenerates by sprouting unless the fire has been severe enough to kill the whole plant, when the restocking is from seed. The other species prominent during the later stages of the burn subsere, before *Calluna* again becomes dominant, are *Agrostis canina*, *Aira praecox*, and *Festuca ovina*, which may form local swards; *Rumex acetosella* interspersed among *Empetrum*; *Erica cinerea*, which quickly becomes dominant on dry areas; *Vaccinium myrtillus* which often occupies recently burned areas, sometimes to the exclusion of *Calluna*; *Juncus squarrosus*, *Potentilla erecta*, etc.

The mixture of *Calluna* and *Eriophorum* on the deepest peat of the highest parts of the plateau, which is a transitional type between heath and blanket bog, is briefly described on p. 703.

Scottish heaths. No recent intensive work, either floristic or ecological, has been done on the heaths of Scotland, and the short account contributed by the late W. G. Smith to *Types of British Vegetation* (pp. 111–16) is still the best available.

W. G. Smith drew a distinction based on "habitat" between "*Calluna* heath" and "*Calluna* moor", but it is very doubtful, as we have seen, if any good vegetational distinction can be made. The former, he says, is developed over sandy or gravelly soil with a surface layer of dry peat, which may in extreme cases reach a depth of 12 in. (30 cm.), as on Hindhead in the south of England (p. 725), but is usually much shallower. These heaths are mainly met with on the eastern side of northern Britain and may occur at any altitude from sea level to about 2000 ft. (*c*. 600 m.). They are generally known as "moors" in common with all other open tracts of country (apart of course from broken rocky surfaces and definitely good grassland used primarily as hill pasture), having an acid peaty humous soil, whether or not deep peat is present and whatever the vegetation. The northern *Calluna* heaths of Smith are however typical "grouse moors" used for the preservation and shooting of the grouse (*Lagopus scoticus*) so far as they do not form part of "deer forests". For this purpose they are regularly burned over every 10 or 15 years to destroy the old "leggy" *Calluna* and make way for new thick growth which is best for feeding and sheltering the grouse. A few sheep may be pastured on some of them, but there is little other human interference. Typical areas occur, as we have seen, on the Cleveland Hills in north-east Yorkshire and there are many in the river basins of the Tay, Dee and Spey in the eastern Highlands. In the western

Highlands and in the southern Uplands of Scotland Callunetum, though developed locally, is much less extensive, and the same is true on the whole of the western sides of England, Wales and Ireland.

Flora. The following generalised list of the more characteristic species from four stations near Blair Atholl in the Tay valley, Perthshire (R. Smith), is taken from W. G. Smith's account in *Types of British Vegetation* (pp. 115–16):

Calluna vulgaris   d

Locally subdominant:

Arctostaphylos uva-ursi	Vaccinium myrtillus
Empetrum nigrum	V. vitis-idaea
Erica tetralix	

Locally abundant:

Erica cinerea	Hypnum spp.
Nardus stricta	

Frequent:

Agrostis tenuis	Festuca ovina
Antennaria dioica	Galium saxatile
Anthoxanthum odoratum	Luzula multiflora
Blechnum spicant	Lycopodium clavatum
Carex dioica	Polygala vulgaris
C. goodenowii	Potentilla erecta
Deschampsia flexuosa	
	Cladonia spp.

Sparse or local:

Genista anglica	Juniperus communis
Melampyrum pratense var.	Trientalis europaea
montanum	

From this list it will be seen that several species are more or less commonly found which do not occur (or occur very rarely) on the southern English heaths, e.g. *Arctostaphylos uva-ursi, Empetrum nigrum, Vaccinium vitis-idaea, Trientalis europaea*; and to these may be added *Listera cordata, Cornus suecica* (rare), *Pyrola* spp. (cf. p. 744).

Burn subsere. No detailed studies of succession have been made on the northern heaths, but it has been noticed that during the first year or two after burning *Cladonia* spp. often form almost the only vegetation. *Erica cinerea* recovers quickly on dry ground, *Vaccinium myrtillus* and *V. vitis-idaea* on steep slopes, *Arctostaphylos uva-ursi* at altitudes towards 2000 ft. (600 m.), the grasses *Agrostis* spp., *Anthoxanthum, Deschampsia flexuosa* and *Festuca ovina* on sandy humus, *Nardus stricta* on moister humous soils. Though the ling typically becomes dominant after a few years, many of these species may maintain themselves in the Callunetum.

On some of the east Scottish coastal sand dunes, as elsewhere in Great Britain (Pl. 126, phot. 309), Callunetum develops, but the successional stages have not been followed in any detail (cf. Chapter XLI, pp. 860–1).

"Heather Moors" of the Pennines. These were originally described from west Yorkshire by Smith and Moss (1903), and by Smith and Rankin (1903), and from the borders of Westmorland and Yorkshire by Lewis (1904). Their accounts were summarised by Lewis and by Moss in *Types of British Vegetation* (1911), and Moss (1913) gave a fuller account from Derbyshire at the southern end of the Pennine range. Crump (1913) studied the water relation of the dominant *Calluna* to the soil on which it grows, and Watson (1932) gives extensive floristic lists. Reasons have already been given (pp. 741–7) for refusing to separate these "heather moors" from "upland heaths", but since they are among the best studied examples a short account of them is given here.

The Pennine range, extending some 140 miles (220 km.) from Northumberland southwards to central Derbyshire, and often called the backbone of northern England, bears some of the most extensive and best known of the English heather moors. In the southern Pennines heather moor may occur as low as 750 ft. (*c.* 250 m.) above the sea, but it usually begins about 1000 ft. (*c.* 300 m.), extending upwards to about 1500 ft. (*c.* 450 m.), or in the northern Pennines and the Wicklow Mountains of eastern Ireland, where the hills are higher, to about 2000 ft. (*c.* 600 m.). Above this level, everywhere in the British Isles, *Calluna* begins to lose its dominance under any conditions of topography and soil, and at about this level the extensive glaciated plateaux bearing the blanket bog or moss formation dealt with in Chapter xxxiv generally occur. Callunetum is the lowest "moorland" zone and occupies the lower, often steeper, slopes of the hill masses (see Figs. 89, 91, 140). In the southern Pennines and in Scotland too, its lower limit often coincides with the upper limit of enclosed grassland; and very often Callunetum adjoins and passes into the bent-fescue grassland dealt with in Chapter xxvi. As Adamson has shown (1918) Callunetum and this grassland have a reciprocal relation, the former passing into the latter under grazing, the latter into the former with cessation of grazing and accumulation of peat.

Water and humus content of soil. According to Crump (1913) and Adamson (1918) the peat of the typical Pennine heather moor "is as a rule so shallow that the heather roots regularly pass through it into the underlying coarse sandy soil that may be conveniently called the 'subpeat'" (Crump, 1913, p. 138). Adamson says that the peat is "rarely over a foot thick and usually only a few inches, and is always much mixed with sand and mineral matter" (1918, p. 100). The roots of the plants "penetrate freely beyond the peat proper into the sandy subpeat", which is "much darkened and discoloured by peat". Crump determined the "coefficient of soil-humidity", i.e. the ratio of water to humus ($W/H = m$), for a great number of peats and subpeats whose actual water and humus contents differed very widely, and found it to vary only between 1·7 and 2·8, with a mean of 2·3, for the peats and between 2·5 and 3·85 with a mean of 3·17 for the subpeats. By deducting the residual water held by the mineral particles of the subpeats he obtained the same mean coefficient ($m = 2\cdot32$) for both. The water content of the soil of the Callunetum is usually a function of its humus content, since clay and silt fractions are generally negligible.

Chemical nature of the peat. According to Pearsall (1938) all samples from the Yorkshire (Pennine) and Lake District "heather moors" "show remarkable similarity, with a pH value near to 3·5, and they are oxidising soils...which are markedly deficient in bases". The extreme range of ten samples from different localities was only 3·41 to 3·63. Nitrates were never present. In "heaths" with quite shallow peat over Triassic sands the pH value was rather more variable but still between 3·2 and 4·0. Surface samples were more acid and had a higher oxidation potential than deeper ones. On a burned and drained heath where *Calluna* had become completely dominant to the exclusion of *Eriophorum* and *Erica tetralix*, formerly present, the pH value was as low as 2·98.[1] These Callunetum soils are very similar to woodland soils on which *Deschampsia flexuosa* is dominant.

Floristic composition. The following list includes the principal species found in the Pennine heather moors (Lewis, 1904; Moss, 1913; Watson 1932):

Agrostis tenuis	f	Lycopodium clavatum	r
Anthoxanthum odoratum	o	L. selago	r
Blechnum spicant	o–a	Melampyrum pratense	o
Carex binervis	o	Molinia caerulea	lf
Deschampsia flexuosa	f	Nardus stricta	o
Empetrum nigrum	f	Polygala serpyllifolia	o
Erica cinerea	l	Potentilla erecta	o
E. tetralix	lf	Pteridium aquilinum	la
Eriophorum vaginatum	lf	Pyrola media	vr
Festuca ovina	f	Rumex acetosella	la
Galium saxatile	f	Salix repens	vr
Genista anglica	r, l	Scirpus caespitosus	o
Juncus bufonius	f	S. pauciflorus	o
J. inflexus	f	Trientalis europaea	vr
J. squarrosus	a	Ulex gallii	la
Lathyrus montanus	o	Vaccinium myrtillus	lsd
Linum catharticum	o	V. vitis-idaea	o
Listera cordata	vr	Veronica serpyllifolia	o
Luzula multiflora	o		

This list of vascular plants includes several species not forming part of the Callunetum, which, especially when typical and closed, is very poor floristically, as in other regions. Thus *Pteridium* is an invader of the edges of the moor from sheltered valleys, *Eriophorum, Scirpus* and *Erica tetralix* belong to bog and are confined to damp or wet places, and *Nardus* and *Molinia* belong to the acid grasslands whose relations to the Callunetum has been discussed on pp. 130 and 500.

[1] Cf. Haines' very low Hindhead values (p. 726).

Watson (1932) gives the following list of bryophytes and lichens from the heather moors of the southern Pennines:

Mosses

Aulacomnium androgynum	–	Hypnum stramineum	o
Bartramia pomiformis	–	Hylocomium loreum	o
Brachythecium purum	f	H. splendens	o
B. rutabulum	f	H. squarrosum	a
B. velutinum	o	H. triquetrum	o
Bryum atropurpureum	–	Leptodontium flexifolium	–
B. capillare	o	Leucobryum glaucum	o
B. erythrocarpum	o	Mnium cuspidatum	o
B. inclinatum	o	M. hornum	f
B. pallens	o	M. serratum	r
B. pendulum	o	M. undulatum	f
B. roseum	r	Oligotrichum hercynicum	f
Campylopus atrovirens	o	Orthodontium	
C. brevipilus	–	heterocarpum	o
C. flexuosus	a	Plagiothecium denticulatum	o
C. fragilis	o	P. elegans	f
C. pyriformis	a	P. sylvaticum	o
Catharinea undulata	f	P. undulatum	o
Ceratodon purpureus	a	Pleuridium subulatum	o
Climacium dendroides	o	Polytrichum aloides	o
Dicranella cerviculata	o	P. alpinum	o
D. heteromalla	o	P. formosum	f
Dicranum bonjeani	–	P. gracile	o
D. scoparium	f	P. juniperinum	f
D. spurium	–	P. piliferum	–
Ditrichum homomallum	–	P. strictum	o
Drepanocladus sendtneri	o	P. urnigerum	o
D. uncinatus	o	Rhacomitrium canescens	–
Eurhynchium myosuroides	o	R. fasciculare	–
E. praelongum	o	R. heterostichum	o
Funaria hygrometrica	o	R. lanuginosum	o
Hypnum cupressiforme	f	Swartzia montana	o
H. cuspidatum	o	Tetraphis pellucida	f
H. var. ericetorum	a	Thuidium tamariscinum	f
H. schreberi	o	Webera nutans	a

Liverworts

Alicularia scalaris	f	Lophocolea bidentata	f
Aplozia crenulata	o	Lophozia atlantica	o
Calypogeia arguta	o	L. attenuata	o
C. fissa	o	L. bicrenata	o
C. trichomanis	f	L. excisa	–
Cephalozia bicuspidata	f	L. floerkii	f
C. connivens	f	L. ventricosa	o
C. media	r	Odontoschisma denudatum	–
Cephaloziella bifida	a	O. sphagni	–
C. starkii	o	Plagiochila asplenioides	o
Diplophyllum albicans	f	Ptilidium ciliare	–
Eucalyx hyalinus	o	Scapania compacta	–
Frullania tamarisci	o	S. curta	o
Lepidozia reptans	f	S. umbrosa	o
L. setacea	o	Sphenolobus minutus	o
Leptoscyphus taylori	f		

Lichens

Acarospora fuscata	–	Coniocybe furfuracea	r
Bacidia umbrina	–	Coriscium viride	r
Baeomyces rufus	o	Gyrophora cylindrica	–
Biatora coarctata	–	G. polyphylla	–
B. granulosa	f	Hypogymnia physodes	o
B. uliginosa	–	Icmadophila ericetorum	–
Bilimbia lignaria	–	Lecanora polytropa	o
B. sabuletorum	o	Lecidia contigua	f
Botrydina vulgaris	o	L. dicksonii	–
Buellia myriocarpa	–	Leptogium lacerum	–
Candelariella vitellina	o	L. sinuatum	–
Cetraria islandica	o	Parmelia fuliginosa	–
Cladina sylvatica	f	P. omphalodes	o
C. uncialis	o	P. saxatilis	f
Cladonia bacillaris	o	Peltidea aphthosa	–
C. caespiticia	–	Peltigera canina	o
C. coccifera	f	P. polydactyla	–
C. degenerans	o	P. rufescens	o
C. fimbriata	o	Pertusaria dealbata	–
C. flabelliformis	o	Physcia hispida	o
C. floerkeana	f	Porina chlorotica	r
C. furcata	o	Rhizocarpon confervoides	–
C. gracilis	f	R. geographicum	o
C. macilenta	o	Sphaerophorus fragilis	r
C. pyxidata	f	S. globosus	r
var. chlorophaea	f	Stereocaulon condensatum	–
C. rangiformis	o	S. coralloides	–
C. squamosa	o	S. denudatum	–
C. subcervicornis	f		

Firing of heather moor. The heather moors of the Pennines, like those of the Cleveland district and of Scotland, are commonly used as grouse moors and are systematically fired every few years. This prevents the effective invasion of trees such as birch and rowan which would occur at the lower altitudes. But at the higher levels the moors are too windswept for good tree growth to occur and the heather moor here represents a climax community. The periodic burning of the heather has the effect of constantly rejuvenating it. According to Moss (1913, p. 178), when it is fired every four years *Calluna* does not grow more than ankle-high, but if eight or ten years elapse before the next burning it becomes knee-deep. In fifteen years or so the plant becomes definitely aged and "leggy" and its flowering capacity is greatly diminished. Twenty or twenty-five years is probably the span of its natural existence, but there are no available observations on the rejuvenation of the community when the bushes are left to complete their natural life.

Burn subsere. After burning, *Rumex acetosella*, *Deschampsia flexuosa* or *Nardus stricta* may first recolonise the burned peat surface. Where present in quantity in the burned community *Vaccinium myrtillus* often asserts itself by sprouting from its underground rhizomes before the heather returns. But the ling commonly sows itself abundantly from seed and soon reasserts its dominance, the bilberry remaining as a lower layer which does not flower below the *Calluna* canopy.

Heather moors of the Wicklow Mountains. From this small mountain complex lying to the south of Dublin, Pethybridge and Praeger (1905, pp. 158–62) describe the Callunetum as occupying the zone between the siliceous grassland of the lower slopes (here mainly dominated by *Ulex gallii*) and the Scirpetum of the summit plateaux, which covers most of the ground above 1500 ft. or 450 m. (see Fig. 89). They emphasise the fact that Callunetum depends on good drainage, and that the thick peat which it often occupies has been formed, not by the ling itself, but by some more hydrophytic plant such as *Sphagnum* or *Scirpus caespitosus*. Where drainage has been increased by turf-cutting *Calluna* has colonised the drying surfaces.

Where the drainage is best, particularly near the lower limits of the heather moor, *Calluna* makes a dense uniform growth 2–3 ft. (*c.* 60–90 cm.) in height, with *Listera cordata*, *Melampyrum pratense*, and a continuous undergrowth of mosses. On its upper limit there is a wide ecotone of *Calluna-Scirpus*, passing above into the pure Scirpetum (Chapter xxxv, pp. 711–13).

The Calluneta of the Wicklow Mountains, like those of Great Britain, are regularly fired for the sake of the grouse, and this burning, together with the strong winds, inhibits tree growth on the moors. The effect of wind may be seen in the few trees of *Sorbus aucuparia*, whose crowns are kept well below the edges of the gullies they inhabit.

Nine samples of Callunetum were examined by Pethybridge and Praeger, and the following species recorded. The number of samples in which a species occurred follows its name:

Calluna vulgaris		d (9)	
Agrostis tenuis	4	Molinia caerulea	2
Anthoxanthum odoratum	1	Nardus stricta	8
Blechnum spicant	4	Narthecium ossifragum	1
Carex binervis	7	Oxalis acetosella	1
C. diversicolor (flacca)	3	Pedicularis sylvatica	1
C. pilulifera	2	Polygala serpyllifolia	1
Deschampsia flexuosa	2	Potentilla erecta	6
Dryopteris dilatata	2	Pteridium aquilinum	2
Empetrum nigrum	3	Scirpus caespitosus	5
Erica cinerea	6	Sieglingia decumbens	1
E. tetralix	1	Sorbus aucuparia	1
Eriophorum angustifolium	1	Succisa pratensis	1
E. vaginatum	2	Ulex gallii	3
Festuca ovina	4	Vaccinium myrtillus	8
Galium saxatile	6		
Juncus effusus	1	Polytrichum sp.	1
J. squarrosus	8	Rhacomitrium lanuginosum	2
Luzula campestris	1	Sphagnum spp.	5
L. sylvatica	5	Cladonia spp.	3
Melampyrum pratense	1		

No fuller list of the bryophytes and lichens is available.

The Mourne Mountains. Armstrong and his co-workers (1930, 1934) have briefly described the vegetation of the Mourne Mountains in north-eastern Ireland in which Callunetum occupies the most extensive areas (Fig. 149), and attains unusual altitudes. According to Armstrong's map (Fig. 149) it covers the summits of 2300 ft. though not those of 2400 ft. and over, which are occupied by alpine or "summit" grassland dominated by *Festuca ovina* and *Rhacomitrium lanuginosum*. But on the southern slope of the highest peak, Slieve Donard (2796 ft. or 850 m.), Callunetum is actually shown (Fig. 150) reaching in one place a height of 2710 ft. or 826 m. This highest Callunetum forms a low dense mat 5–8 cm. high and with very few other species.[1] The principal associates of *Calluna* at the higher levels are *Erica cinerea* and *Carex pilulifera*, and "on certain sections of the mountain side *Erica cinerea* is definitely the dominant plant". The basins and cols between the summits are occupied by extensive tracts of Molinietum and Scirpetum caespitosi (Fig. 149).

Bilberry moor (*Vaccinietum myrtilli*)

Vaccinium myrtillus (Bilberry, Whortleberry, Wimberry, or, in Lowland Scots, Blaeberry—compare the American "Blueberry") is a local associate of *Calluna* on the southern English heaths, but is there confined to the somewhat moister areas, being absent altogether from many of the drier heaths. In the west and north it is far more abundant, and becomes a constant constituent of the Callunetum, often preceding the ling in succession and frequently forming a subordinate layer of the vegetation. As we go north, in Scandinavia, and as we ascend the higher mountains in Scotland it replaces *Calluna* as a dominant species, and is evidently able to withstand more rigorous climatic conditions.

Highland Vaccinieta. Robert Smith (1900) was the first to describe the Vaccinietum in the Highlands of Perthshire. On the mountains composed of rocks which produce the "poorer" soils, and on which raw humus or acid peat tends to accumulate, *Calluna* is commonly dominant up to about 2000 ft. (*c.* 600 m.) and tends to be replaced by *Vaccinium* above that altitude, the Vaccinietum, either nearly pure or still accompanied by *Calluna*, occupying the mountain slopes up to about 3000 ft. (*c.* 900 m.). Robert Smith gives a list of thirty-five vascular plants found on the "heather moors" at lower altitudes, occurring also in this high-level Calluno-Vaccinietum or pure Vaccinietum, and ascending to altitudes varying from 2350 to 3600 ft. (*c.* 715–1100 m.). The following species are almost confined to this high-level community:

Betula nana	Lycopodium annotinum
Cornus suecica	Rubus chamaemorus
Loiseleuria procumbens	Vaccinium uliginosum

[1] Similar "alpine mats" of Callunetum have been described from high altitudes in other regions, e.g. by Crampton from Caithness and Sutherland in northern Scotland.

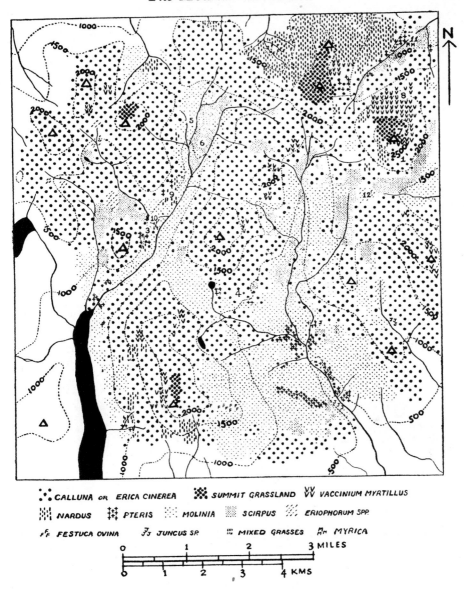

Fig. 149. Vegetation of the Mourne Mountains (N.E. Ireland)

Callunetum occupies about half the area, covering the lower summits up to between 2300 and 2400 ft. and ascending above 2500 ft. on Slieve Donard in the north-west corner of the map. Vaccinietum occurs "on the steep faces of old boulder screes". The higher summits are covered with grassland dominated by *Rhacomitrium lanuginosum* and *Festuca ovina*. The stream valleys and flatter cols are occupied by Molinietum and Scirpetum caespitosi. Pteridietum is confined to sheltered valleys, and Eriophoretum and Nardetum are quite local. From Armstrong, Ingold and Vear, 1934.

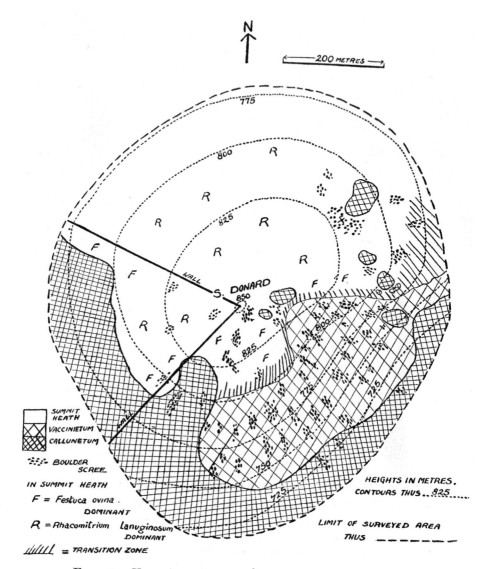

FIG. 150. VEGETATION OF THE SUMMIT OF SLIEVE DONARD,
MOURNE MOUNTAINS (IRELAND)

The actual summit (2800 ft.) and down to the 2600 ft. (800 m.) contour on the northern side
is covered with "*Rhacomitrium* heath" (see pp. 787–8), with *Festuca ovina* very abundant,
Empetrum nigrum abundant and *Salix herbacea* frequent. This is really an "arctic-alpine"
community (see Chapter XXXVIII). On the southern side *Festuca* is dominant. The boulder
scree to the south-east is occupied by Vaccinietum myrtilli, and below is Callunetum which
covers most of the high ground of this mountain group. Contours in metres. From Armstrong,
Calvert and Ingold, 1930.

All of these have a claim to be considered "arctic-alpine" species, and the first three definitely belong to this category (compare Chapter XXXVIII, pp. 780–1).

Pennine Vaccinieta. Smith and Moss (1903) and Smith and Rankin (1903) recognised substantially the same displacement of *Calluna* by *Vaccinium myrtillus* at the higher levels in the southern Pennines of west Yorkshire, where Vaccinietum occupies the highest summits of the much

"Vaccinium summits" and "edges"

lower hills (1400–1700 ft., *c.* 420–510 m.). These rocky boulder-strewn "Vaccinium summits" are composed of rocks with peaty humus between them, and the dominance of the bilberry is apparently determined by the better drainage and the greater exposure as compared with the peat-covered heather moor of the slopes below. The flora includes a small selection of the species of the Callunetum and lacks the characteristic arctic-alpine species of the high-level Scottish Vaccinietum.

Similar Vaccinietum occupies the very steep rocky escarpment slopes falling from the cotton-grass mosses of many of the Pennine plateaux (Fig. 140), but here the flora is somewhat richer, doubtless because of the greater shelter. Moss (1913, p. 182) gives the following list of vascular plants from such a "*Vaccinium* edge":

Vaccinium myrtillus			d
Arctostaphylos uva-ursi	la	Galium saxatile	la
Calluna vulgaris	lsd	Pteridium aquilinum	la
Deschampsia flexuosa	la	Rumex acetosella	o
Empetrum nigrum	la	Vaccinium vitis-idaea	lsd
Erica cinerea	la		

So good is the shelter that not only the bracken but even occasional individuals of hawthorn, rowan, birch and sessile oak have succeeded in establishing themselves.

Two rather sandy peats from "*Vaccinium* edges" at altitudes of 450 and 500 m. respectively showed pH values of 3·26 and 3·19. The soils are highly oxidising and nitrates are absent (Pearsall, 1938).

Northern Pennines. From the northern Pennines Lewis (1904) described, at about 2000 ft. (*c.* 600 m.), a "rocky slope with shallow peat in which *Calluna* tends to disappear, *Rubus chamaemorus* and *Eriophorum* are altogether absent, and the vegetation is made up of *Vaccinium myrtillus* and *V. vitis-idaea*, with *Juncus squarrosus* and a small quantity of *Nardus*". This seems to be essentially a "*Vaccinium* edge".

Mourne Vaccinieta. Armstrong, Ingold and Vear (1934) record Vaccinietum from the Mourne Mountains on steep boulder scree at 1500–2500 ft. also with *Blechnum, Deschampsia* and *Galium* and in other places *Calluna* and *Empetrum*, as well as *Anthoxanthum, Festuca ovina*, etc. (Figs. 149, 150).

Wicklow Vaccinieta. Pethybridge and Praeger (1905, p. 162) describe quite similar Vaccinieta "on well-drained rocky slopes" in the Wicklow Mountains, varying greatly in elevation "from a few hundred feet on Bray Head to over 2000 ft. on the hills". The vegetation of one of these "*Vaccinium* edges" (1600–1800 ft., c. 480–550 m.) contains a similar, though somewhat different, selection of species of the Callunetum (four species common to the two lists, excluding the dominant):

Vaccinium myrtillus   d

Blechnum spicant	Luzula sylvatica
Calluna vulgaris	Melampyrum pratense
Deschampsia flexuosa	Oxalis acetosella
Galium saxatile	Vaccinium vitis-idaea

Vaccinietum succeeding "Moss". On the plateau of "the Peak" in Derbyshire (southern Pennines), and on similar areas a little to the north (Fig. 140), Moss described the gradual conversion of an extensive moss, originally dominated by *Eriophorum vaginatum*, into a Vaccinietum, consequent on the gradual drainage and drying of the wet cotton-grass peat by the cutting back of streams into the edges of the Eriophoretum. This is described in Chapter XXXIV (pp. 705–6).

Mixture of *Vaccinium* with *Calluna* and *Eriophorum*. Very extensive mixed communities in which bilberry is associated with heather and cotton-grass are described by Lewis from the northern Pennines in a region where extensive peat erosion is proceeding. While it is not clear that these are exactly comparable with the Vaccinietum described by Moss on the Derbyshire Peak, it seems likely that here also *Calluna* and *Vaccinium* are colonising drying *Eriophorum* bogs, which still exist on these high-lying slopes and plateaux in immediate proximity to the mixed communities described. It is to be noted that while *Vaccinium myrtillus* is often mixed with *Calluna*, and both of these with *Eriophorum*, the bilberry does not seem in this region to appear as a co-dominant with the cotton-grass alone. This suggests that the relations of the three plants to drainage are in sequence—*Eriophorum, Calluna, Vaccinium*—a conclusion borne out by their general moorland habitats—so that a drying moss will tend to be colonised first by heather and then by bilberry. In the areas described by Moss it may be conjectured that drainage of the *Eriophorum* peat is so rapid that the bilberry is able to enter at once. The facts described by Adamson (1918) harmonise well with this conclusion. But there may well be other factors, not yet clear, connected with the autecology of the three species and helping to determine their distribution.

The peat bearing mixtures of *Calluna* and *Eriophorum vaginatum* shows somewhat higher pH values than that of pure Calluneta but they are below 4·0. The oxidation-reduction potential fluctuates about the critical range (Pearsall, 1938).

DISTRIBUTION OF THE HEATH FORMATION IN THE BRITISH ISLES

Though *Calluna vulgaris* is a ubiquitous species, being recorded from every vice-county in Great Britain and from every county in Ireland, Callunetum is not extensively developed on the western side of Ireland, Wales and Scotland, being confined to well-drained ground on poor soil which is not heavily pastured, where wind or frequent burning prevents the development of trees, and to acid peat which is drying as a result of natural or artificial drainage. This comparative scarcity of Callunetum in the west is probably due to the very wet climate, which favours the development of the bog or moss formation on all ground which is not exceptionally well drained.

Upland heaths. On the eastern side of Ireland and Wales, for example on the Wicklow Mountains, the Mourne Mountains, and on the Welsh Marches; and of Great Britain, as in the Eastern Highlands and on the Cleveland moors, upland heath dominated by *Calluna* is far more extensive, both on steep and on gentle slopes, and even on plateaux where the soil is permeable. These Calluneta are the typical grouse moors.

Lowland heaths. These are scattered throughout the English lowlands wherever there is sandy soil poor in bases and acid in reaction which has not been planted or cultivated—much more abundantly in the south and east where such soils are most prevalent. Many lowland heaths, except where they are exposed to violent winds, are always being colonised by birch, by pine (in the neighbourhood of plantations), less freely by oak, and locally by beech; but these are as constantly destroyed by felling, pasturing and burning. Many lowland heaths are still unfenced commonland and are pastured to a slight extent by the commoners, others are used as rabbit warrens, some are simply wasteland, a good deal of which is now being afforested. The following are some of the principal areas of English lowland heath:

(1) *East Anglian heaths*. These occupy considerable stretches of flat sandy country on the Pliocene crag of north-east Norfolk and south-east Suffolk, and especially on the overlying glacial sands and gravels there and in Essex. The Breckland heaths in south-west Norfolk and north-west Suffolk are developed on post-glacial sands and on leached sandy glacial till. Here however there are varying amounts of lime in the soil—the chalky boulder clay lying close below—and not only grass heath, but grassland rich in bases, sometimes approximating to chalk grassland, is often present.[1] Where rabbits are in excess Callunetum is converted into bent-fescue grassland. Colonisation by trees is not great except near plantations.

(2) *South-eastern (Lower Cretaceous and Wealden) heaths*. The heaths of

[1] See p. 511.

Kent, southern Surrey, and Sussex are extensive. They are developed on Lower Greensand (Cretaceous), and on the Ashdown Sand (Wealden) which forms the "Forest Ridge" in the centre of the Weald. Much of this land is occupied by "oak-birch heath", birchwood, and subspontaneous pinewood, some by oakwood, and locally there are small areas of beechwood.

(3) *Heaths of the London basin.* The Lower London Tertiaries (Reading Beds, etc.) on the edge of the London basin (together with overlying Plateau Gravels and Valley Gravels) and the Bagshot Sand, largely developed to the south-west of London, overlying the main formation of London Clay, also bear extensive heaths, those of the Bagshot Sands, especially, being much overgrown with subspontaneous pine.

(4) *Heaths of the Hampshire basin.* This, like the London basin, is a broad syncline of Tertiary rocks consisting of alternating sands and clays. The low plateau formed by these was covered in late Tertiary times by extensive sheets of river and estuarine gravels (Plateau Gravels) which have since been dissected by river systems forming shallow valleys. These sands and gravels bear wide stretches of heath, mainly to the west of Southampton Water in the New Forest area and in south Dorset. Much of this heathland is certainly very old: Domesday Book (eleventh century) and Leland (sixteenth century) mention the great "bruaria" of this region. The Callunetum alternates with dry oakwood, especially in the New Forest area. The soil of some of these Calluneta is very sterile—apparently toxic—and will not support tree growth, while the *Calluna* itself is poorly developed.

(5) *South-western heaths.* In west Dorset, south Somerset, Devon and Cornwall, heaths occur on non-calcareous rocks of very various age which form permeable soils. These are characterised by several south-western ("Atlantic southern") species. Thus large areas of the heath on the serpentine rock of the Lizard peninsula are dominated by *Erica vagans*,[1] a south-west European plant occurring nowhere else in the British Isles; and in the wet heaths of both Dorset and Cornwall *Erica ciliaris* occurs[2] and also *Pinguicula lusitanica*, both Atlantic southern species, the last of wider distribution in the west.

The heaths of the higher south-western hills—Exmoor, Dartmoor and Bodmin Moor on Devonian grits and on granite—are upland heaths and not extensively developed. Blanket bog and acid grassland are the typical communities of these high-lying moors.

(6) *Midland heaths.* On the Jurassic and Triassic rocks of the midlands heaths are not so common as on the Cretaceous, Tertiary and post-Tertiary sands and gravels of the south and east, but the Bunter Sandstone bears typical Callunetum in Sherwood Forest in Nottinghamshire. Delamere Forest in Cheshire, where the Bunter Sandstone is largely covered by recent deposits of alluvium and peat, has little Callunetum. The Lower Greensand

[1] Pl. 127, phot. 314.   [2] Pl. 127, phots. 312–13.

PLATE 127

Phot. 312. *Erica ciliaris* and *Molinia caerulea*. Wet heath near Perranwell, Cornwall.
Mrs Cowles.

Phot. 313. *Erica ciliaris*, closer view.
J Massart.

Phot. 314. *Erica vagans* and *Ulex gallii* on the
Lizard peninsula. *J. Massart.*

SPECIES OF SOUTH-WESTERN ENGLISH HEATHS

is well developed in Bedfordshire (south-east midlands) and bears Callunetum, but here it is very largely planted with conifers, especially Scots pine, which does exceedingly well.

STATUS OF THE HEATH FORMATION

The vegetational status of the Heath Formation requires some consideration. Primarily it is a west European formation, interdigitating so to speak with bog and forest, replacing the former on better drained soil, and the latter on the lighter and poorer soils, where pasturing or burning prevents the colonisation of trees. It can only be counted as a *climatic* formation on exposed coasts and on exposed mountain slopes between 300 and 600 m.,[1] that is, in situations where the wind factor is definitely inimical to tree growth, and where this is combined with good drainage and acid soil, either derived from acidic rocks or where the climate favours rapid leaching and surface peat formation. In these situations heath is certainly a climax, for, although the upland heaths of northern England and Scotland are practically always used as "grouse moors" and are fired periodically, there is no evidence that such heath, where it lies above the forest limit, would be superseded by any other type of vegetation if it were left entirely untouched.

The lowland heaths, so widespread in the south and east of England, are for the most part to be regarded as a stage in the succession to forest. In *Types of British Vegetation* (pp. 99–103) the heath formation was considered, following Graebner, as a result of the degeneration of oak forest on sandy soils which became podsolised, and some heaths have probably taken origin in this way, either through podsolisation of the soil, or more likely by felling, burning and pasturing. But many of them may well have been subject to burning or pasturing, or both, from the earliest times, and perhaps have never borne forest at all. Apart from some areas of Bagshot Sand, one of which (on Wareham Heath in Dorset) is now being investigated, there are few, if any, southern English lowland heath which will not support tree growth, at least of birch and pine, though the soil of some is probably too poor and dry for the good growth of oak. In general, however, they may be regarded as subclimax or "deflected climax" to dry oak or beech forest, belonging to the succession: heath → birch (→ pine) → oak → beech. It has been suggested that the development may really be cyclical, the beech, on these soils, creating soil conditions which render it incapable of regeneration, so that the forest ultimately dies and is replaced by heath, to begin the cycle anew (pp. 355, 415).

In regard to the development of heath itself, xeric subseres have been described (pp. 729–34, 751, 752, 756). Heath also develops on stabilised sand dunes (Pl. 126, phot. 309), but this prisere has not been ecologically investigated. It has been argued (pp. 734–5, 737) that the evidence for a hydrosere

[1] Higher than this in Scotland and north-east Ireland.

leading to heath is far from conclusive, though the possibility cannot be excluded. But the occupation of wet acid peat by Callunetum, as distinct from the abundant presence of *Calluna*, seems to depend, so far as the positive evidence goes, either on drainage or on change to a drier climate.

SUMMARY OF PEAT COMMUNITIES

It will be convenient to summarise here the seral relationships of the communities of peat and peaty soil described in the last six chapters.

Subaquatic peats. Peat may be formed under water by aquatic and reedswamp vegetation and Pearsall demonstrated that different conditions favour the colonisation of a lake floor by particular aquatic communities which form peat of different kinds; these seem to be analogous to the peats formed by the subaerial communities of fen and moss. But very little is known about British subaqueous peats, except that they have relatively high pH values (above 5·0) and are always reducing in character.

Ionic content of waters. We can draw a sharp distinction between the highly acidic waters of moorland pools, very poor in basic ions (or at least in calcium) and in nitrates, which are colonised by aquatic Sphagna and a few other water plants tolerant of such conditions and from which reedswamp is often absent, and the alkaline waters of a calcareous fen basin in which luxuriant and varied aquatic vegetation and reedswamp are present and typical fen is formed. Between the two come a series of waters, neutral or somewhat acid, which have never been given serious study. Most of them are fringed with some kind of reedswamp, and it is certain that the different reedswamp dominants show different degrees of toleration towards acidity and lack of calcium and nitrates. The "basic ratio", i.e. of potassium (and sodium) to calcium (and magnesium), may also be important. Until further study has been given to these matters there must remain a wide gap in our knowledge of the British hydroseres.

Reedswamp. Of the three commonest tall reedswamp dominants *Scirpus lacustris* and the two species of *Typha* (*T. latifolia* and *T. angustifolia*) are confined to reedswamp, *Scirpus* often occupying its outer zone in fairly deep water, while the commonest of all, *Phragmites communis*, maintains itself in fen. Of others, which may dominate the transition to fen, *Glyceria maxima* (*aquatica*) and *Phalaris arundinacea* seem to be characteristic of silting, *Cladium mariscus* is most luxuriant in alkaline, but not silting, waters, while some Carices and Equiseta, e.g. *Carex inflata* and *Equisetum fluviatile* (*limosum*), though by no means all or most, are tolerant of more acidic waters.

Phragmites communis is by far the most widespread and abundant of all the reedswamp-fen species, and is widely tolerant of different conditions. It forms reedswamp in various depths of water and is often abundant, or even dominant, in fen. By means of its strong, horizontal, deep-lying and wide-spreading rhizomes the common reed can freely invade wet soil (which may be relatively dry upon the surface), sending up its shoots among other vegetation. It is not always clear whether its presence in fen is due to survival from a reedswamp phase in which it has been dominant, or to secondary invasion. Nevertheless it is rather remarkably absent from many places which seem quite suitable for it.

Fen. *Cladium mariscus* is a very local plant in the British Isles, but it can tolerate very different conditions of climate and soil reaction, occurring alike in the East Anglian fens and in the west Irish bogs. It is a characteristic pioneer and dominant in the calcareous fen basins of East Anglia, forming a very pure community owing to its power of excluding other species.

In more mixed fens *Juncus subnodulosus* and *Carex panicea* are two of the most abundant species. In small calcareous basins in the south midlands *Juncus subnodulosus* sometimes forms reedswamp, maintaining itself as an abundant species in the fen, but followed as a dominant by *Schoenus nigricans*, a species characteristic of such habitats in most of England, though in the south-west and in Ireland it lives in and sometimes dominates acid bogs: it has recently been suggested that this depends upon salt spray. In northern Ireland the dominants of the "lower fen" are very various.

Succession in fen. It is only at Wicken Fen that exact observations have been made on the critical level for the colonisation of woody plants, and this is so close above the summer water level that there would seem little room for fen development between reedswamp and carr in the natural prisere. The decisive factors for the production of extensive fen may be (1) deep winter flooding as in north Ireland, or (2) lack of seed parents for the initiation of woody vegetation. Many existing fens are however certainly maintained by cutting, or in the drier fens, grazing. Under natural conditions the first shrubs of the fenwood or carr colonise Cladietum or other fen community while it is still very wet, and sometimes even reedswamp.

Cutting Cladietum not only prevents the development of carr but, when it is heavy and repeated at short intervals, handicaps and may eventually exterminate the dominant. At Wicken under these conditions *Molinia caerulea* enters the community and may become dominant, continuing to build up the peat level so as to produce a much drier type of fen than is formed by primitive carr. This is a good case of "deflected succession". If cutting is stopped the fen is at once colonised by bushes. *Molinia* is a species of great habitat range, occurring widely in fen, in wet heath, and in *relatively* dry bog, besides dominating extensive areas of wet upland grassland. It is a real peat former, though *Molinia* peat is seldom of any great depth.

Carr. Perhaps the commonest pioneer of the woody vegetation of fen is *Salix atrocinerea*. At Wicken this sallow is accompanied or preceded by *Frangula alnus*, which is the most abundant species in young carr, and followed by *Rhamnus catharticus* and *Viburnum opulus*, in Norfolk also by

Alnus glutinosa, and *Betula pubescens*. *Fraxinus excelsior* is often a frequent or abundant constituent of developing carr. *Crataegus monogyna* and *Ligustrum vulgare* also occur, and in some fens other constituents of calcareous woodland, such as *Viburnum lantana* and *Daphne laureola*, as well as calcicolous herbs like *Paris quadrifolia* and *Mercurialis perennis* and the climber *Tamus communis*.

Fully formed scrub carr at Wicken (where trees are practically absent) is dominated mainly by *Rhamnus catharticus*, locally by *Viburnum opulus* and *Salix atrocinerea*, or a mixture of species. Almost pure sallow carr is often met with on fens, but the sallow seems usually to give way to other species, sometimes dying in large numbers from a cause not yet ascertained. In Norfolk Alnetum is the typical ultimate carr, but Betuletum also occurs, and alder and birch are frequently mixed, together with ash, in other marsh and fen woods.

It seems probable that there would be a natural succession from fen carr to oakwood in this country under conditions which permitted of the gradual raising of the soil level by continued accumulation of humus, but it is difficult to demonstrate because of the almost universal interference with the soil and vegetation as soon as land conditions begin to be established.

Moss or bog. In sharp contrast with typical alkaline fen we have the moss or bog formation characteristic of highly acidic waters poor in basic ions, especially calcium.

The establishment of bog vegetation is a development of fen alternative to the progress to scrub and forest indicated above. It begins with the appearance of species of *Sphagnum* on the tussocks of the fen plants when they are raised somewhat above the alkaline ground water. The Sphagna are those of relatively high mineral requirements, such as *S. squarrosum* and *S. subsecundum*, which are more tolerant of alkaline solutions and do not themselves produce so much acid as other species that occur later in the succession. These are accompanied by a selection of such plants as *Eriophorum angustifolium*, *Pinguicula vulgaris*, *Polytrichum commune* and species of *Drosera*. The *p*H ranges of these have not been worked out— like the Sphagna themselves some are more, others less, oxyphilous. In eastern and central England, where the normal succession is from fen to scrub or woodland, the "acidic evolution" of fen usually seems to stop at the formation of small colonies of *Sphagnum* with their associated plants. This is probably correlated with the relatively dry climate, which may prevent the bog moss from growing far above the water table. In English fens *Sphagnum* colonisation and bush colonisation may be seen side by side, but the closing of the shrub canopy eventually kills out the bog moss, though some species may maintain themselves for a time in open carr.

Raised moss or bog. It is otherwise in Wales and western England, in parts of Scotland, and in central Ireland. In the moister climate of these regions, though tree colonisation of fen and carr formation may also occur,

Sphagnum is able to build upon fen a *raised moss* or *bog*, entirely obliterating the fen vegetation and forming a lens-shaped cushion of acid peat several metres thick and many hundreds of metres in diameter. The development of a typical Irish raised bog, as recorded in the peat, is fully described on pp. 686–96.

There is a regular succession of species of *Sphagnum* from the semi-aquatic species, of relatively high mineral requirement, which originally colonised the fen at the base of the bog, to highly acid species of low mineral requirements in the bog itself. These last, independently in each of the elementary units of which the bog consists, show repeated cyclic successions, from species of high to species of low water requirements, as "hollows" give place to "hummocks"—the "regeneration cycle" of Osvald (pp. 687–8).

But though the development of raised bog depends essentially on the growth of different species of *Sphagnum*, its vegetation contains other and more conspicuous elements, flowering plants which determine the physiognomy, the tone, of the whole bog, and form a considerable fraction of the whole mass of plant material present. Of these *Scirpus caespitosus* and *Eriophorum vaginatum* are the most abundant and prominent, and *Calluna vulgaris* is nearly as conspicuous. The remaining species, except *Narthecium ossifragum* and *Rhynchospora alba*, which are the phanerogamic dominants of the "hollows", are of little vegetational importance.

There is no evidence in Ireland that trees ever colonise raised bog, except rather sporadically the well-drained marginal slope, where occasional isolated pines and birches may be seen. The slope normally bears Callunetum, with *Pteridium* and *Ulex* locally dominant. Drained bog peat also normally develops Callunetum.

Blanket bog. In the very wet climates of the extreme west of Scotland and Ireland another type of bog occurs, which may be called *blanket bog*. This is indeed the climatic climax formation of these regions, covering the general surface of the country, except on rocky outcrops and on the steeper slopes. Essentially the same general type of bog also covers the high plateaux, many about 2000 ft. (*c.* 600 m.), of the English, Scottish and Irish mountains, which also possess a wet climate.

Sphagnum rarely forms the essential basis of blanket bog as it does of raised bog (extensive Sphagneta occur only in one place in the Pennines). Hummocks of bog moss, of similar construction to those of raised bog,[1] may be formed locally (and sometimes abundantly), especially in the lowland blanket bogs of the west, but the dominants of blanket bog are vascular plants. The most widespread dominants are *Scirpus caespitosus* (west Ireland, west Scotland, Wicklow Mountains), *Eriophorum vaginatum* (Pennines), and *E. angustifolium* (northern Pennines), in that order, with *Schoenus nigricans* more local, and *Molinia caerulea* in drier places. *Rhynchospora alba*, *Narthecium ossifragum* or *Erica tetralix* may be locally

[1] Often capped by *Rhacomitrium lanuginosum* or *Leucobryum glaucum*.

dominant. Very often however blanket bogs are not dominated by single species.

There is no evidence and no likelihood that blanket bog is *generally* built up on fen, though doubtless it had this origin in certain calcareous basins in the west of Ireland. The plants of the blanket bog probably colonised the general surface of the country when the climate became sufficiently wet, whatever the previous vegetation. In many parts of the west of Ireland, it is clear from the continuous layers of pine stumps at the base of the peat that here bog succeeded pine forest in many places. The early stages of such successions are unknown, but from observations on current successions, and from our knowledge of what happens in the north of Europe, we may conjecture that they have included the alga *Zygogonium* as well as *Sphagnum* itself. The soil on which blanket bog has developed, both in the lowlands and on the high plateaux, is very commonly glacial clay—the ground moraines left behind on the recession of the Pleistocene ice-sheets. Much of the blanket bog may have originated in the innumerable pools of acidic water that must have been present in local depressions of this ground moraine, colonised by aquatic species of *Sphagnum*, whose growth filled up the pools; but the details of such successions have not been studied.

Callunetum. The occurrence of *Calluna* in blanket bog is very widespread, practically universal, but it does not attain anything like luxuriant growth or abundant flowering except on drier knolls or the tops of local hummocks. A succession to Callunetum has not been observed except where the peat has been drained, and there it is a regular occurrence. It is probable that under natural conditions change to a drier climate is necessary for the supersession of the bog vegetation by Callunetum.

The Heath formation itself, as we have seen, generally occurs on a permeable mineral soil poor in bases, in drier climates developing only a shallow layer of surface raw humus. In damper climates it (or rather its accompanying lichens) forms a certain depth of acid peat, rarely exceeding 12 in. (30 cm.). Where Callunetum occurs on deeper peat this has been formed by some other community which has been superseded by the heather as a result of drainage or of the incidence of a drier climate.

The lowland Calluneta of the south of England show xeroseral succession, and it is not clear that wet heath can pass into Callunetum without the agency of erosion. Wet heath does not differ essentially from bog or moss except in the much lesser frequency of the Cyperaceae (*Scirpus, Eriophorum*, etc.), which are abundant or dominant in the latter. The upland Calluneta have been fully discussed, and from that discussion the conclusions emerge that while the *habitats* of *Calluna* are very varied, its *dominance* in the British climates depends upon relatively well-drained soils, and that it does not come to dominate the peat of moss or bog except through increased drainage or change of climate. Where erosion leads to rapid drying of the peat especially at relatively high altitudes on the Pennine plateaux, it is *Vaccinium myrtillus* which occupies the Vaccinietum ground. And at the higher altitudes on the high plateaux above 2000 ft. in the Scottish Highlands, the same species, in company with *V. vitis-idaea*, *V. uliginosum* and *Arctostaphylos*, replaces *Calluna* as the general dominant.

Synopsis of peat vegetation

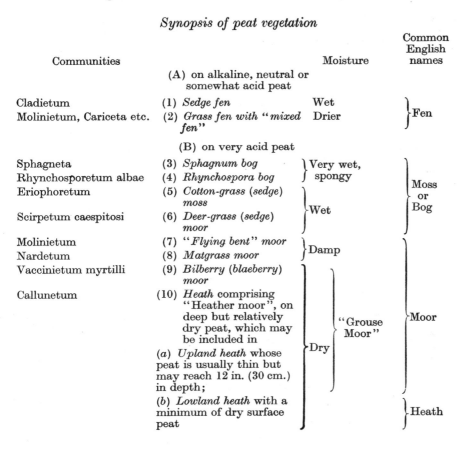

Communities		Moisture	Common English names
	(A) on alkaline, neutral or somewhat acid peat		
Cladietum	(1) *Sedge fen*	Wet	Fen
Molinietum, Cariceta etc.	(2) *Grass fen with "mixed fen"*	Drier	
	(B) on very acid peat		
Sphagneta	(3) *Sphagnum bog*	Very wet, spongy	Moss or Bog
Rhynchosporetum albae	(4) *Rhynchospora bog*		
Eriophoretum	(5) *Cotton-grass (sedge) moss*	Wet	
Scirpetum caespitosi	(6) *Deer-grass (sedge) moor*		
Molinietum	(7) *"Flying bent" moor*	Damp	Moor
Nardetum	(8) *Matgrass moor*		
Vaccinietum myrtilli	(9) *Bilberry (blaeberry) moor*	Dry	"Grouse Moor"
Callunetum	(10) *Heath* comprising "Heather moor", on deep but relatively dry peat, which may be included in		
	(a) *Upland heath* whose peat is usually thin but may reach 12 in. (30 cm.) in depth;		
	(b) *Lowland heath* with a minimum of dry surface peat		Heath

REFERENCES

ADAMSON, R. S. On the relationships of some associations of the Southern Pennines. *J. Ecol.* 6, 97–109. 1918.

ARMSTRONG, J. I., CALVERT, J. and INGOLD, C. T. The ecology of the mountains of Mourne with special reference to Slieve Donard. *Proc. Roy. Irish Acad.* 39, 440–52. 1930.

ARMSTRONG, J. I., INGOLD, C. T. and VEAR, K. C. Vegetation map of the Mourne Mountains, Co. Down, Ireland. *J. Ecol.* 22, 439–44. 1934.

CRUMP, W. B. The coefficient of humidity: a new method of expressing the soil moisture. *New Phytol.* 12, 125–47. 1913.

ELGEE, F. The vegetation of the eastern moorlands of Yorkshire. *J. Ecol.* 2, 1–18. 1914.

LEWIS, F. J. Geographical distribution of the vegetation of the basins of the rivers Eden, Tees, Wear and Tyne. Parts I and II. *Geogr. J.* **23**. 1904.

LEWIS, F. J. and Moss, C. E. "Heather Moor" in *Types of British Vegetation*, pp. 275–9. Cambridge, 1911.

Moss, C. E. *Vegetation of the Peak district*. Cambridge, 1913.

PEARSALL, W. H. The soil complex in relation to plant communities. III. Moorlands and bogs. *J. Ecol.* **26**, 298–315. 1938.

PETHYBRIDGE, G. H. and PRAEGER, R. LL. The vegetation of the district lying south of Dublin. *Proc. Roy. Irish Acad.* **25**. 1905.

SMITH, R. Botanical survey of Scotland. *Scot. Geogr. Mag.* 1900.

SMITH, W. G. "Scottish Heaths", in *Types of British Vegetation*, pp. 113–16. Cambridge, 1911.

SMITH, W. G. and Moss, C. E. Geographical distribution of vegetation in Yorkshire. Part I. Leeds and Halifax district. *Geogr. J.* **22**. 1903.

SMITH, W. G. and RANKIN, W. M. Geographical distribution of vegetation in Yorkshire. Part II. Harrogate and Skipton district. *Geogr. J.* **22**. 1903.

TANSLEY, A. G. In *Types of British Vegetation*, Chapter IV, "The Heath formation". Cambridge, 1911.

WATSON, W. The bryophytes and lichens of moorland. *J. Ecol.* **20**, 284–313. 1932.

Part VIII

MOUNTAIN VEGETATION

Chapter XXXVIII

THE UPLAND AND MOUNTAIN HABITATS

MONTANE AND ARCTIC-ALPINE VEGETATION. THE SCREE SUCCESSION. COMMUNITIES OF EXPOSED ARCTIC-ALPINE HABITATS

The highest hills in the south-east of England scarcely ever exceed 1000 ft. (304 m.) in height, and up to this altitude the vegetation shows no perceptible change. In the south-west there are considerable areas over 1000 ft. and some (Dartmoor) over 1500 ft. (457 m.) and up to 2000 ft. (610 m.) in altitude. These however are high-lying plateaux, and corresponding with the damp south-western climate they have a relatively high rainfall and often bear blanket bog (see Chapter XXXIV). Genuine mountain habitats are only to be found in parts of Ireland, Wales, the north of England, and especially in the Scottish Highlands.

Upper limit of forest. Except in the most exposed situations and on the wet waterlogged plateaux the former upper limit of forest must have commonly reached altitudes of 1500 and in places 2000 ft. where the mountain masses were of adequate height; and the existing vegetation of the hill slopes up to these levels has replaced forest, mainly as the result of grazing. The grassland and heath communities which now occupy these habitats have already been dealt with (Chapters XXVI and XXXVII). So far as flora is concerned there are a few species which may be called "montane" because while they are in no sense high mountain plants they are rarely found at altitudes of much less than 1000 ft.; but there are not many plant communities specially characteristic of these intermediate altitudes. Nardetum, upland Callunetum and Vaccinietum may however be considered as belonging to this category.

Montane vegetation

As we ascend to higher levels several important climatic factors show progressive change. Mean temperatures are lower, rainfall and humidity higher, snow lies much longer, and exposure to wind, except in sheltered places, is much greater. In this country very little attention has been given to the study of the factors which condition the characteristic mountain habitat in relation to vegetation, though their combined effect in determining the so-called arctic-alpine communities is sufficiently clear. In spite of the general severity of the mountain climate its total flora is extremely rich and the variety of small communities very great. This is because the mountain habitats are very varied and show many different combinations of factors, some of which are very favourable to plant life.

Talus and scree. There is one physiographic factor characteristic of craggy hills and mountains—namely the effect of gravity on the weathering debris of hard precipitous rocks (*crags*)—which produces a highly characteristic plant habitat, namely the accumulations of talus detached from crags as the result of alternate heating and cooling of the parent rock and of the freezing and thawing of water in rock fissures. The fragments so detached accumulate in a slope (*scree*) below the crag at the angle of repose, and as fresh rock fragments fall they disturb the equilibrium of the mass of talus so that there is a constant tendency for the stones to slide down the slope. Screes may be formed at any altitude, but they become more numerous and extensive at the higher levels where the parent crags are commoner, and are best considered as one type of mountain habitat, though when developed at low or moderate altitudes they may be entirely lacking in "arctic-alpine" species.

The scree succession. The vegetation of British screes has not received a great deal of attention, but Leach (1930) has published a good preliminary account of some non-calcareous screes in the Lake District and around Snowdon. In this he describes the leading factors of the scree habitat and the various successions that can be traced.

Screes differ much in their degree of stability as well as in the size of the fragments into which the parent rock weathers. Leach found that the most unstable type, most difficult of colonisation by plants, is that composed of tabular fragments of slate about 10–20 cm. in length and breadth by 1–4 cm. in thickness, while screes formed of more or less isodiametric fragments are much more stable. "Block screes" he does not consider.

Lithophytes such as *Andreaea petrophila*[1] and *Rhacomitrium fasciculare*, **Lithophilous bryophytes** which may colonise the actual surfaces of the rock fragments, play little or no active part in the sere, since they do not form a suitable nidus for the colonisation of higher plants. Mosses may however actually disintegrate the rock surface, which is apparently rendered porous by the action of the closely applied rhizoids of *Rhacomitrium heterostichum* and *Dicranella heteromalla*. The first really **Pioneers** effective stage in the sere is formed by chomophytes,[2] at first mosses and liverworts, followed by ferns, which settle in the spaces between the stones where they find small accumulations of fine detritus. The pioneer bryophytes in these situations are *Diplophyllum albicans*, *Rhacomitrium lanuginosum* and *Rh. heterostichum*, which occur on practically every non-calcareous scree examined between the altitudes of 800 and 2000 ft. (240 and 610 m.). In the cushions formed by these bryophytes, particularly *Diplophyllum*, the parsley fern (*Crypto* **The parsley fern** *gramma crispa*) finds enough water-retaining humus to enable its prothallus and young sporophyte to get a footing, and it is the principal pioneer vascular plant on all the more extensive siliceous screes examined. To enable the parsley fern to continue its growth a position of some stability is required; and this is generally provided by the

[1] Still conspicuous at 2000 ft. (610 m.).　　　　[2] See p. 787.

occurrence of a larger block which resists the flow of the surrounding talus, diverting the sliding fragments to each side. In the shelter of such a block, i.e. on its lower side, *Cryptogramma* can grow from year to year with the aid of the constantly increasing accumulation of humus supplied by the decay of its annually deciduous fronds. Below every large plant of this

Other ferns species there is a mass of root-bound humus composed entirely of the debris of dead fronds (Pl. 128, phot. 317). Other ferns may occur in the same situations:

Athyrium filix-femina	Dryopteris filix-mas
Blechnum spicant	D. montana

but these are not so uniformly successful as *Cryptogramma crispa* in providing year by year for their own future requirements and thus acting as pioneer vascular plants.

Eventually *Cryptogramma* produces a mass of humus more than sufficient

Bryophytes and Flowering plants for its own needs, and then numerous bryophytes and some flowering plants are able to settle down round the pioneer fern.

The following species occur in this situation:

Agrostis canina	Festuca ovina and f. vivipara
A. tenuis	Galium saxatile
Anthoxanthum odoratum	Oxalis acetosella
Deschampsia flexuosa	Viola riviniana

Campylopus flexuosus	Hylocomium splendens
C. fragilis	H. squarrosum
Catharinea undulata	Plagiothecium elegans
Dicranella heteromalla	Polytrichum alpinum
Dicranum fucescens	P. commune
D. scoparium	P. formosum
Eurhynchium pumilum	P. gracile
*Hypnum cupressiforme var.	P. juniperinum
ericetorum	P. piliferum
H. cuspidatum	Webera nutans
H. schreberi	
Hylocomium loreum	Lophozia floerkii

* Still conspicuous at 2000 ft. (610 m.) on Lliwedd (Snowdon).

The amount of surface covered by bryophytes varies with the general humidity of the habitat. The number of species present decreases with altitude. Thus at 1000 ft. there were seventeen conspicuous species, at 1750 ft. eight, and at 2000 ft. six.

Formation of vertical strips. If the protecting block remains *in situ* the area of vegetation so arising tends to extend downwards from the site originally colonised, and the longer the vegetation remains the more permanent it is likely to become, because of the fixing and stabilising effect on the substratum of the continuous formation of root-bound soil between the rock fragments of the scree. The vertical strips of vegetation thus produced (Fig. 151, and Pl. 128, phots. 315, 316) are often dominated by *Calluna* and show the composition of a typical heath

The heath community community. Besides those mentioned above as succeeding *Cryptogramma* the following species may be found in this heath community:

Calluna vulgaris	Polygala serpyllifolia
Digitalis purpurea	Potentilla erecta
Erica cinerea	Solidago virgaurea
E. tetralix	Teucrium scorodonia
Euphrasia officinalis (agg.)	Thymus serpyllum
Lycopodium selago	

FIG. 151. TABULAR SCREE, SEMI-DIAGRAMMATIC SKETCH

Some of the vertical strips of vegetation are stabilised by protecting blocks at their upper ends, others (more permanent) are developed below areas at the top of the scree which are relatively quiescent. From Leach, 1925.

Many such patches and strips are however transient, being carried away by the general movement of the scree.

Besides the strips originating from this type of colonisation below large protecting blocks, others arise in places where the fragments of rock falling from the crags above are only accumulating slowly, so that there is comparative stability at the top of the scree. Such strips extend downwards from the foot of the crags to the bottom of the scree slope, though the lower ends may be overwhelmed and destroyed by the lateral spreading of fans of

PLATE 128

Phot. 315. Strip vegetation (heath) of scree on south slope of Causey Pike (Lake District) with the Birkrigg ks behind. Alt. 1100–1400 ft. From *W. Leach* (1930).

Phot. 316. Screes on south-west slope of Grasmoor above Crummock Water, showing vertical strips of Callunetum. Alt. 1500 ft. From *W. Leach* (1930).

Phot. 317. Large parsley fern (*Cryptogramma crispa*) th *Festuca ovina* on Hindscarth scree, alt. 1700 ft. om *W. Leach* (1930).

Phot. 318. *Vaccinium myrtillus* invading Grasmoor scree by rhizome growth. Alt. 1500 ft. From *W. Leach* (1930).

SCREE VEGETATION

unstable material from intervening strips of actively moving scree (Fig. 151). The vertical strips of vegetation are usually at a lower level than the scree on each side (Fig. 151, section AB), because they are protected from the piling up of fragments that occurs in a moving scree which is constantly being added to from above.

Destruction of seral stages. The complete succession from *Diplophyllum* and *Rhacomitrium* through *Cryptogramma* to Callunetum is rarely completed because the colonised areas are so often destroyed in one stage or another, either by movement of the scree as a whole or by the piling up and overflowing of the rock fragments, which obliterate the plant colonies. Thus an active scree shows fragments of seral communities, representing various stages of the succession, scattered upon its surface, and the more active the scree the sparser and younger the colonies.

Stable screes. The more stable screes, on the other hand, show more extensive and more advanced stages of progressive colonisation, and these connect with more gentle slopes and with areas of mountain-top detritus (see p. 787) where formation and disintegration of the rock fragments occur *in situ*. Here soil, often clayey, accumulates more rapidly between the fragments of rock, and *Agrostis tenuis*, *Alchemilla alpina* and *Festuca ovina*, accompanied by mosses such as *Polytrichum nanum* and *P. piliferum*, act as pioneers, in addition to *Diplophyllum*, *Rhacomitrium* and *Cryptogramma*. The roots of the grasses, and to some extent the moss rhizoids, bind the soil and prevent its being washed away by heavy rain.

Rhizomatous and creeping plants invade the edges of scree from
Marginal invasion neighbouring areas, either an adjacent hillside covered with vegetation, or an already fully stabilised scree. The most important of these marginal invaders are bracken (*Pteridium aquilinum*) and bilberry (*Vaccinium myrtillus*).[1] On the edge of the scree humus collects in the deeper and more stable layers of the substratum, and this is penetrated by the roots and rhizomes of the marginal plants which progressively colonise the margin of the scree. *Dryopteris linnaeana* and *Vaccinium vitis-idaea* are less frequent. The bilberry and crowberry (*Empetrum nigrum*) with their upright woody shoots here act as pioneers and have a markedly stabilising effect on the slowly moving stones. *Hedera helix* and *Silene maritima*, which creep over the surface, are occasionally met with.

Climax vegetation on screes. On completely stabilised scree up to an altitude of about 1500 ft. (450 m.) wood of *Quercus sessiliflora*, the climatic climax, may develop, and many of the hillside Querceta both in Wales and the Lake District undoubtedly occupy the sites of ancient scree (Pl. 128, phot. 315), just as Fraxinetum may be developed on calcareous scree (Chapter XXI, pp. 436–40). Woodland may develop as a result of the final weathering down of the crags causing cessation of the supply of falling stones.

[1] Pl. 128, phot. 318.

ARCTIC-ALPINE VEGETATION

Comparatively little work has been published in recent years on the higher mountain vegetation of the British Isles—the two most important contributions being Watson's account of the bryophytes and lichens of arctic-alpine vegetation (1925) and Price Evans' study of the vegetation of Cader Idris (1932). Both of these authors adopt the classification of the arctic-alpine plant communities proposed in 1911 (in *Types of British Vegetation*) by the late W. G. Smith, whose account is indeed so excellent as a preliminary treatment that the present chapter will be based upon it and considerable passages reproduced.

The arctic-alpine zone. "The recognition of an arctic-alpine zone of vegetation is based on the fact that in ascending the higher mountains one finds that at some altitude the general tone of the vegetation alters, and that new species characteristic of higher altitudes appear, forming plant communities distinct from the lowland ones. The species also present various growth-forms adapted to conditions ranging from the barrenness of windswept slopes and summits to the comparative shelter and generally favourable environment afforded by rock-crannies amongst precipitous crags and ravines. In the central Highlands of Scotland the contour line of 2000 ft. (*c.* 600 m.) indicates roughly the lower limit of this zone: above this the vegetation is characterised by an increase of such plants as saxifrages, dwarf willows and other low-growing perennials, generally associated with characteristic mosses and lichens. Such woodlands as still remain in the country cease either below or near this limit. On the subalpine ericaceous moorland the change is marked by decrease of *Calluna* and increase of *Vaccinium*, while the grasslands become poorer in species belonging to such families as Leguminosae and Umbelliferae, and assume a new physiognomy through increasing frequency of arctic-alpine plants, such as *Alchemilla alpina*, viviparous grasses, and arctic species or varieties of *Carex, Juncus, Luzula, Draba, Cerastium, Potentilla*, etc.

"On the Atlantic side of Scotland and Ireland a similar vegetation occurs at much lower altitudes, almost down to sea-level. These plant communities include many of the same species grouped in a manner comparable to that of the communities of the higher mountains. It is thus evident that in defining the arctic-alpine vegetation, too much stress should not be laid on zones of altitude expressed numerically; other factors must be sought which may explain the limited distribution of this vegetation.

"The arctic-alpine vegetation presents well-marked features both in its plant communities and in its characteristic growth forms. It can thus be characterised not only on a floristic, but also on a strictly ecological basis, especially as sub-alpine species occurring in the arctic-alpine zone assume the growth-forms of low-growing cushions, rosettes, or mats, more or less similar to those of the arctic-alpine species proper" (W. G. Smith, 1911, pp. 288–90).

Arctic-alpine species. The "arctic-alpine species proper" are those which are more or less confined in Britain to high altitudes and were called "Highland" by H. C. Watson (*Compendium* to the *Cybele Britannica*, 1868, Introduction, p. 28), and count between one and two hundred angiosperms, according to the number of arctic-alpine forms or ecotypes of lowland aggregates which we recognise as "species". These plants are characteristic of high northern latitudes in arctic and subarctic Europe, many of them occurring at sea-level in the high arctic and becoming restricted to higher and higher altitudes as we pass south from these regions. In central Scotland, as we have seen, they do not occur in any quantity below about 2000 ft. (*c.* 600 m.) and are much more abundant above 3000 ft. (*c.* 900 m.), while in the Swiss Alps many of the same species occur only in alpine communities above about 8000 ft. (*c.* 2400 m.). In western Scotland and western Ireland, however, several of them descend to sea-level.

Life forms of arctic-alpines. Both the arctic and the alpine angiosperms, as groups, have, as Raunkiaer showed (1908), a percentage of species of the chamaephytic life form[1] markedly exceeding that which obtains in the flora of most other climates, and the higher the latitude or altitude the greater the percentage of chamaephytes. Indeed the few species which inhabit the more remote arctic islands and those occurring above 4000 m. in the Alps are more than half chamaephytes, the rest being hemicryptophytes.[2] Of the eleven species of vascular plants occurring above 1000 m. (*c.* 3280 ft.) in the Clova region of the Scottish Grampians three are chamaephytes, seven are hemicryptophytes and one is a geophyte, i.e. 27 per cent are chamaephytes, as compared with 9 per cent in the flora of the world as a whole (Raunkiaer, 1908; Willis and Burkill, 1901–4). Between 800 and 900 m. (*c.* 2625–2950 ft.) while 60 per cent of the seventy-two species are hemicryptophytes there are already 22 per cent of chamaephytes, a very marked rise from lower altitudes.

Chamae-phytes (margin note)

The relatively large number of chamaephytes in the arctic-alpine vegetation is undoubtedly connected with the fact that the ground is covered with snow during most of the winter, in some places during the whole winter and the first half of summer,[3] so that the temperature just above the soil surface remains fairly constant round about 0° C., and the vegetation

---

[1] Chamaephytes are plants whose perennating buds are borne between the soil surface and a height of 25 cm. (10 in.) above it.

[2] Plants with perennating buds in or on the surface layer of soil.

[3] The summit of Ben Nevis in the western Scottish Highlands at an elevation of 4409 ft. (1344 m.)—the highest point in the British Isles—was only free from snow in August and September during 10 years' observation. The snow, which began to lie in October, steadily increased during the winter, reaching a mean maximum of 78 in. (198 cm.) at the beginning of May and not disappearing entirely till after the middle of July (Buchan and Omond, 1905).

is not subjected to the rapid alternations of frost and thaw which occur during winter in the lowlands of western Europe where the soil is exposed. The low-growing forms of the plants (the vegetative parts often forming rosettes or cushions) ensures that the vegetation is covered with snow during the cold period and may also be related to the violence of the winds that often blow at high altitudes during the growing period. The velocity of the wind is very much less close to the ground, and the close setting of the shoots and leaves gives a great deal of mutual protection. When the snow melts the surface of the soil warms up quickly under the rays of the sun, especially on southern slopes, and the rise in temperature of the soil surface and of the air immediately above it stimulates rapid development of the perennating buds into leaves and flowers. This is very necessary in a climate where the growing season is extremely short because the snow lies very late and fresh snowfalls occur in the early autumn.

Bryophytes and Lichens. Another marked feature of arctic-alpine vegetation is the prominence of mosses and especially of lichens. There are a large number of species of these groups which occur in, and many of which are peculiar to, the arctic-alpine zone, but even more striking is their massive conspicuousness in the vegetation. Everywhere they are locally dominant at the higher altitudes, and some dominate considerable communities, such as the "*Rhacomitrium* heath", especially on exposed summits and plateaux.

During the summer, extremes, and especially rapid fluctuations, of temperature and of humidity are much greater than in the lowlands, and mosses and lichens are particularly well adapted to endure such conditions of alternate drying and wetting, heating and cooling. Most lichens and many mosses are able to withstand almost complete desiccation, absorbing water again as soon as it is available and immediately resuming active life. The low-growing and tufted mat- or cushion-like habit of most mosses also gives them the maximum protection from wind. Furthermore, as W. Watson especially has shown, the leaves of many species of moss have structural characters which tend to check evaporation from their leaves. Thus most of these non-vascular plants are extremely well protected against killing by drought; but for the reproductive processes of bryophytes, unlike those of the higher plants, external water is essential in the sexual stage and dry air in the stage of spore distribution. In fact some species very rarely fruit, reproducing themselves almost entirely by vegetative means.

Though the habitats of the exposed upper slopes and the summits of mountains are frequently very dry, and the vegetation is correspondingly xerophytic, it must not be forgotten that there are, on many mountains, both very wet and also very well-protected habitats with abundant soil water. It is in these last that, side by side with arctic-alpine species, various lowland plants (or arctic-alpine ecotypes of lowland plants) may flourish very luxuriantly, though they often do not flower.

Effect of rock and soil. The effect of the nature of the rock (which conditions the type of erosion) and of the soil which it produces on the kind of vegetation developed on the higher Scottish mountains is well marked. On the acidic rocks of which the majority of the mountains are composed, sub-alpine moorland with regular acid peat formation extends up to a high level, e.g. 3000 ft. on the Clova-Canlochan plateau. This type of moorland, dominated by *Vaccinium myrtillus*, though often with *Calluna* associated (Calluno-Vaccinietum) has been described in Chapter xxxvii (pp. 761–2). The flora is poor and typical of sub-alpine moorland, including very few arctic-alpine and those mostly arctic species, such as *Loiseleuria procumbens*, *Betula nana* and *Cornus suecica*. But on exposed knolls, where the drainage is good and peat cannot form, a few other arctic-alpines occur, such as *Alchemilla alpina*, *Carex rigida*, *Gnaphalium supinum* and *Salix herbacea*. Even on the summits which rise from such moorlands to a much higher level (in the neighbourhood of 4000 ft. or 1200 m.) the arctic-alpine flora is relatively poor.

Richer flora of basic rocks. Sharply contrasted with these floristically poor mountains are those composed of easily weathering rocks rich in mineral salts. Actual limestone is very rare in the higher Scottish mountains but some of the schists and other metamorphic and volcanic rocks are relatively rich in bases (calcium, magnesium and potassium). It is on the slopes, particularly the southern slopes, especially if they are irrigated by numerous springs, in the corries (steep-sided valleys), and on the rock ledges of such mountains that the arctic-alpine flora is seen in its greatest profusion. The two typical Scottish areas are the Breadalbane Mountains in the central Highlands of Perthshire, of which Ben Lawers is the best known example, and the Clova region on the borders of Forfarshire and Aberdeenshire in the eastern Highlands; though here, as we have seen, the elevated plateau bears peat moor and most of the numerous arctic-alpine species are confined to the slopes and corries. The Cairngorm group, composed mainly of granitic rocks, on the borders of Aberdeen, Banff and Inverness, though it has by far the most extensive area of high mountain land, including some of the highest summits, is not so rich as these two, though considerably richer than the western Highlands.

The same contrast between the flora and vegetation of acidic rocks and of those rich in basic salts is seen in the higher English and Welsh mountains, for example in the Lake District, in Snowdonia and on Cader Idris (see pp. 786, 808).

ARCTIC-ALPINE GRASSLAND

On the old, stabilised, scree-covered slopes, especially the southern slopes, of those mountains whose rocks produce a more basic soil, generally "where the individual summits begin to be differentiated from the more continuous undulating slopes of the valleys", at some altitude between 2000 and

3000 ft. (about 600–900 m.) there is developed a characteristic arctic-alpine grassland, or at least a vegetation in which grasses are prominent and not dominated by dwarf shrubs.

By its fresh bright green colour this differs markedly in appearance from the dingy, brownish green, peaty grassland often dominated by *Nardus*, and developed on the flatter morainic terraces below. The soil of arctic-alpine grassland is derived by direct weathering of the basic rocks of the mountain, and occurs in continuous sheets between and over the overgrown boulders and outcropping rocks (Pl. 130, phot. 322). Raw humus formation is at a minimum and peat is absent, except locally. "The habitat is kept moist throughout the greater part of the year by numerous springs emerging at various levels, many of them marked by swelling cushions of bright green mosses. This water is mainly derived [on Ben Lawers] from schistose rocks comparatively rich in calcium, magnesium and potassium, and owing to the steepness of the slopes it drains away and does not become stagnant and acid. An essential condition of the habitat is that the soil is stable and little disturbed by surface erosion" (W. G. Smith). In these respects this grassland resembles the limestone grasslands at lower altitudes and contains many of the same species, though it is characterised by an assemblage of arctic-alpine plants. It may certainly be considered as a climax community, developed above the natural forest zone and not grazed to any considerable extent nor interfered with by artificial drainage.

Floristic composition. The floristic composition of a rich arctic-alpine grassland, like that of Ben Lawers, is very varied. There is no general dominant, but the two commonest local dominants are *Festuca ovina*, forma *vivipara* and *Alchemilla alpina* (both arctic-alpine plants), which often cover considerable areas.[1] The community has not been analysed in detail, and we can only give a general list of species mostly recorded by Robert Smith (1900) and taken from W. G. Smith's account (1911, pp. 300–1). Species limited to wet stream banks are omitted.

Abundant and locally dominant lowland species

Agrostis tenuis ⎫	
Anthoxanthum odoratum ⎪	general and abundant
Deschampsia flexuosa[2] ⎬	
Festuca ovina (agg.) ⎭	
Carex goodenowii (forma) ⎫	
Molinia caerulea ⎬	locally dominant[2]
Nardus stricta ⎭	

"Highland" (arctic-alpine) species

Alchemilla alpina ⎫	
Festuca ovina f. vivipara ⎭	abundant, locally dominant

[1] Pl. 130, phot. 322.

[2] *Nardus, Molinia, Carex goodenowii* (forma) and *Deschampsia flexuosa* are evidently species of peaty soil and must be regarded as local intrusions determined by local conditions and not belonging to the community described.

Carex capillaris
C. rigida
Cerastium alpinum
Luzula spicata
Lycopodium alpinum
L. selago

Phleum alpinum
Poa alpina
Potentilla crantzii
Sagina saginoides
Selaginella selaginoides
Sibbaldia procumbeus

Lowland species

Achillea millefolium
Alchemilla vulgaris (agg.)
Anemone nemorosa
Antennaria dioica
Avena pratensis
Bellis perennis
Blechnum spicant
Botrychium lunaria
Campanula rotundifolia
Carex binervis
C. stellulata
C. flava (forma)
C. pilulifera
Cerastium vulgatum
Deschampsia caespitosa
Euphrasia officinalis (agg.)
Galium saxatile
Heracleum sphondylium
Hypericum pulchrum
Juncus squarrosus
Lathyrus montanus
Leontodon autumnalis
Linum catharticum

Lotus corniculatus
Luzula multiflora var. congesta
Lycopodium annotinum
Melampyrum pratense var.
   montanum
Orchis maculata
Oxalis acetosella
Plantago lanceolata
Polygala serpyllifolia
Potentilla erecta
Ranunculus acris
Rumex acetosa
R. acetosella
Sagina procumbens
Scirpus caespitosus
Succisa pratensis
Taraxacum officinale
Thymus serpyllum
Trifolium repens
Vaccinium myrtillus (sparse and
   dwarf)
Veronica serpyllifolia
Viola lutea f. amoena

Similar types of high altitude grassland have been described from the slopes of other mountains, but over rocks producing soil poorer in mineral salts they are much less rich floristically, and contain but few arctic-alpine species. Thus Price Evans (1932) gives lists from five different sites of "dry grassland on a porous substratum" on Cader Idris in Wales. **Festucetum ovinae of Cader Idris. Dry facies.** This grassland occurs on "parts of the summit plateau, and on drained ledges of the great escarpment, but it reaches its greatest development on the lower slopes where the high summits begin to be differentiated from the moorland". These sites, which vary from 1600 ft. (488 m.) to 2700 ft. (823 m.) in altitude, are dominated or partly dominated by *Festuca ovina*. They are mostly situated at somewhat lower absolute altitudes than the corresponding grassland described from Ben Lawers (the summit of Cader Idris is only 2927 ft. while Ben Lawers reaches 3984 ft.) and the conditions of the habitat are clearly not so favourable. The list of species is therefore much shorter and there are only one or two which can be considered arctic-alpines, but there are very few which are not recorded as occurring in the Ben Lawers grassland. It is clearly the same type of community. The following is a composite list from the five sites:

Festuca ovina		d or co-d	
Agrostis spp.	o	Luzula campestris	r–o
Anthoxanthum odoratum	o	*Lycopodium alpinum	f–a
Blechnum spicant	r	L. clavatum	r–la
Campanula rotundifolia	r–o	*L. selago	o–f
Carex pilulifera	r–f	Nardus stricta	r–f
Cryptogramma crispa	l	Polygala vulgaris	o
Deschampsia flexuosa	r	Potentilla erecta	o–f
Empetrum nigrum	o–a	Vaccinium myrtillus	f–co–d
Galium saxatile	a	V. vitis-idaea	o–a
Juncus squarrosus	r–o	Viola sp.	o

Species of *Cladonia* are locally abundant or sub-dominant on some of the sites, and *Rhacomitrium lanuginosum* varies from "frequent" to "co-dominant".[1] Four other species of lichen were noted:

*Cerania vermicularis	Cetraria glauca
*Cetraria aculeata forma hispida	C. islandica

* "Highland" or arctic-alpine species.

On the lower slopes of the northern face of Cader Idris, but still at a relatively high altitude (1500–1800 ft., or 457–550 m.), in the neighbourhood of springs and flushes, under a band of calciferous rocks, there are small detached patches of mixed grassland locally dominated by *Festuca ovina*, *Agrostis tenuis* or *Poa annua* accompanied by a number of lowland species more than half of which are also recorded from Ben Lawers arctic-alpine grassland, and some of which are present as arctic-alpine varieties or forms. Among these are the following, besides the local dominants just mentioned:

Lower Festucetum. Moist facies

Achillea millefolium	f	Prunella vulgaris	f
Bellis perennis	la	Ranunculus acris	o
Campanula rotundifolia	l	R. repens	o
Cerastium vulgatum	o	Rumex acetosa	o–f
Cirsium palustre	la	R. acetosella	o–f
Dactylis glomerata	o	Sagina procumbens	o
Euphrasia officinalis	la	Sieglingia decumbens	o
Ficaria verna	o	Stellaria media	o
Leontodon autumnalis	l	Thymus serpyllum	o
Oxalis acetosella	o–f	Trifolium repens	la
Plantago lanceolata	f	Veronica officinalis	o
P. major	o	Viola sp.	o

Out of a total of fifty-nine species of vascular plants recorded from the Ben Lawers grassland and forty-four from Cader Idris, twenty-nine are common to both.

[1] Compare also the Festucetum ovinae with *Rhacomitrium* on the summit of Slieve Donard (p. 758 and Fig. 150).

These two communities on Cader Idris, taken together, may fairly be regarded then as the equivalent of the Ben Lawers grassland, the dry and the moist facies being separated on Cader Idris, which also has the arctic-alpine element very feebly represented. The lower is transitional, as Price Evans says, to the *Agrostis-Festuca* grassland of lower altitudes (Chapter XXVI).

UPPER ARCTIC-ALPINE ZONE

Above the level of the continuous high-level moorland or arctic-alpine grassland, i.e. at an altitude of about 3000 ft. (900 m.) or more on the higher mountains, we come to the most typical arctic-alpine communities, consisting very largely of "highland" species. The habitats may be broadly distinguished as "exposed" and "sheltered". The exposed higher slopes, for

Exposed habitats

Mountain-top detritus

example the northern slopes of Ben Lawers, and the summits, are covered with so-called "mountain-top detritus", a waste of loose rocks and rock debris which has accumulated *in situ* from the continuous action of frost and wind on the solid rock of the higher mountain slopes and summits, and is often of great depth. This is the most extreme habitat and supports only an open community of mosses and lichens with a few associated phanerogams, which, on the higher mountains, are almost exclusively arctic-alpine species; or a closed Rhacomitrietum (often called a "*Rhacomitrium* heath") with fewer "highland" species of flowering plants and more heath plants which depend on the accumulation of raw acid humus by the dominant moss.

In strong contrast to the exposed habitats are the *corries* with their sheltered hollows, fissures and ledges containing pockets of mineral debris

Protected habitats

and soil, with little humus, but moist and well aerated. These are much the most favourable habitats at the high altitudes and support by far the greater number of the total list of arctic-alpine species. W. G. Smith groups the plants of these habitats together as

Chomophytes

chomophytes—a term due to Oettli, from χῶμα a heap or pile of earth. Along with the numerous arctic-alpine species the fissures and ledges of the corries also hold a number of lowland plants. This vegetation is dealt with in the next chapter.

FORMATION OF MOUNTAIN-TOP DETRITUS

The succession. "In the earlier phases [of succession] lichens form crusts and xerophytic mosses gradually cover the rock detritus and bridge the gaps, till, by their growth and decay, cushions of humus become available for the growth of flowering plants. At the same time fine soil collects in the pockets amongst the stones.

"The formation includes two extreme phases. There is an open plant community on a substratum where stones and boulders are still visible, but are covered with crusts of lichens and a partially discontinuous carpet of moss and lichen (moss-lichen open community).

Moss-lichen associes This leads by gradations to the *Rhacomitrium* heath in which the substratum is often completely hidden by masses of woolly fringe-moss (*Rhacomitrium*) which frequently form a thick layer of humus.

Rhacomitrium heath This latter is the final or closed stage of the formation, but at the lower altitudes it may advance a stage further and become a community of ericaceous shrubs. The final stage has however a precarious existence on the summit ridges, as the moss carpet may frequently be observed completely torn up by wind so that the rocky floor is exposed, and the succession begins again. No hard and fast line can be drawn between the extreme communities, and all transitional stages between them may frequently be seen within a limited area (Pl. 129).

"The formation coincides topographically with the 'alpine plateau' of Robert Smith. The two most marked communities are described in the Faröes by Ostenfeld as the 'alpine formation on the rocky flat' and the '*Grimmia (Rhacomitrium)* heath'; the species given by Ostenfeld correspond closely with those recorded on the Scottish mountains. The *Rhacomitrium* community was recognised by R. Smith (1900):—'above 3000 ft. (c. 910 m.), and even on bare exposed places at lower elevations, such as the summits of hills, the ground is usually stony and sparsely covered with vegetation. Mosses and lichens dominate: in particular, the woolly fringe-moss (*Rhacomitrium lanuginosum*) which on many of the steep mountain rubbles forms the peat on which the alpine humus plants develop.'[1]

Status of the community. "The arctic-alpine formation of mountain-top detritus is primarily climatic and represents the extreme limit of plant-life on an extremely porous substratum in places greatly exposed to wind, subject to extremes of temperature for the greater part of the year and probably under snow for a considerable period. On the one hand, it may be regarded as an outpost of vegetation represented in the extreme case by the first lithophytes on bare rock, and progressing towards a condition of permanent covering. On the other hand, passing upwards from below it appears as the remnant of the grass slope and the ericaceous moorland of the lower arctic-alpine zone, with all species eliminated which cannot adapt themselves successfully to the more extreme ecological conditions. At the same time the vegetation is recruited as regards 'highland' species, e.g. *Silene acaulis*, *Minuartia (Arenaria) sedoides*, *Saxifraga oppositifolia* on

---

[1] Pethybridge and Praeger (1904, p. 168) record *Rhacomitrium* "associations" from Killakee and other summits in Ireland below 2000 ft.; two forms of this community are recognised by them, one in which this moss forms high bosses or cushions on wet boggy moorland and another in which *Rhacomitrium* forms a flatter carpet over granite debris. The former seems to be identical with the *Rhacomitrium* hummocks of blanket bog (pp. 716, 718), but the latter is analogous to that described in the text, though arctic-alpine species play no part.

PLATE 129

Phot. 319. Mountain top detritus on summit of Cross Fell (northern Pennines), alt. 2900 ft. (c. 880 m.). *Rhacomitrium lanuginosum* between the boulders. *J. Massart.*

Phot. 320. Near the summit of Cross Fell. *Carex rigida, Rhacomitrium lanuginosum. J. Massart.*

Phot. 321. "*Rhacomitrium* heath", with open associes of mosses and lichens in foreground and distance. Summit of Mynydd Moel, 2700 ft. (c. 820 m.), Cader Idris. From *E. Price Evans* (1932).

VEGETATION OF MOUNTAIN TOP DETRITUS

Lawers, from the more sheltered corries where these species find their most favourable habitat.

Habitat conditions. "The surface is dry and well-drained, either because of its porous character or from steepness of slope, but the frequency of rain, snow or mist makes it periodically wet. There is little or no soil layer, so that the competition of the closed moorland communities is excluded. The plant covering is distinctly xerophilous in response to frequent dry periods as a result of wind action, less frequently to insolation and lack of precipitation. The growth forms are those indicated by Warming as characteristic of the arctic mat-vegetation and 'Felsenflur', the moss tundra or heath, and the lichen tundra or heath. Most of the species of flowering plants are deeply rooted amongst the stones and boulders in pockets of soil which are deeper than a superficial glance might suggest. Advantage is taken of any shelter afforded by boulders, so that depressions are more rapidly occupied than the more exposed ridges. An observation (supplied by Mr W. E. Evans) illustrates this; while ascending the Cairngorms, the stony waste appeared devoid of green plants, but on looking down the slope from above there was a distinct green tint, due mainly to *Juncus trifidus* growing in the shelter afforded by the slightly raised edges of a series of terraces of rock debris.

Life forms. "The more characteristic growth [life] forms on the drier substrata are as follows: cushion plants (*Silene acaulis, Minuartia* (*Arenaria*) *sedoides*), low mats of decumbent intertwined branches (*Loiseleuria procumbens, Sibbaldia procumbens, Empetrum nigrum, Saxifraga oppositifolia* which also has a cushion form), mats formed by rhizomes and shoots intertwined just below the surface (*Luzula spicata, Carex rigida*), mats with rhizomes deeply buried (dwarf *Vaccinium myrtillus, Salix herbacea*), rosettes (dwarf *Ranunculus acris, Festuca ovina* f. *vivipara*). These growth forms give much protection, and this is further increased by hairs and other protective adaptations. Most of the species are well adapted for vegetative propagation and several are viviparous.

Recruitment. "Where the moss-lichen associes is adjacent to exposed rock-faces and corries with a rich flora (as on Ben Lawers) the list of species is considerably increased. This indicates a great wastage of plant life under the extreme conditions of this formation, a loss which can only be replaced from habitats more favourably situated. While the closed *Rhacomitrium* association would appear to increase the shelter for other species, it is noteworthy that the proportion of 'highland' species is generally less in it than on the more open stony waste" (W. G. Smith, 1911, pp. 310–13).

Flora of moss-lichen associes. This is the pioneer open associes on mountain-top detritus developed on summits exposed to extreme variations of temperature and humidity. While the general physiognomy is extremely uniform the micro-habitats differ very much according to whether they are moist or dry. The associes consists of a great assemblage of species and

varieties of bryophytes and lichens, which include not only a large number
of strictly arctic-alpine forms, but also many species that are common at
low altitudes, as for example the following very widely distributed species:
*Polytrichum piliferum, Dicranum scoparium, Dicranella heteromalla, Eurhyn-
chium myosuroides, Hypnum cupressiforme, Hylocomium splendens, H. triquetrum, Cephalozia bicuspidata, Peltigera canina, Cladonia sylvatica.* These
are accompanied by a much smaller number of scattered vascular plants,
nearly all arctic-alpine species.

The lists of mosses (70), liverworts (39) and lichens (172) given by Watson
(1925) in his comprehensive study of this vegetation are reproduced in full
below. Asterisks are attached to the names of species or varieties which
are strictly arctic-alpine:

Mosses

*Andreaea alpina (moister places)	o
A. crassinervia (usu. on rock)	f
A. nivalis (moister places)	l
A. petrophila	a
*var. alpestris (rocks)	o
Blindia acuta	f
*Brachythecium glaciale	r and l
*B. plicatum	r and l
B. rutabulum	o
Bryum pseudotriquetrum (moist)	o
Ceratodon purpureus	o
*Conostomum boreale	f
Dichodontium pellucidum (moist)	o
*var. fagimontanum (moist)	o
Dicranella heteromalla	o
var. interrupta	o
Dicranoweisia crispula	lf
Dicranum falcatum	lf
D. fuscescens	a
*var. congestum	f
*D. molle (usu. on ledges)	r
D. schisti	lf
D. scoparium	o
var. spadiceum	o
D. starkei (often on rock)	o
*Ditrichum zonatum	f
*var. scabrifolium	
*Eurhynchium cirrosum	r and l
E. myosuroides	o
E. swartzii	o
Hylocomium loreum	a
*H. pyrenaicum	l
H. rugosum	l

Hylocomium splendens	a
H. squarrosum	o
H. triquetrum	o
Hypnum callichroum	la
*H. bambergeri	o
H. cupressiforme	f
var. ericetorum	f
*H. hamulosum	f
H. incurvatum	r
H. molluscum	lf
*H. procerrimum	l and r
*H. revolutum	r
H. schreberi	la
Oedipodium griffithianum (usu. rock nooks)	l
Oligotrichum hercynicum	a
Orthothecium rufescens	o
Plagiothecium denticulatum	f
var. obtusifolium	la
P. elegans	o
*P. muhlenbeckii	l
*Polytrichum aloides	o
*P. alpinum	a
P. commune	o
P. juniperinum	o
P. piliferum	a
*P. sexangulare	la
Pterigynandrium filiforme	o
Rhacomitrium canescens	o
R. heterostichum	o
var. gracilescens	a
R. lanuginosum	a
Swartzia montana	f
*var. compacta	f
Webera albicans (moist)	o
W. elongata	o
*W. ludwigii (moist)	o
W. nutans	a

Liverworts

*Alicularia breidleri	lf	Gymnomitrium crenulatum	o
A. scalaris	a	*G. obtusum	o
*Anthelia julacea	a	*G. varians	la
*A. juratzkana	a	Leptoscyphus taylori	o
Bazzania triangularis	o	*Lophozia alpestris	f
B. tricrenata	o	L. floerkii	o
*Cephalozia ambigua	l	L. incisa (damp pl.)	o
C. bicuspidata	f	L. quinquedentata	o
Cephaloziella byssacea	f	L. ventricosa	o
Diplophyllum albicans	a	*Marsupella condensata	la
*D. taxifolium	r	*M. ustulata	o
Eucalyx subellipticus	o	Plagiochila asplenioides	f
Gymnocolea inflata	o	var. minor	lf
f. compacta (drier pl.)	o	*Pleuroclada albescens	lf
f. laxa (moist gr.)	o	Ptilidium ciliare	f
*Gymnomitrium adustum	o	Scapania dentata (moist pl.)	o
*G. alpinum	o	S. irrigua (small form)	o
*G. concinnatum	la	S. undulata (moist pl.)	o
*G. corallioides	o	*Sphenolobus politus	o
*G. crassifolium	la		

Lichens

Alectoria bicolor	lf	*C. nivalis	r
*A. divergens	r	*Cladonia amaurocraea	r
*A. ochroleuca	o	*C. bellidiflora	la
*A. sarmentosa	r	C. cervicornis	a
*var. cincinnata	r	var. subcervicornis	a
*Arthopyrenia bryospila	r	C. coccifera	o
Baeomyces roseus	o	C. crispata	lf
B. rufus	o	C. deformis	o
Bacidia flavovirescens	o	C. degenerans	o
*Biatorella fossarum	l	var. phyllophora	la
*Biatorina contristans	la	*C. destricta	lf
B. cumulata	l	C. flabelliformis	o
Bilimbia lignaria	o	C. floerkeana f. trachypoda	la
B. melaena	la	C. furcata	a
B. sabulosa		var. pinnata	a
*var. montana		C. gracilescens	lf
B. sabuletorum	o	C. gracilis	la
var. simplicior	r	C. pyxidata	o
*B. rhexoblephara	l	C. rangiferina	
*Buellia alpicola (on stones)	l	C. squamosa	o
*B. badioatra (on stones)	l	C. sylvatica	f
var. atrobadia (on stones)	l	C. uncialis	a
B. myriocarpa (on stones)	o	f. adunca	a
*B. pulchella (rock nooks)		f. turgescens	f
*Cerania vermicularis	la	C. verticillata	o
Cetraria aculeata	f	*Collema ceraniscum	r and l
var. alpina	o	Collema tenax	o
C. crispa	o	*Dacampia hookeri	la
*C. cucullata	r	Dermatocarpon cinereum	la
*C. hiascens	r	*var. cartilagineum	lf
C. islandica	a	D. hepaticum	o
*forma platyna	o	D. lachneum	o

*Gyalecta foveolaris o
Gyrophora cylindrica a
*G. erosa (rock) r
G. polyphylla o
*G. torrefracta (rock) r
Haematomma ventosum
(rock) la
Icmadophila ericetorum o
*Lecania curvescens vr
Lecanora badia (rocks) f
*L. epibryon l
*L. geminipara vr
L. polytropa a
L. tartarea f
  *var. frigida la
  *var. gonatodes
L. subtartarea o
*L. upsaliensis r
Lecidea aglaea o
*L. alpestris l and r
*L. arctica la
L. atrofusca (on moss) o
L. auriculata (rocks and
stones) lf
  var. diducens (on rocks
  and stones)
*L. berengeriana l
  *var. lecanodes
*L. breadalbanensis r
L. coarctata o
  var. glebulosa o
  var. elacista o
L. confluens (rocks and
stones) f
*L. consentiens (rocks and
stones) r
L. contigua (rocks and
stones) f
L. contigua v. flavi-
cunda (on rocks and
stones) f
  var. platycarpa (on rocks
  and stones) f
L. crustulata (rocks and
stones) f
*L. cuprea
L. demissa f
*L. deparcula (rocks and
stones) r
L. dicksonii (rocks and
stones) lf
L. epiphorbia r
L. fuscoatra (rocks
and stones) o

L. fuscocinerea (rocks and
stones) l
L. granulosa o
  *var. escharoides la
L. griseoatra (rocks and
stones) o
L. kochiana (rocks and
stones) o
L. lapicida (rocks and
stones) o
*L. limosa la
L. lithophila (rocks and
stones) o
*L. nigroglomerata vr
*L. pycnocarpa (rocks and
stones) l
*L. rhizobola r
L. sanguinaria o
  *var. affinis
L. sorediza (rocks and
stones) o
L. sublatypea (rock) o
*L. tabidula (rocks and
stones) r
*L. tesselata (rocks and
stones) o
L. uliginosa a
*L. vernalis
Leptogium lacerum o
  var. pulvinatum o
Pannaria brunnea f
*P. hookeri l
*Parmelia alpicola (often on
quartz) r
*P. pubescens a
  *var. reticulata
P. saxatilis o
P. tristis (usu. on rock) la
*P. vittata r
*Parmeliella lepidota r
Peltidea aphthosa o
Peltigera canina (small
forms) o
P. polydactyla o
P. rufescens o
*Pertusaria bryontha l
*P. dactylina l
*P. glomerata l
*Pertusaria oculata la
*P. xanthostoma l
*Placynthium delicatulum r
*Platysma fahlunense o
*P. hepatizon la
*P. polyschizum o

*Polyblastia gelatinosa	r	*Solorina bispora	r
P. nigritella	r	*S. crocea	la
*P. sendtneri	r	S. spongiosa	r
*Porina furvescens	r	Sphaerophorus fragilis	o
Rhizocarpon confervoides		S. corallioides	a
(on stones)	la	f. congestus (drier)	a
R. geographicum (on		*Stereocaulon alpinum	f
stones)	a	S. corallioides	f
var. atrovirens (on		S. evolutum	a
stones)	o	S. denudatum	f
R. oederi (on stones)	lf	*S. tomentosum	r
*R. postumum	r	*Thelopsis melathelia	r
*Schizoma lichinodeum	la	*Varicellaria microsticta	l and r

Adaptations to habitat. Watson notes that the species are preponderantly xerophytic, but that many which are common to the lowland and the arctic-alpine habitats show variations in structure which he thinks indicate that they are living in moister conditions on the mountain tops. Thus the hair points of the leaves of *Polytrichum piliferum* and *Rhacomitrium canescens*, well developed in the typical lowland habitat, are often less conspicuous at high levels, and in *Rhacomitrium* may be almost absent. Similar variations are found in the lowlands when the plant grows in ditches. Again *Dichodontium pellucidum, Bryum pseudotriquetrum, Sphenolobus politus, Scapania dentata* and *S. undulata*, usually found in more constantly wet situations at lower altitudes, are able to exist in this exposed community because they can absorb water through the whole of their surfaces.

Plagiothecium denticulatum and *Hypnum schreberi*, on the other hand, usually flourishing best in damp and somewhat shaded situations, have intensified xeromorphic characters in this high-lying habitat.

Vascular flora. The following vascular plants are listed by W. G. Smith (1911) as members of this associes:

*Alchemilla alpina	*Lycopodium selago
*Arabis petraea	*Minuartia (arenaria) sedoides
Deschampsia flexuosa (dwarf)	Nardus stricta
*Draba incana	*Polygonum viviparum
Empetrum nigrum	Ranunculus acris (dwarf)
*Festuca ovina f. vivipara	*Salix herbacea
*Gnaphalium supinum	*Saxifraga oppositifolia
*Juncus trifidus (dwarf)	*S. stellaris (dwarf)
*Loiseleuria procumbens	*Silene acaulis
*Luzula arcuata	Solidago virgaurea
*L. spicata (dwarf)	*Sibbaldia procumbens
*Lychnis alpina (Clova)	

* Arctic-alpine species.

Watson (1925) records *Rhytisma salicinum* on *Salix herbacea* on Ben Doran at 3000 ft., and *Ticothecium erraticum* on the crustaceous lichens.

The moss-lichen associes on Cader Idris. On the summit plateau of Cader Idris, which lies roughly between 2500 and 2900 ft. (762 and 883 m.),

Price Evans (1932, pp. 32–3) describes this associes as "a meagre vegetation consisting chiefly of mosses and lichens" with a few flowering plants. He ascribes its poverty not only to the severe climatic conditions but also to the infertility of the granophyre of which the detritus is composed.

Rhacomitrium lanuginosum is the most abundant and characteristic plant, and associated with it are:

Dicranella heteromalla		Polytrichum piliferum	a
var. interrupta	—	Scapania gracilis	—
Lophozia floerkii	—		
Baeomyces roseus	—	Cladonia furcata	f
*Cerania vermicularis	—	C. gracilis	f
*Cetraria aculeata f. hispida	—	C. rangiformis	f
C. glauca f.	f	C. subcervicornis	a
C. islandica	f	C. uncialis	f
Cladonia cervicornis	f		

Of rock mosses and lichens on the stones of the debris there are:

Andreaea petrophila	Rhizocarpon geographicum
Gyrophora cylindrica	R. oederi
Lecidea contigua f. flavicunda	Rhacomitrium gracilescens
L. dicksonii	*Parmelia pubescens

and of vascular plants only the following:

Campanula rotundifolia	r	*Lycopodium alpinum	a
Empetrum nigrum		*L. selago	r–o
(prostrate)	o	Potentilla erecta	a
Festuca ovina	a	Vaccinium myrtillus	f
Galium saxatile	f	V. vitis-idaea	o

* "Highland" or arctic-alpine species.

The flora of the detritus on the summit of this Welsh mountain is very meagre and the flowering plants are nearly all "lowland" forms, arctic-alpine species being represented only by two species of *Lycopodium*. Nevertheless, the vegetation is certainly a fragmentary example of the same community as that which occurs on the tops of the floristically rich Scottish mountains.

Flora of *Rhacomitrium* heath. As W. G. Smith pointed out (see pp. 787–8) this community arises where a species of *Rhacomitrium* (usually *R. lanuginosum*), already abundant in the moss-lichen associes, has succeeded in establishing continuous dominance, at the same time forming a nidus of raw humus open to the colonisation of such ericaceous plants as can tolerate the extreme conditions of exposure. The bryophytes and lichens are naturally greatly reduced from the enormous numbers occurring in the open associes, owing to the general dominance of *Rhacomitrium*, but, as Watson says, many species of the open associes may still occur in the closed "heath". Though a few species and varieties recorded by Watson for the latter are not included in the lists for the former community, he informs me (*in litt.*) that it is doubtful if there are any bryophytes or lichens in the *Rhacomitrium* heath which may not occur in the open community. Liverworts are rare or absent in the Rhacomitrietum.

The following are recorded:

Bryophytes

Rhacomitrium lanuginosum   d

Dicranum fucescens	o	Rhacomitrium canescens	
D. uncinatum	r	var. ericoides	o
Hylocomium loreum	o	R. heterostichum	
H. splendens	o	var. gracilescens	la
Hypnum cupressiforme	o	Polytrichum alpinum	o
H. schreberi	o		
Leptoscyphus taylori	o	Ptilidium ciliare	o

Lichens

Alectoria nigricans	la	Cladonia flabelliformis	o
*A. ochroleuca	o	C. furcata	la
Bilimbia melaena	r	C. gracilis	o
*Cerania vermicularis	o	C. rangiferina	a
Cetraria aculeata	o	C. sylvatica	a
f. subnigrescens	o	C. squamosa	o
*var. alpina	o	C. uncialis	a
C. crispa	la	f. obtusata	a
C. islandica	la	f. turgescens	a
*Cladonia bellidiflora	o	Galera hypnorum	r
C. deformis	la	Icmadophila aeruginosa	r
C. degenerans		Lecanora tartarea	o
f. pleolepidea	o	Peltigera canina	o
*C. destricta	l		

The following is an amended list of the vascular plants given by W. G. Smith (1911) for the *Rhacomitrium* heath:

Arctostaphylos uva-ursi	Empetrum nigrum
*Arctous (Arctostaphylos) alpina	Euphrasia officinalis
*Astragalus alpinus	Galium saxatile
Calluna vulgaris	*Lycopodium alpinum
Campanula rotundifolia	Potentilla erecta
*Carex rigida	Rumex acetosa
*Cerastium alpinum	*Salix herbacea
*C. arcticum	Vaccinium myrtillus
*Cornus suecica	*V. uliginosum
Deschampsia flexuosa	V. vitis-idaea

For the *Rhacomitrium* heath on the summit plateau of Cader Idris on thin black peat with a *p*H value of 5·6 (Pl. 129, phot. 321), Price Evans gives:

Rhacomitrium lanuginosum   d

*Cetraria aculeata f. hispida	o	Cladonia uncialis	f
C. islandica	la		
Cladonia furcata	f	Hylocomium loreum	o
C. gracilis	f	Hypnum schreberi	o
C. rangiferina	f	Polytrichum piliferum	o–f
C. sylvatica	f		

Campanula rotundifolia	r–o	*Lycopodium alpinum	la
Carex pilulifera	o	*L. selago	o
Empetrum nigrum (dwarf)	r–o	Potentilla erecta	o
Festuca ovina	o–f	Vaccinium myrtillus (dwarf)	o–f
Galium saxatile	a	V. vitis-idaea	o

The species of *Cladonia* and *Cetraria* are sometimes co-dominant with *Rhacomitrium*. On exposed parts of the summit plateau "*Rhacomitrium* moor" occurs. This is a community dominated by the same moss, but moister and with more peat formation. It has several of the same species as *Rhacomitrium* heath but without the Cladoniae, and with *Juncus squarrosus*, *Luzula maxima*, *Nardus stricta*, *Eriophorum vaginatum*, *Plagiothecium* sp., *Sphagnum* sp. and *Polytrichum commune* (f–ld).

The summit plateaux of many British mountains have the same type of vegetation but an extremely poor flora. *Rhacomitrium lanuginosum* and *Carex rigida* are fairly constant species (Pl. 129, phots. 319, 320).

REFERENCES

BUCHAN, A. and OMOND, R. T. The meteorology of the Ben Nevis observatories. Part III. *Trans. Roy. Soc. Edinb.* **43**. 1905.

LEACH, W. A preliminary account of the vegetation of some non-calcareous British screes (Gerölle). *J. Ecol.* **18**, 321–32. 1930.

PRICE EVANS, E. Cader Idris: a study of certain plant communities in south-west Merionethshire. *J. Ecol.* **20**, 1–52. 1932.

RAUNKIAER, C. Livsformernes Statistik som Grundlag for biologisk Plantegeografi. *Bot. Tidskr.* **29**, 1908. Translated as Chapter IV of *The Life forms of Plants and Statistical Plant Geography*. Oxford, 1934.

SMITH, R. Botanical Survey of Scotland. Part II. North Perthshire. *Scot. Geogr. Mag.* 1900.

SMITH, W. G. "Arctic-alpine Vegetation", Chapter XIII of *Types of British Vegetation*, 1911, pp. 288–329.

WATSON, W. The Bryophytes and Lichens of Arctic-alpine Vegetation. *J. Ecol.* **13**, 1–26. 1925.

WILLIS, J. C. and BURKILL, I. H. The Phanerogamic Flora of the Clova Mountains in special relation to Flower Biology. *Trans. Bot. Soc. Edinb.* **22**, 109–25. 1901–4.

Chapter XXXIX

ARCTIC-ALPINE VEGETATION (*continued*)

"SNOW-PATCH" COMMUNITIES. VEGETATION OF PROTECTED HABITATS. LITHOPHYTES AND CHOMOPHYTES

"SNOW-PATCH" COMMUNITIES

Effect of snow lie. The length of time for which snow lies during the year has an important effect in differentiating arctic-alpine plant communities. Snow, as we have seen, is very effective in protecting the low-growing vegetation during winter. This can be well observed in places where the snow can get no permanent lodgment because it is constantly removed by the wind. Such places are either quite barren or support a scanty population of the most resistant species. On the other hand, where snow lies very late, so that the soil is exposed for a short time only during the growing season, vegetation is necessarily much restricted, and in extreme cases, where the soil is uncovered on the average for a few weeks or even for a few days only, plants may not have time to develop at all. This can be very well seen in late August in the Scandinavian mountains. Some patches of soil where the last snow is only just melting are obviously quite barren and will necessarily remain so during the two or three weeks before snow falls again in September. Other places, where the last snow disappeared in July, support a few scattered plants.

In situations where the snow lies late, but the soil is exposed long enough for a well-developed vegetation to flourish, these peculiar conditions may determine a special vegetation. "The chief factors are patches of fine silty soil derived from dust accumulated in the snow, and an abundant supply of cold snow-water by which the soil is saturated: then as drying proceeds, there results a slippery slime, and finally, in summer, a dry cracked surface" (W. G. Smith, 1911). The Swiss botanists have long recognised the peculiar vegetation of such areas ("Schneeflecken" and "Schneetälchen") in the Alps. Our knowledge of the Scottish representatives is mainly due to W. G. Smith (1912) and Watson (1925).

Like the arctic-alpine communities previously described the snow-patch vegetation is at first mainly composed of bryophytes and lichens, and a liverwort, *Anthelia juratzkana*, is the dominant pioneer. According to Watson (1925) three of Rübel's five Swiss snow-patch communities are represented in the British Isles—Anthelietum, Polytrichetum sexangularis and Salicetum herbaceae. These apparently stand in successional relationship, the pioneer *Anthelia* forming a crust on the soil surface, which may then be colonised by *Polytrichum* (*P. sexangulare* and *P. alpinum*), and later by vascular plants, of which the most constant and sometimes the dominant is *Salix herbacea.*

The following is a collective list of species recorded from these communities, mostly taken from Watson (1925):

Lichens

Alectoria bicolor	o	Cetraria crispa	la
*Cerania vermicularis	la	C. islandica	a
Cetraria aculeata	—	*Lecidea consentiens	la
*var. alpina	—	*Solorina bispora	la

Liverworts

Alicularia breidleri	—	Gymnomitrium crassifolium	la
A. scalaris	a	G. varians	la
*Anthelia julacea	a	Lophozia alpestris	—
*A. juratzkana	ld	L. floerkii	—
Cephalozia bicuspidata	o	L. ventricosa	o
Diplophyllum albicans	a	*Marsupella condensata	la
D. taxifolium	—	*M. ustulata	o
*Gymnomitrium adustum	o	*Moerckia blytti	la
*G. alpinum	o	*Pleuroclada albescens	ld
G. concinnatum	—	Ptilidium ciliare	—
G. coralloides	—	Scapania undulata	o

Mosses

*Brachythecium glaciale	o	Oligotrichum hercynicum	la
*B. plicatum	o	*Polytrichum alpinum	o
*Bryum arcticum	o	*P. sexangulare	ld
*B. muhlenbeckii	—	Rhacomitrium fasciculare	o
Dicranum falcatum	—	Webera commutata	f
D. schisti	—	W. cucullata	o
D. starkei	—	W. nutans	a

* Arctic-alpine species.

Watson remarks that the whitish appearance of *Pleuroclada albescens,* from which its name is derived, is due to the moribund condition of the plant after the snow has melted. Underneath the snow-patch, if sufficient light and air have access, this liverwort is quite green and active.

The flowering plants recorded from Scottish snow-patch vegetation are few, all but one arctic-alpines:

*Alchemilla alpina	*Juncus triglumis
*Cochlearia micacea	*Loiseleuria procumbens
Galium saxatile	*Phleum alpinum
*Gnaphalium supinum	*Salix herbacea
*Juncus biglumis	*Saxifraga stellaris

VEGETATION OF RELATIVELY PROTECTED HABITATS

In habitats protected from the more violent winds conditions are naturally much more favourable for vegetation in general and though the plant habitats based on mountain-top detritus are far more extensive they possess many fewer species of flowering plants than the sheltered corries, ledges and fissures, which are by far the richest collecting grounds for the rarer species of all classes. Here the conditions are very varied, owing to extreme fragmentation of the topography; and in place of extensive and more or less uniform communities we have small groups of plants constantly varying from place to place. The dynamic factors of the habitat—constant erosion by frost, wind and water—bring about continual change, so that the vegetation furnishes typical examples of "migratory" communities. From another point of view these represent early stages of succession which never develop into climax vegetation.

Lithophytes, chomophytes and chasmophytes. The stations available for plant growth are of two kinds: exposed rock surfaces inhabited by *lithophytes*, and the piles and pockets of debris resulting from erosion and supporting the so-called *chomophytes*. On the Scottish mountains the arctic-alpine chomophyte communities reach their greatest development in the corries of the mica-schists. Plants inhabiting rock fissures are often called *chasmophytes*, but here they will not be separated from chomophytes, since the two groups pass into one another. These habitats with several of the species they bear are shown in Pls. 130–132.

Succession. "The succession begins on the bare rock faces with lithophytes, coatings of algae and lichens, followed by typical rock-mosses (*Andreaea*, etc.); flowering plants rarely occur in this phase. As weathering proceeds, and soil accumulates in fissures, the chomophytes become established. Xerophilous species (*Minuartia* (*Arenaria*) *sedoides, Silene acaulis, Saxifraga oppositifolia*) here precede the mesophytes, which will ultimately occupy the deeper deposits of fine soil and humus wherever they are accumulated. Later still, as the ledges become more or less obliterated by soil accumulations, and as the crevices become filled up, there is invasion by surface-rooting and mat-forming plants ('exochomophytes' of Oettli) from the closed plant communities of the grassland or heath of the lower arctic-alpine or sub-alpine zone. This last phase is well seen on many Scottish hills (e.g. Ben Lawers), where the lower terraces of the corrie area are covered with closed grassy swards, although they are clearly parts of the same system of rock exposures as the ledges and ravines bearing open communities at a higher level. There is reason to believe that if erosion of the rocks ceased, the whole system of rock-ledges and corries would ultimately become closed plant communities, and, as is the case now on the closed terraces, the characteristic arctic-alpine chomophytes would be largely exterminated" (W. G. Smith, 1911, pp. 319–20).

Screes. "The same phases may be traced on screes, beginning with lithophytes and ending with a closed vegetation. Mosses play a larger part in the intermediate phases, they become established on the blocks, and extend over the narrower cavities, and by their growth and decay provide a humus covering, but frequency of drought prevents any but xerophilous flowering plants from becoming established. Within the larger interstices, vegetation is restricted to liverworts, mosses and such species as can live in deficient light, but as soil and humus accumulate the conditions become not unlike those of the chasmophytes or shade chomophytes of fissures in the rock face."

Humidity. "Other factors, besides the progressive disintegration of rock, obviously play an important part in modifying the vegetation, especially the supply of moisture, and its conservation for plants through shelter from wind. Atmospheric humidity is probably quite as essential for many species as soil moisture. Lithophytes and other plants on the rock faces depend mainly on aerial sources for moisture, and are therefore subject to recurrent drought. On the ledges and in crevices there is frequently telluric water available, more constant in supply and richer in food materials. Sometimes, as on Ben Lawers, this supply is so ample that throughout the summer water continues to run over extensive ledges and forms the sources of numerous streamlets; this is the habitat for the community of hydrophilous chomophytes referred to later. Another aspect of water-supply is the temporary supply of cold water from melting snow which in early summer can be seen trickling down the rocks in streamlets or spreading over ledges that later become comparatively dry. Here we have a 'snow-valley' (Schneetälchen) and with it the occurrence of certain species of plants, e.g. *Saxifraga rivularis, Veronica alpina, Arabis petraea*" (W. G. Smith, 1911, pp. 320–1).

Lithophytes. "The first plants to find a footing on the bare rock are algae and lichens, the former requiring a damper substratum than the latter in order to become noticeable constituents of the vegetation, though *Trentepohlia* gives reddish-orange hue to many fairly dry rocks. The structure of some lichens renders them specially able to act as lithophytes. Collemoid lichens seldom occur except when the rocks are almost constantly damp and are rarely very dry or very wet. Crustaceous lichens are very abundant on dry rocks, and those which are common are widely distributed species frequent at lower elevations, e.g. *Lecanora parella, L. tartarea, L. atra, L. polytropa, L. contigua, L. sorediza, L. crustulata, Rhizocarpon confervoides* and *R. geographicum....*

"A slight amount of erosion enables such lichens as *Dermatocarpon miniatum*, and species of *Gyrophora, Umbilicaria, Stereocaulon*[1] and *Sphaerophorus* to establish themselves on exposed rock faces, and these are accompanied or followed by mosses, species of *Andreaea* and *Grimmia* being the earliest representatives, different species of the same genus varying in their powers of establishment. For example the very rare moss *Blindia caespiticia* occurs in some of the small crevices, its near relative, *B. acuta*, being scattered over the face of the same rock. Some pleurocarpous mosses appear later, and then phanerogamic chomophytes become well established in fissures or on rock ledges. In this succession hepatics are rare at first—gradually

---

[1] Pl. 132, phot. 332.

becoming more abundant till on ledges they often form the chief constituents. The opposite is true of lichens" (Watson, 1925, p. 13).
Lichens.

On rocks in earlier stages of erosion

*Acarospora admissa	—	L. picea	r
A. discreta	—	L. polytropa	a
*A. peliocypha	—	*var. alpigena	—
*Aspicilia alpina	r	L. subtartarea	a
*A. chrysophana	r	L. tartarea	a
A. cinerea	o	Lecidea albocoerulescens	—
*A. cinereorufescens		*L. armeniaca	—
*f. diamarta (moist)	r	L. auriculata	—
*A. depressa	o	var. diducens	—
A. gibbosa	o	L. cinerascens	—
*A. leucophyma	r	*L. commaculens	—
*A. pelobotrya	r	L. confluens	—
*A. superiuscula	r	L. contigua	a
Bacidia flavovirescens var.	—	var. flavicunda	a
alpina	o	var. platycarpa	—
Biatorella simplex	o	*L. contiguella	—
Biatorina biformigera	—	L. crustulata	—
B. candida	—	L. dealbatula	—
*B. confusior	—	*L. deparcula	r
*B. contristans	—	L. dicksonii	la
*B. cumulata	—	L. fuscoatra	—
*B. rhypodiza	r	L. goniophila	—
*Buellia alpicola	r	L. griseoatra	—
*B. badioatra	r	L. kochiana	—
*var. atrobadia	r	L. lapicida	—
B. colludens	—	L. leucophea	—
*B. deludens	r	L. lithophila	—
*Callopisma siebenhaariana	r	L. mesotropoides	r
Coenogonium ebeneum		*L. nigroglomerata	—
(damp)	—	L. plana	—
Dermatocarpon miniatum		L. polycarpa	—
var. complicatum (moister)	—	*L. pycnocarpa	—
D. hepaticum	—	L. rivulosa	la
*Gyrophora arctica	—	L. sanguinaria	—
G. cylindrica	a	L. sorediza	—
*var. delisei	—	*L. subgyratula	—
*G. erosa	—	L. subkochiana	—
*G. hyperborea	—	L. sublatypea	—
*G. leiocarpa	—	*L. tabidula	—
G. polyphylla	—	*L. tessellata	—
G. polyrrhiza	—	*L. umbonella	—
G. proboscoidea	—	Microglaena bread-	
*G. torrefracta	—	albanensis	—
Haematomma ventosum	a	M. corrosa	—
*Lecanora austera	r	*var. nericensis	—
L. badia	f	Microthelia exerrans	—
*L. frustulosa	—	*Parmelia alpicola	—
L. intricata	f	P. conspersa	f
*var. leptacina	—	var. stenophylla	la
L. parella	a	*P. encausta	la

* Arctic-alpine species and varieties (and so throughout the lists).

Parmelia omphalodes	f	P. theleodes	—
*P. pubescens	a	Porina chlorotica	o
Pertusaria ceuthocarpa	—	Racodium rupestre (damp)	—
P. concreta	—	Rhizocarpon confervoides	a
P. dealbata	—	*R. geminatum	r
P. lactea	—	R. geographicum	a
P. monogyna	—	R. obscuratum	o
*Placodium elegans	o	R. oederi	la
*Platysma fahlunense	—	R. petreum	la
*P. hepatizon	f	*R. plicatilis	r
*P. polyschizum	—	*Squamaria chrysoleuca	r
Polyblastia fuscoargillacea	—	S. gelida	f
P. intercedens	—	Thelidium papulare	o
*P. scotinospora		Umbilicaria pustulata	—

On rocks in later stages of erosion

Cetraria aculeata (usually on ground)	a	Ephebe lanata	o
		*Europsis granatina	—
C. islandica (usually on ground)	a	*Leptogium glebulentum (moist)	—
C. nivalis	lf	L. rhyparodes (wet rocks)	lf
C. odontella	—	Peltidea aphthosa (usually on ground)	—
*Cladonia bellidiflora	—		
C. cervicornis	—	Peltigera canina (usually on ground)	—
f. stipata	la		
var. subcervicornis	a	P. polydactyla (usually on ground)	—
C. furcata	—		
C. pyxidata	—	P. rufescens (usually on ground)	—
C. rangiferina	—		
C. squamosa	—	*Psorotichia furfurella	r
C. sylvatica	—	*Pterygium pannariellum	—
C. uncialis	—	*Pyrenopsis furfurea (damp)	r
C. verticillata	—	P. fuscatula	—
*Collema ceraniscum	r	*P. haematopsis	—
C. granuliferum	—	*P. homoeopsis	—
C. multifidum	—	*Schizoma lichinodeum	lf
f. marginale	—	*Solorina crocea [1]	—
C. tenax	o	Sphaerophorus compressus	—
Coriscium viride (usually on ledges)	—	S. corallioides	a
		f. congestus	a
Ephebe hispidula	—	Thermutis velutina (damp)	r

Liverworts. These plants are rarely present on rocks unless they are wet or shaded or with a soil-cap, but they are often abundant on rocky ledges and may there form conspicuous members of the community.

A. On rocks in the early stages of erosion

*Chandonanthus setiformis	*Radula lindbergii var. germana
Frullania tamarisci	*Sphenolobus saxicolus
*Gymnomitrium obtusum	

[1] Pl. 132, phot. 333.

B. *On rocks which are moist, shaded or partly covered with a thin soil cap*

Alicularia scalaris
*Anthelia julacea
*A. juratzana
Diplophyllum albicans
*D. taxifolium
*Gymnomitrium alpinum
*G. concinnatum
  *var. intermedium

*Gymnomitrium coralloides
G. crenulatum
Herberta adunca
Lophozia floerkii
*L. quadrifolia
*Marsupella sparsifolia
Plagiochila asplenioides var. minor
Ptilidium ciliare

Mosses.

A. *On rocks or boulders in the earlier stages of erosion*

*Andreaea alpina
A. crassinervia
*A. nivalis
A. petrophila
  *var. alpestris
  var. acuminata
A. rothii
Blindia acuta
*B. caespiticia
*Campylopus schwarzii
Dichodontium pellucidum (moist)
  *var. fagimontanum
*Dicranum elongatum
*D. falcatum

D. fulvellum (clefts)
D. longifolium
D. schisti
Grimmia apocarpa
  *var. alpicola (wet)
  var. pumila
  var. rivularis
*G. atrata (wet)
G. conferta
G. doniana
G. funalis
G. torquata
G. trichopylla

B. *In cracks, on soil caps or on wet rock faces, etc.*

Anoectangium compactum
Bartramia oederi
*Brachythecium glaciale
B. glareosum
*B. plicatum
B. rutabulum
*Bryum arcticum
B. pseudotriquetrum
Ceratodon purpureus
*Conostomum boreale
Cynodontium gracilescens
*C. virens
*C. wahlenbergii
Dicranella heteromalla
Dicranoweisia crispula
Dicranum fuscescens
  *var. congestum
D. scoparium
D. starkei
Encalypta ciliata
*E. commutata
  *var. imberbis
E. rhabdocarpa
*Eurhynchium cirrhosum
E. myosuroides
Fissidens osmundoides

*Grimmia atrata (wet)
*G. alpestris
*G. unicolor
*Hylocomium pyrenaicum
Hypnum bambergeri
H. callichroum
H. cupressiforme
*Hypnum halleri
*H. hamulosum
*H. sulcatum
*Mnium lycopodioides
M. orthorrhynchum
M. serratum
*M. spinosum
Myurella apiculata
M. julacea
Orthochecium rufescens
*Plagiobryum demissum
Plagiothecium denticulatum
  var. obtusifolium
*P. muhlenbeckii
P. pulchellum
*Polytrichum alpinum
*Pseudoleskea atrovirens
*P. catenulata
*P. patens

*Pseudoleskea striata var. saxicola	Swartzia montana
Pterigynandrium filiforme	Trichostomum tortuosum
Rhacomitrium fasciculare	*var. fragilifolium
R. heterostichum	Webera commutata
var. gracilescens	W. nutans
R. lanuginosum	Zygodon lapponicus
R. ramulosum	

Vascular plants. Terrestrial vascular plants are never strict lithophytes, at least in this country, since they must have *some* soil or humus in which to root. In the foregoing lists of bryophytes and lichens, however, many forms are included which are equally unable to colonise bare rock surfaces, but require at least a thin layer of disintegrated mineral or organic material. A distinction is drawn in Watson's lists between the species living on rocks "in the earlier stages of erosion", corresponding roughly to lithophytes proper, and those which settle in cracks, on thin soil over the tops of rocks, or on wet rocks where a thin film of organic substance is provided by algae. It is obvious that no hard and fast line can be drawn between such habitats and those providing enough soil to support the typical chomophytes of arctic-alpine vegetation. W. G. Smith (1911, p. 322) gives a list of thirty flowering plants and ferns forming "open communities on exposed rock faces", very nearly all arctic-alpine species, the habitat being an

on open ledges

"open ledge without much soil" (the habitat of *Erigeron alpinus* can be seen from Pl. 130, phot. 324), but these, he says, generally have "a single rootstock extending into a fissure, within which a branching root system anchors the plant firmly"; and this is obviously a character which renders the plant independent of *surface* soil. The following species are included:

*Alchemilla alpina[1]	*Minuartia sedoides[4]
*Arabis petraea	*Oxytropis uralensis (rare[2])
*Asplenium viride	*Potentilla crantzii
*Astragalus alpinus (rare[2])	*Sagina nivalis
*Bartsia alpina (rare[2])	*Salix reticulata[5]
Campanula rotundifolia	*Saxifraga cernua[6]
*Cerastium alpinum	*S. hypnoides
Cystopteris fragilis	*S. nivalis
*Draba incana	*S. oppositifolia
*D. rupestris	*Sedum rosea[7]
*Dryas octopetala	*Sibbaldia procumbens[8]
*Erigeron alpinus (rare[2])[3]	*Silene acaulis
*Juncus trifidus	Thymus serpyllum
*Luzula spicata	*Woodsia alpina
*Minuartia rubella	

[1] Pl. 131, phot. 328.

[2] "Rare" means rare within the habitat: the majority of the species are "rare" in the ordinary sense because their habitats are very restricted.

[3] Pl. 130, phot. 324. [4] Pl. 131, phot. 328. [5] Pl. 130, phot. 323.

Pl. 132, phot. 330. [7] Pl. 130, phot. 325. [8] Pl. 132, phot. 331.

PLATE 130

Phot. 323. *Salix reticulata* and *Rhacomitrium lanuginosum* on a dry ledge near the summit of Ben Lawers, alt. c. 3500 ft. (1060 m.). *F. F. Laidlaw.*

Phot. 326. Hydrophilous chomophytes—*Webera albicans* var. *glacialis*, *Chrysosplenium alternifolium*. *F. F. Laidlaw.*

Phot. 325. *Sedum rosea (rhodiola)* and *Alchemilla alpina* on steep rocks, alt. c. 3600 ft. *J. Massart.*

Phot. 322. Arctic-alpine grassland on Ben Lawers, alt. 2700 ft. (c. 820 m.). *Alchemilla alpina, Festuca ovina f. vivipara*, etc. Abundance of mosses and lichens on boulders.

Phot. 324. *Erigeron alpinus* and *Festuca ovina* f. *vivipara* on an open ledge, alt. c. 3500 ft. *F. F. Laidlaw.*

ARCTIC-ALPINE VEGETATION OF BEN LAWERS

PLATE 131

Phot. 327. *Athyrium alpestre, Cystopteris fragilis, Polystichum lonchitis, Oxyria digyna, Saxifraga stellaris, Oxalis acetosella*, etc. At top of cleft, more exposed, *Alchemilla alpina, Festuca ovina* f. *vivipara, Cerastium alpinum*, etc. *N. F. G. Cruttwell.*

Phot. 328. *Minuartia sedoides, Alchemilla alpina. J. Massart.*

Phot. 329. *Oxyria digyna, Saxifraga aizoides, Nardus stricta. J. Massart.*

ARCTIC ALPINES (BEN LAWERS)

CHOMOPHYTES

Rock ledges offer very varied habitats, according to shelter, exposure and insolation, the slope and width of the ledge, and whether the surface is relatively dry or constantly irrigated. More or less mineral soil and humus will accumulate upon the ledge according to the conditions. Watson gives the following lists of bryophytes and lichens from this type of habitat. Though numerous in species they do not so greatly outnumber the vascular plants as in the more exposed habitats dealt with in the last chapter. Watson remarks that the bryophytic constituents are often very varied, though sometimes pure masses of one species occur. Foliose liverworts are generally abundant, but lichens and Sphagna are rare.

Mosses

Andreaea petrophila	—	Mnium hornum	o
Bartramia ithyphylla	o	*M. lycopodioides	—
B. oederi	o	M. orthorrhynchum	—
Bryum pseudotriquetrum		M. punctatum (moist)	lf
(moist)	f	var. elatum	lf
*Campylopus schimperi	—	*M. spinosum	o
Dicranum fulvellum	—	*Myurella apiculata	r
D. fuscescens	o	*M. julacea	o
*D. molle	lf	Oedipodium griffithianum	f
D. scoparium	o	Oligotrichum hercynicum	—
*Eurhynchium cirrhosum	lf	Plagiothecium denticulatum	—
E. myosuroides	o	var. obtusifolium	—
E. swartzii	o	P. elegans	—
Grimmia doniana	—	*P. muhlenbeckii	—
Heterocladium squarro-		P. pulchellum	—
sulum		*Polytrichum alpinum	o
Hylocomium loreum		P. commune	—
H. splendens	o	P. piliferum	—
H. squarrosum	o	*Pseudoleskea atrovirens	—
H. rugosum	—	Pterigynandrium filiforme	o
H. triquetrum	o	Rhacomitrium fasciculare	—
Hypnum callichroum	—	R. heterostichum	—
H. cupressiforme	—	R. lanuginosum	a–d
*H. halleri	—	Sphagnum acutifolium	r
*H. hamulosum	—	S. molluscum	o
H. molluscum	—	Webera elongata	—
H. schreberi	—	W. nutans	—

Liverworts

Alicularia scalaris	lf	*C. pleniceps	—
Anastrepta orcadensis	lf	Diplophyllum albicans	a
Anastrophyllum donianum	—	*D. taxifolium	l and r.
Bazzania triangularis	—	*Gymnomitrium concin-	
B. tricrenata	f	natum	—
B. trilobata	—	Herberta adunca	f
Blepharostoma tricho-		H. hutschinsiae	f
phyllum (moist)	o	Jamesoniella carringtonii	lf
Cephalozia bicuspidata		Leptoscyphus taylori	la
(moist)	o	Lophozia floerkii	o

Lophozia hatcheri	—	Pleurozia purpurea	lf
*L. heterocolpa	r	Preissia quadrata	o
*L. incisa	f	Ptilidium ciliare	o
*L. lycopodioides	—	Reboulia hemispherica	lf
*L. obtusa	r	Scapania aequiloba	o
L. quadriloba	r	S. aspera	o
L. quinquedentata	f	S. irrigua	o
L. ventricosa	o	*var. alpina	r
Marsupella emarginata	—	*S. nimbosa	r
Metzgeria furcata	o	*S. ornithopodioides	o
Pellia neesiana	o	*Sphenolobus politus	—
Plagiochila asplenioides	f	*var. medelpadicus	r

Lichens

Biatorina candida	o	Lecidea decipiens	o
*Buellia pulchella	r	L. demissa	o
Cetraria islandica	lf	L. lurida	o
Cladonia cervicornis	o	*L. rubiformis	r and l
var. stipata	lf	*Schizoma lichinodeum	r
Coriscium viride	o	Solorina saccata	o
Dermatocarpon lachneum	o	Sphaerophorus coralloides	o
*Lecidea cupreiformis	r		

Vascular plants

Smith (1911, p. 323) gives the following eight species from dry, well-drained rock-ledges, of southern exposure, with abundance of soil and relative freedom from competition:

Botrychium lunaria†	*Potentilla crantzii
*Gentiana nivalis†	*Rhinanthus borealis
Linum catharticum	*Veronica saxatilis†
*Myosotis alpestris	Viola lutea

† Three species recorded together in various stations.

And the following eighteen from ledges with shelter and deep soil, the community nearly closed:

*Alchemilla alpestris	Melandrium dioicum (rubrum)
Angelica sylvestris	Orchis maculata
*Carex atrata	*Poa alpina, etc.
*Deschampsia alpina	Pyrola minor
Dryopteris dilatata	P. rotundifolia
Galium boreale	P. secunda
Geranium sylvaticum	Rubus saxatilis
*Hieracium alpinum, etc.	Rumex acetosa
Luzula sylvatica	Trollius europaeus

Here under the better conditions we have a much larger proportion of "lowland" species.

Shade chomophytes. Where light is deficient, as in wide and deep fissures in the rock face, or in hollows among block screes, there is also very good shelter, exceptionally high atmospheric humidity, and soil moisture. Among the mica schists such cavities have earthy or rocky walls, moist and bare except where covered by mats of mosses and liverworts. Here and

PLATE 132

Phot. 330. *Cochlearia micacea, Sagina linnaei, Saxifraga cernua. J. Massart.*

Phot. 331. *Sibbaldia procumbens, Gnaphalium supinum, Euphrasia scotica, Galium saxatile. J. Massart.*

Phot. 332. *Stereocaulon alpinum. J. Massart.*

Phot. 333. *Solorina crocea. J. Massar'*

ARCTIC-ALPINES (BEN LAWERS)

there grow a few flowering plants, more or less pale and etiolated. On shaded rocks, where water oozes down through cushions of algae and mosses, flowering plants are more abundant. The deeper fissures are evidently buried in snow early in the winter, but the plants obtain water trickling under the snow before it melts altogether (W. G. Smith).

The following characteristic species are listed by Smith (1911, pp. 324–5) and W. Watson (1925):

Vascular plants

Adoxa moschatellina
*Agropyron donianum
Anemone nemorosa
*Athyrium alpestre[1]
*A. flexile
Cardamine flexuosa
C. hirsuta
*Carex atrata
Cystopteris fragilis[1]
*C. montana
Chrysosplenium alternifolium
C. oppositifolium
Dryopteris linnaeana

Galium saxatile
Geranium robertianum
*Lactuca alpina
*Luzula spicata
Oxalis acetosella[1]
*Oxyria digyna[1]
*Polygonum viviparum
*Polystichum lonchitis[1]
Potentilla erecta (f.)
*Salix lapponum
*Saussurea alpina
*Saxifraga stellaris[1]
Taraxacum officinale (f.)

In this habitat more woodland species are mixed with the arctic-alpines. Watson gives the following list of bryophytes mainly from the hollows of block screes above 3000 ft. on the east side of Ben Lawers:

Mosses

Anoectangium compactum   o
Brachythecium glareosum
*B. plicatum
B. rutabulum
Dicranum fuscescens   f
Eurhynchium confertum
   (among boulders)   f
E. myosuroides
E. praelongum
E. swartzii
Fissidens osmundioides   o
Grimmia patens   f
Hylocomium loreum
*H. pyrenaicum   lf
H. squarrosum
H. umbratum   lf
Hypnum callichroum   f
H. cupressiforme
H. falcatum (well-shaded or
   moist places among
   boulders)
H. molluscum   a
*H. procerrimum
*Mnium lycopodioides
M. orthorrhynchum

M. punctatum var. elatum
   (well-shaded or moist
   places among boulders)
*M. spinosum   f
Plagiobryum zierii
Plagiothecium denticulatum   f
*P. muhlenbeckii
P. pulchellum
*Polytrichum alpinum   f
*Pseudoleskea atrovirens   f
Rhacomitrium canescens
   var. ericoides
R. heterostichum
R. lanuginosum
*R. sudeticum
*Thuidium philiberti var.
   pseudotamarisci
T. tamariscinium   o
*Timmia norvegica (moist
   places among rocks)
Trichostomum tortuosum   o
Webera annotina
W. elongata (also mossy
   boulders)
W. nutans

[1] Pl. 131, phot. 327.

The following are more usually found in small nooks of the rocks

Bartramia ithyphylla	Polytrichum aloides
Dicranum fulvellum	Rhabdoweisia crenulata
Ditrichum zonatum	R. denticulata
*Grimmia elongata	R. fugax
Myurella julacea	*Saelania caesia
Oedipodium griffithianum	Zygodon lapponicus

Hepatics

Alicularia scalaris	L. ventricosa
Aneura pinguis	*Marchantia polymorpha var.
Anastrepta orcadensis	alpestris f
Bazzania tricrenata	Pellia neesiana
Cephalozia bicuspidata	Plagiochila asolenioides
Diplophyllum albicans	var. minor, f. laxa (in
*D. taxifolium	damper places)
*Gymnomitrium concinnatum	Preissia quadrata
*var. intermedium	Ptilidium ciliare
Lophozia floerkii	Scapania curta
L. hatcheri	S. dentata
*L. lycopodioides	var. ambigua (damper
L. obtusa	places)
Lophozia quinquedentata f	

Lichens

Lichens are naturally not characteristic but the following may be found occasionally:

Cladonia furcata	Peltidea aphthosa
C. sylvatica	Peltigera canina, etc.
Collema multifidum, etc.	Racodium rupestre
Pannaria brunnea	*Solorina bispora

Cader Idris. On the calciferous "pillow lavas" (upper and lower basic volcanic series) of Cader Idris, Price Evans (1932) records a rich **Basic rocks** flora including a fair number of arctic-alpine species. His lists give "representative" species and do not profess to be exhaustive, but they are interesting to compare with the general lists given above which are mainly taken from the Scottish Highlands. The lists below are composite, including the upper and lower bands of "pillow lava" (1500–1700 ft. and 1800–2400 ft., *c.* 450–520 and 550–730 m.) and do not distinguish between the types of habitat:

Lichens

si	Lecanora parella	f	si	Pertusaria dealbata	o
	L. subtartarea	a		Rhizocarpon petraeum	f
	L. tartarea	a	si	Sphaerophorus globosus	o
	Lecidea contigua f. calcarea	o–f		Stictina fuliginosa	la
	Peltigera canina	o	ca	Solorina saccata	f

Liverworts

	Aneura pinguis	o		Frullania tamarisci	f
	Anthelia julacea	a	ca	Reboulia hemisphaerica	f
	Fissidens adiantoides	f–a	si	Scapania dentata	o

Mosses

	Bartramia pomiformis	f	ca	Neckera crispa	f–a
ca	Hypnum commutatum	f		Plagiobryum zierii	f
	Mnium undulatum (shade)	f		Polytrichum urnigerum	o

Vascular plants

ca	Adiantum capillus-veneris	vr		H. pulchrum	o
	Adoxa moschatellina	o		Hymenophyllum peltatum	f
	Alchemilla alpestris	f		Lathyrus montanus	o
	Anemone nemorosa	o		Linum catharticum	o
	Antennaria dioica	o		Luzula maxima	o
ca	Arabis hirsuta	o		*Lycopodium selago	o
	Asplendum ruta-muraria	r–o		L. clavatum	o
	A. trichomanes	f		Meconopsis cambrica	o–f
*ca	A. viride	f		Molinia caerulea	o–f
si	Blechnum spicant	o		Oxalis acetosella	o
si	Calluna vulgaris (lower alt.)	f	*ca	Oxyria digyna	r–la
	Campanula rotundifolia	o	ca	Pimpinella saxifraga	o, l
	Cardamine hirsuta	o		Pinguicula vulgaris	o
	Carex flava	o		Polygala vulgaris	o
	C. pulicaris	o		Polypodium vulgare	o
	Chrysanthemum leucan-			Primula vulgaris	o
	themum	o		Ranunculus repens	o
	Chrysosplenium oppositi-			*Rhinanthus cristagalli,	
	folium	la		forma	r
si	Cryptogramma crispa	o		*Saxifraga hypnoides	la
	Cystopteris fragilis			*S. stellaris	o–f
	*ca. f. dentata	f		Sedum anglicum	o
si	Deschampsia flexuosa	o		*S. rosea	r–la
si	Digitalis purpurea	o		Scirpus caespitosus	o
	Dryopteris filix-mas	o		Selaginella selaginoides	o–f
	D. linnaeana	o		Solidago cambrica	o–f
si	D. phegopteris	o–f		S. virgaurea	o
	Epilobium angustifolium	la		Succisa pratensis	o–f
	E. palustre	o		Taraxacum officinale	o
	Euphrasia officinalis	o	ca	Thalictrum minus (agg.)	lf
	*Festuca ovina f. vivipara	o		Thymus serpyllum	f
	Ficaria verna	o		Urtica dioica	o
	Filipendula ulmaria	o		Valeriana dioica	o
	Geum rivale	o		Veronica officinalis	o–f
	Geranium robertianum	o		Viola sp.	o
	Hieracium sylvaticum	o	si	Vaccinium myrtillus	
	Hypericum humifusum	o		(higher alt.)	f

* Arctic-alpine species, ca, calcicolous, si, silicicolous (calcifuge).

Price Evans points out that a few species (ca) in these lists are calcicolous, but the majority are "indifferent". The rich flora must be attributed
to the generally favourable conditions for vegetation on these lava soils,
conditions of which the high base status is one. There are however a certain
number of "calcifuge" species (si) and the presence of these he attributes to
the formation of acid humus on some of the rock ledges and also to the
soil produced by acidic beds interstratified with the calciferous lavas.

Acid rocks. On the granophyre escarpment (the great northern cliff) of Cader Idris the available calcium is low. Much of the surface is bare rock and the variety of species is much less than on the basic lavas, but the high altitude (up to 2900 ft. = 884 m.) determines the presence of a few arctic-alpine forms, notably the mountain forms of the sea pink, sea campion and sea plantain.

The following are some characteristic species:

Agrostis stolonifera	Oxalis acetosella
Alchemilla alpina	Pinguicula vulgaris
*Armeria maritima, forma	*Plantago maritima, forma
Athyrium filix-femina	Potentilla erecta
Blechnum spicant	*Saxifraga stellaris
Campanula rotundifolia	Scirpus caespitosus
Carex flava	*Sedum rosea (rhodiola)
Cryptogramma crispa	*Silene maritima, forma
Deschampsia flexuosa	Solidago cambrica
Dryopteris phegopteris	Sorbus aucuparia (1·2 m. high, up
Galium saxatile	to 700 m.)
*Lycopodium selago	Succisa pratensis
Luzula maxima	Thymus serpyllum
Molinia caerulea	Viola riviniana
Narthecium ossifragum	

On the ledges dwarf *Calluna* is frequent at lower levels and *Vaccinium myrtillus* at higher. *Festuca ovina* forma *vivipara* and *Rhacomitrium lanuginosum* are generally abundant, with *Polytrichum commune* and *Sphagnum* in the wetter places.

The following are abundant on the rock faces:

Andreaea petrophila	Lecidea dicksonii
Lecidea contigua	Rhizocarpon confervoides

Gyrophora cylindrica and *Parmelia pubescens*, which are frequent on the summit plateau, are occasional on the rocks at the base of the escarpment, i.e. so low as 550 m., and cushions of *Anthelia julacea, Scapania* sp. and *Tetraplodon mnioides* are abundant on wet surfaces in the gullies, the last named in association with *Peziza rutilans* saprophytic on the decayed bones of sheep.

Hydrophilous chomophytes. Wet habitats are common and varied on many of the high mountains. Springs emerge at quite high altitudes and form numerous small shallow runlets before they converge to form the "burns" or mountain streams at lower levels. In the areas of schist, as on Ben Lawers, series of terraces are formed, and excess of water from the numerous springs, not yet confined to definite channels, flows gently over these. Such ground is covered with a bright green, wet carpet of mosses, liverworts and algae, with a certain number of flowering plants. On Ben Lawers the lower corries show the main development of this vegetation, but on a smaller scale it occurs also at higher levels. The substratum of the vegetation is fairly uniform and is formed mainly of flakes of mica schist and other rocks caught among larger stones and boulders. Thus the habitat is constantly wet and at the same time well aerated and supplied with abundant nutrient salts from the moving spring water. Along the lower stream flats, on the other hand, where drainage is impeded, more stagnant conditions prevail, and dull green bog or marsh is formed, dominated by *Sphagnum* and by rushes and sedges.

Societies of bryophytes. Watson (1925, p. 19) distinguishes six societies dominated by bryophytes:

(1) *Philonotis fontana* society (described by Ostenfeld in *Botany of the Faroes*, 1908, and mentioned by W. G. Smith, 1911). The dominant is often pure, but may be mixed with species of *Hypnum, Bryum, Mnium, Sphagnum* and *Scapania*. This forms the typical bright green wet carpet irrigated by slowly moving spring water and referred to above.

(2) **Sphenolobus politus* society in places similar to (1):

*Harpanthus flotowianus	*Scapania obliqua
Hypnum sarmentosum	Sphagnum inundatum
Pellia neesiana	

(3) **Scapania obliqua* society in a spring at 3000 ft. on Ben Lui, associated with:

Aulacomnium palustre	Scapania dentata
Bryum pseudotriquetrum	S. uliginosa
Fissidens adiantoides	S. undulata
Hypnum vernicosum	

(4) *Hypnum trifarium* society on wet boggy slopes, sometimes associated abundantly with *H. exannulatum, revolvens, sarmentosum* and *stramineum*.

(5) *Marsupella* society (*M. aquatica* and *M. pearsoni*) in water flowing over stony ground.

(6) *Aplozia* society in a boggy spring on Ben Lawers over 3000 ft.:

Aplozia cordifolia	a	Hypnum falcatum	a
A. atrovirens		H. revolvens	a
A. riparia		Scapania dentata	o
Alicularia scalaris	f	S. irrigua	
Aneura pinguis	o	S. undulata	
Bryum pseudotriquetrum	a	Sphagnum cymbifolium	a
Fissidens osmundoides	o		

Such bryophytic societies require further study before they can be properly defined and their relations to habitat established.

Flora. Watson (1925) records the following hydrophilous bryophytes from the arctic-alpine region. For notes on their habitats and associates his paper should be consulted:

Mosses

Aulacomnium palustre	f	B. pallens
Blindia acuta		Cinclidium stygium
Bryum alpinum		*Cynodontium virens

*Cynodontium wahlenbergii	H. uncinatum
Dichodontium pellucidum	H. undulatum
Dicranella squarrosa	H. vernicosum o
Fissidens osmundoides	Meesia trichodes
Grimmia apocarpa var.	Mnium cinclidoides r
rivularis	M. punctatum var. elatum a
*H. arcticum	M. subgobusum
H. commutatum a	Orthothecium rufescens
H. cuspidatum	Philonotis adpressa
*H. decipiens	P. calcarea
H. exannulatum a	P. fontana
var. orthophyllum	*Polytrichum alpinum
H. falcatum a	P. commune
H. fluitans o	*P. sexangulare
H. lycopodioides o	Splachnum vasculosum
*H. molle	*Tayloria lingulata
H. ochraceum	*Thuidium philiberti var.
H. revolvens a	pseudotamarisci
H. sarmentosum f–a	*Timmia norvegica
H. scorpioides	Webera albicans
H. stellatum	*var. glacialis (Pl. 130, phot. 326)
H. stramineum a	W. elongata
H. trifarium la	W. ludwigii

Liverworts

Alicularia scalaris	*L. kunzeana r
Aneura pinguis	L. lycopodioides f
*Anthelia julacea var.	L. muelleri r
gracilis	*L. wenzelii r
*Aplozia atrovirens var.	Marsupella aquatica a
sphaerocarpoidea	M. emarginata
A. cordifolia a	M. pearsoni a
A. riparia f–a	Pellia epiphylla o
*A. sphaerocarpa var. nana	P. neesiana a
Blepharostoma tricho-	Ptilidium ciliare o
phyllum o	*Scapania crassiretis
Cephalozia bicuspidata a	S. dentata a
Chiloscyphus polyanthus	var. ambigua a
var. fragilis	S. irrigua
*Eremonotus myriocarpus	*S. obliqua lf
Eucalyx obovatus o	*S. paludosa r
Gymnocolea inflata o	S. subalpina
forma nigricans a	S. uliginosa a
*Harpanthus flotowianus lf	S. undulata a
*Hygrobiella laxifolia	Sphenolobus politus a
Lophozia bantriensis la	

Lichens on wet rocks. The following lichens may occur on stones in streams or on wet rocks by stream sides (Watson, 1925):

Aspicilia lacustris	E. pubescens
Collemodium fluviatile	Lecidea contigua
Dermatocarpon miniatum	forma hydrophila
var. complicatum	Polyblastia theleodes
Ephebe hispidula	*P. inumbrata

*Psorotrichia furfurella
*Pyrenopsis phylliscella
Staurothele clopima
Verrucaria aethiobola

V. laevata
V. margacea
V. submersa

Representative vascular hydrophytes. W. G. Smith gives the following flowering plants as "representative species" occurring in the hydrophilous moss communities referred to above, or in other hydrophytic communities: one or two are added from Watson's paper:

Highland species	*Lowland species*
Carex saxatilis	Alchemilla vulgaris (agg.)
C. vaginata	Caltha palustris var. minor
Cerastium alpinum	Carex curta
C. cerastioides	C. dioica
Epilobium alsinifolium	C. stellulata
E. anagallidifolium	Chrysosplenium oppositifolium
Juncus biglumis	Cochlearia officinalis (agg.)
J. triglumis	Equisetum hyemale
Oxyria digyna	Montia fontana
Polygonum viviparum	Narthecium ossifragum
Sagina saginoides	Pinguicula vulgaris
Saxifraga aizoides	Sedum villosum
S. stellaris	Stellaria uliginosa
Thalictrum alpinum	Veronica humifusum
Tofieldia palustris	
Veronica alpina	

REFERENCES

PRICE EVANS, E. Cader Idris: a study of certain plant communities in south-west Merionethshire. *J. Ecol.* **20**, 1–52. 1932.

SMITH, W. G. Arctic-alpine vegetation. Chapter XIII in *Types of British Vegetation.* 1911.

SMITH, W. G. Anthelia: an arctic-alpine plant association. *Scott. Bot. Rev.* pp. 81–9. 1912.

WATSON, W. The bryophytes and lichens of arctic-alpine vegetation. *J. Ecol.* **13**, 1–26. 1925.

Part IX

MARITIME AND SUBMARITIME VEGETATION

Chapter XL

INTRODUCTORY. THE SALT-MARSH FORMATION

Maritime vegetation. It is convenient to distinguish *maritime vegetation*, that is land vegetation affected more or less profoundly by the neighbourhood of the sea, from *marine vegetation*, living in the sea. This of course cannot be an absolute distinction, because vegetation which is covered with sea water at high tide, and exposed to the air at low tide is clearly intermediate between the two. Of this "tidal" vegetation the seaweed communities of the intertidal zone of rocky coasts are reckoned as "marine" and are not dealt with in this book, while the "salt-marsh" communities of more or less protected mud or sand flats, dominated mainly by flowering plants, though marine algae are present and are often important constituents dominating some of the communities, are included in maritime vegetation.

The habitats. Maritime vegetation is here taken to include the communities which occupy the following habitats:

(1) The sand and mud of the intertidal zone where these are protected from the more violent wave action (*salt-marsh formation*) (Chap. XL).

(2) The uppermost zone of the sea shore on exposed coasts, barely reached by the spring tides (*seashore community*) (Chap. XLI).

(3) Coastal sand dunes (*sand-dune formation*) (Chap. XLI).

(4) Coastal shingle beaches (Chap. XLII).

(5) Brackish marshes (Chap. XLIII, p. 896).

(6) Sea cliffs and coastal rocks out of reach of the tides, but exposed to spray (Chap. XLIII, p. 897).

(7) Maritime and submaritime grasslands, above the limit of high tide but affected by salt spray (Chap. XLIII, p. 899).

The characteristic plant formation (1) inhabiting the sand and mud of the intertidal zone is the nearest to marine vegetation proper because it is exposed to immersion in salt water at some high tides. It is essentially composed of halophytes and includes many salt-water algae. This formation is based on marine alluvium and occurs only where it is protected from the erosive action of waves and the scour of swift currents and tide races, particularly where the flat shores of an estuary are protected by headlands, shingle banks or sand dunes. The intertidal zone on coasts formed of hard rock and exposed to the waves is inhabited by communities of marine algae, zoned according to their degree of tolerance of exposure to the air and also of wave action. These seaweeds are firmly fixed to their rock substratum, and depend for their mineral nutrition entirely on what they can absorb from the sea water while they are covered by the tide. Salt-marsh plants, on the other hand, are rooted in mud or sand and, like rooted fresh-water aquatics and land plants, get their water and mineral nutrients through their root-systems.

Sandy or muddy sea shores between tide marks on *exposed* coasts are destitute of higher plants because of their instability under wave action:[1] it is the uppermost zone alone, reached only by the highest spring tides, that often bears a scattered population of rooted phanerogams—*littoral* plants proper—and often known as the "beach" or "sea shore" community (2).

Other maritime plant communities, less exposed to the direct effect of the tides but showing more or less maritime influence, are (3) coastal sand-dune vegetation, which is not composed of halophytes and passes over into typical land vegetation; (4) shingle-beach vegetation, inhabiting a local and special habitat, and having much in common with that of sand dunes, but including halophytic species; (5) brackish marshes in the upper reaches of estuaries or wherever salt water, diluted with fresh water, has access; (6) the vegetation of sea cliffs and coastal rocks out of reach of the tide, but exposed to drenching by spray and therefore largely halophytic; and finally (7) grasslands close to the sea and subject to the effect of sea spray driven on to the land by the wind, which show a mixture of halophytic and non-halophytic species.

The types of maritime community just enumerated are not only constantly found in proximity because they all depend on maritime influence, they are also frequently interdependent. Thus the measure of protection required for the development of salt marsh is often provided by shingle spits (cf. pp. 569–73) or dune complexes, while sand dunes themselves are often formed on a basis of shingle beach. These relations are illustrated on Plates 133 and 134 as well as in several of the photographs reproduced on other plates in this and the following chapters.[2]

Plate 133 shows the parallel dune ridges of the Headland at Blakeney Point which are largely built on pre-existing parallel shingle banks. The continuation of one of these banks as a naked shingle spit is seen in phot. 335 curving round and protecting a large salt marsh which is traversed by smaller and older spits. This marsh was doubtless formed in successive compartments, each developed after the formation of the spit which established the condition of protection. All the compartments of the marsh now drain into a common channel shown on the right of the photograph. Plate 135, phot. 338, is a view of this marsh from a north-western bay and shows three old worn-down spits, two of which nearly shut off the marsh compartment seen in the foreground.

Plate 134, phot. 336, shows a series of much older spits with their enclosed mature marshes farther up the estuary, and collectively known as "The

---

[1] Pl. 133, phots. 334–5.

[2] E.g. Pl. 135; Pl. 136, phot. 341; Pl. 147, phot. 373; Pl. 158, phot. 405.

PLATE 133

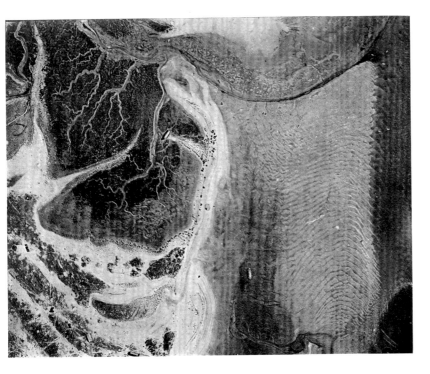

Phot. 335. Part of the *Salicornia-Pelvetia* salt marsh behind the Headland. The marsh is enclosed and protected by an old lateral shingle spit running across the centre of the picture. Two minor spits ending freely in the marsh are seen (centre and right) at the top of the picture. The spits are fringed with bushes of *Suaeda fruticosa* (black dots). Tidal access to the marsh is by the channel on the right. Ribbed sandy mud of the estuary at bottom. *Major J. C. Griffiths* (alt. 7000 ft.), 1921.

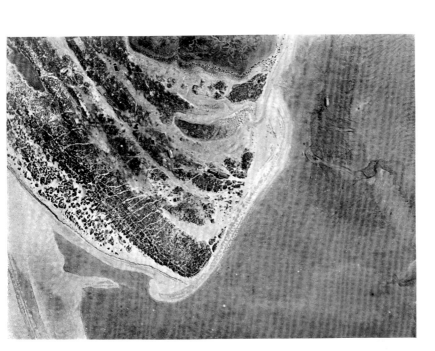

Phot. 334. The Headland at Blakeney Point showing the parallel lines of old shingle beach overlaid by sand dunes, with lows between. The outer-most dunes are young and unconsolidated. The white lines crossing the main dune ridges are human foot tracks. Mottled and ribbed sand and mud of the estuary occupy the bottom of the picture. The continuation of the main shingle bank leading to the Far Point is seen in the top left-hand corner. *Major J. C. Griffiths* (alt. 7000 ft.), 1921.

BLAKENEY POINT FROM THE AIR

PLATE 134

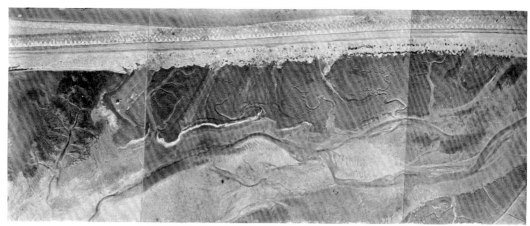

Phot. 336. At the top of the picture is the edge of the sea with mottled sand and lines of drift marking different high tide levels; then the main shingle bank with dark *Suaeda fruticosa* (dots and irregular lines) and other vegetation (fainter). Below is the series of salt marshes in compartments, bounded and protected by the lateral spits or "hooks" and drained by sinuous branching channels up which the tide gains access. Below is the channel of the River Glaven bordered by expanses of estuarine mud. The length of the photograph is about 1¼ miles (2 km.). *Major J. C. Griffiths* (alt. 7000 ft.), 1921.

Phot. 337. Central portion of the Marams on a larger scale. The three irregular zones of *Suaeda fruticosa* on the main bank, the deltas of shingle invading the marshes and the *Suaeda* bushes on the laterals can be distinguished. *Major J. C. Griffiths* (1921).

"THE MARAMS" AT BLAKENEY POINT FROM THE AIR

PLATE 135

ιot. 338. Salt marshes protected and partly enclosed by lateral shingle spits, which are attached to the dune-
covered main spit (left). Two run across the centre of the picture, and the proximal end of a third is in the foreground.
ll three are fringed with *Suaeda fruticosa* and end freely in the marsh, which is dark coloured and dominated by
licornia with unattached *Pelvetia* as an understorey. Waters of the estuary to the right behind. The photograph
taken from a north-western bay of the marsh shown in Pl. 133, phot. 335, just out of the top of the air photograph.
. *W. Oliver.*

Phot. 339. The other side of the marsh shown in phot. 338. On the right is an old lateral spit covered with dunes
much worn down and abundant *Senecio jacobaea*. The marsh is fringed with *Suaeda fruticosa* on marsh soil covered
with shingle, severely rabbit-eaten on the side towards the spit. Dune-covered main spit (separating the marsh from
the open sea) in the distance. *F. W. Oliver.*

MARITIME VEGETATION AT BLAKENEY POINT (SAND DUNE, SHINGLE AND SALT MARSH)

PLATE 136

Phot. 340. Centre of the marsh shown on Pl. 135, with drainage channels and pans. Dune-covered main shingle b
in the distance. *F. W. Oliver.*

Phot. 341. Old lateral shingle spit at "the Marams", curving round a salt marsh. Zone of *Suaeda fruticosa* on sl
of the spit: above (right foreground) is the zone of *Festuca rubra* (see p. 889). Other *Suaeda*-covered spits enclos
salt marshes beyond. *Mrs Cowles* (1911).

MARITIME VEGETATION AT BLAKENEY POINT (SALT MARSH AND SHINGLE SPITS)

Marams". Here the spits are seen to be "laterals" arising from the main shingle bank shown at the top of the picture (see Chapter XLII, pp. 870–1). Pl. 136, phot. 341, shows one of these lateral spits curving round a salt-marsh compartment.

(1) THE SALT-MARSH FORMATION

In tidal estuaries into which sand and mud are carried by the tide and laid down on the flat shores, where wave action is at a minimum, a luxuriant phanerogamic vegetation, with associated algae, is developed. This vegetation is generally known as *salt marsh*, its flora, like the factors of its habitat, is extremely distinct, and it constitutes a well-defined plant formation. The master factor which differentiates this from other formations is the salt water which bathes the whole plant body during the periodic immersions and forms the soil solution (though varying in concentration) at all times. Hence the plants are said to be *halophytes* or *halophilous*, many of them growing well only when supplied with salt water, though others can do without it. Halophytes occur not only in coastal salt-marsh formations but also inland, wherever there is a high concentration of salt in the soil, for example round existing salt lakes or on the sites of old lake basins in arid regions, or again in the neighbourhood of salt springs.

Relation to tides. The salt-marsh formation is periodically immersed by the tide. It extends upwards from levels which are just about reached by high water of neap tides (tides of least range) and are covered by all the tides of intermediate range, to the highest levels of the marsh, which are only reached by high water of the highest spring tides (i.e. the tides of greatest range). At low water of spring tides the tide recedes far below the level of the lowest salt-marsh community and uncovers stretches of bare sand or mud and also the fields of "sea wrack", "grass wrack" or "sea grass" (*Zostera*), a marine monocotyledon with flexible, band-shaped leaves, which sometimes occupy this position. The Zosteretum must be reckoned as belonging to marine vegetation for *Zostera* is one of the small group of marine angiosperms, and has a different habit and economy from the salt-marsh plants; but Zosteretum, where present, may be considered as the starting point of the halosere since it exists on similar substrata and shows transitions to the Salicornietum.

Different levels of the marsh are thus covered by the sea water for very different lengths of time, the lowest zones for long periods, the higher for much shorter ones, while the highest of all are visited only by a few tides, usually twice a year, around the equinoxes.

Salt content of ground water. Besides undergoing periodic immersions in sea water the plants of the salt marsh have a ground water which is normally salt, though the amount of salt it contains may vary at different periods from a percentage higher than that of sea water (about 3·3) to quite a low percentage approximating to that of fresh water. The variation in the saltness of the ground water is due to the interaction of various factors—immersion by the tide, the rate of evaporation when the tide is down, and the washing out of salt by rain during the periods of emersion. One of the characteristic features of most salt-marsh plants is their succulence, caused by the presence of abundant "water tissue" whose cells are typically large, thin walled, and swollen with cell sap. The halophytes absorb large amounts of salt, which remains in their tissues, and the concentration of osmotically active solutes in their cell sap is necessarily brought more or less into equilibrium with the medium surrounding their roots, the salt ground water, and, during the periods of immersion, with the sea water of the tide which covers the plants. It has been shown that the osmotic pressure of halophytic cells is not only much higher than that of mesophytes, a character which is shared by the cells of xerophytes, but that it changes in response to the concentration of the surrounding medium.

Zonation and succession. As always where some decisive ecological factor undergoes a regular spatial change in intensity ("gradient") the salt-marsh formation is very definitely zoned. These zones undoubtedly correspond, in a general way, with the salt-marsh succession (*halosere*) passing from pioneer communities, on wetter, more mobile mud or sand exposed to the air for shorter periods, to those on drier more stable substrata which are exposed for longer periods, the highest being very rarely immersed by the tide. The cause of the succession is the rise of level brought about by continuous tidal silting. As soon as the substratum becomes sufficiently stable for colonisation by plants the silt is trapped between them and the general level raised, with increasing rapidity at first, as the density of the vegetation increases (F. J. Richards, 1934), then more slowly, as the surface is less and less often covered by the silt-bringing tide. At the highest levels of the marsh silt deposition is almost negligible and any further rise in level depends mainly on humus accumulation.

Ecological factors. The most important factor differentiating the zones of the marsh is probably the relative lengths of time during which they are submerged or exposed to the air. Other factors are the nature of the soil and its water content, determining aeration and depending on drainage. Besides these physical factors there is of course the competition between the plants of the later, closed, communities.

Submersion and exposure. It is not generally realised for how long a time even the lower salt-marsh communities are exposed to the air. This is well brought out in V. J. Chapman's very thorough studies (1934, 1938) of the times of exposure of the salt marshes at Scolt Head island on the north coast of Norfolk. He shows that at the bottom of the Zosteretum, i.e. well below the level of the salt marsh proper, the ratio of hours of submergence to hours of emergence per month (calculated for the period 31 July to 9 October) is 0·628, representing a submersion of 282·4 out of 732 hr., while at the top of the marsh this *S/E* ratio is 0·004, representing a total submersion of only 3 hr. per month. The whole extent of the salt-marsh vegetation is thus submerged by the tide for much less than half the time, and the upper levels for very small fractions only.

Effects on different zones. Further, the duration of submergence varies a good deal at different levels according to the season. The upper communities of the marsh, which Chapman classes together as Group X (see Fig. 152), are most frequently submerged at the spring equinox, that is during the period of seedling growth when a good supply of water is essential, and at the autumn equinox, the period of fruiting, when the tides can distribute the seeds. During the summer there is a long uninterrupted period of exposure, when, in the absence of rain, the surface soil may become very dry, so that the plants of the X communities have to endure drought, besides having to tolerate brief periods of immersion in sea water around the equinoxes. The most frequent submergences of the lower communities, which Chapman separates as Group Z, occur in midwinter and at midsummer, i.e. during the period of maximum growth, so that the plants are unlikely to suffer from water deficiency at a time when those of the X communities may be wilting or dying from drought. Thus the Z communities are probably limited upwards by the inability of the species of which they are composed to withstand drying, while those of the X communities are limited downwards by their inability to tolerate frequent immersions in sea water during their period of maximum growth. Most of the communities of the salt marsh belong either to Group X or to Group Z, very few crossing the line between them. These few intermediate communities Chapman classes as Group Y.

SALT-MARSH COMMUNITIES

Variation in detail. The general zonation and succession of communities in salt marsh is very uniform, but different marshes show very considerable variation in detail, owing to differences in the substratum (sand and mud) and also in physiography and topography. A salt marsh rarely shows a uniform gentle slope with a series of zones of uniform width: it is nearly always dissected by creeks of varying depth and is very often developed in bays as individual marshes of varying size and varying relation to other physiographic features such as sand dunes and shingle beaches; and all these differences involve variations in detailed development.

A very usual arrangement on the east coast of England, where shingle "spits" are common (see Fig. 168, p. 872), is the development of separate marshes on the estuarine mud or sand of the bays enclosed by lateral shingle banks attached to the inner side of a main shingle spit which runs between the open sea and the estuary.[1] These laterals extend inwards at various angles from the main bank and end freely in the estuary, and the

[1] Cf. Pl. 134.

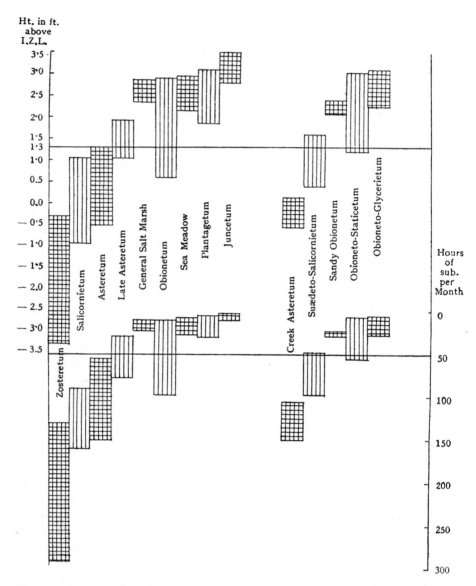

FIG. 152. RANGES OF SALT MARSH COMMUNITIES AT SCOLT HEAD, NORFOLK

The upper part of the figure shows the ranges in vertical heights. The critical height, separating the X communities (above the horizontal line 1·3) from the Z communities below, is 8·3 ft. above Ordnance Datum (I.Z.L. = 7 ft. above O.D.). The lower part of the figure shows the range in hours of submergence per month. 50 hours is the critical number separating the X communities (above) from the Z communities (below). The communities in the right hand group do not belong to the primary series, but are partly determined by local soil conditions. From Chapman, 1934.

marshes they enclose commonly form a progressive series, the most recently arisen nearest the extremity of the main spit. Such marshes also differ according to the nature of the substratum and according to whether the bay has a wide or a narrow opening.

The most various phases of retrogression and regeneration of salt marsh are to be found, especially on the edges of creeks and also on the seaward face of an old marsh which has reached a relatively high level and has then been eaten away by the tides owing to a change in currents. Here there is often a vertical crumbling cliff or escarpment, several feet high, at the top of which the old salt marsh persists, with soft mud below formed from the debris of the cliff, on which much younger communities are developed.

Individual salt-marsh species sometimes appear at different stages of the succession owing to causes not fully understood. For example at Scolt Head on the north coast of Norfolk, *Glyceria maritima*, one of the most widespread and ubiquitous salt-marsh plants, and generally entering the succession very early, does not appear till a later stage than in most marshes; and at the neighbouring Blakeney Point it is only prominent in upper marsh communities.

General zonation. The following is the general zonation, though one or more of the zones may be absent:

(1)* Algal communities beginning below the level reached by high water of neap tides. This zone may be occupied by a Zosteretum, associated with the same algae.

(1) Salicornietum herbaceae.

(1*a*) Spartinetum townsendii.

(2*a*) Glycerietum maritimae.

(2*b*) Asteretum tripolii.

(3) Limonietum vulgaris.

(4) Armerietum vulgaris.

(4*a*) Obionetum portulacoidis.

(4*b*) Suaedetum fruticosae.

(5) Festucetum rubrae (sandy marshes).

(6) Juncetum maritimi.

Fig. 153 shows the general zonation in the sandy salt marsh at Ynyslas on the shore of the Dovey estuary in Cardiganshire. Here (3), (4*a*) and (4*b*) are (exceptionally) absent.

(1*) *Algal communities and Zosteretum.* The first colonists of the still mobile sand or mud are algae, most commonly species of *Rhizoclonium*, *Ulothrix*, *Chaetomorpha*, *Vaucheria*, *Monostroma* and *Enteromorpha* among the green algae (of which the first and last genera are the most constant and abundant), and in some muddy estuaries a rich flora of diatoms and blue-green algae (Fig. 154) as described by Carter (1932). In sandy estuaries the algal flora is much poorer (Fig. 155). The filamentous forms undoubtedly play a considerable part in binding the loose mud, and probably serve as a good nidus for the colonisation of *Salicornia*. The brown seaweed *Fucus vesiculosus* var. *evesiculosus* has also been recorded from this zone (Chapman, 1934).

It is at this level too that a Zosteretum may be developed. This is dominated in some localities by *Zostera marina* (Philip, 1936), in others by

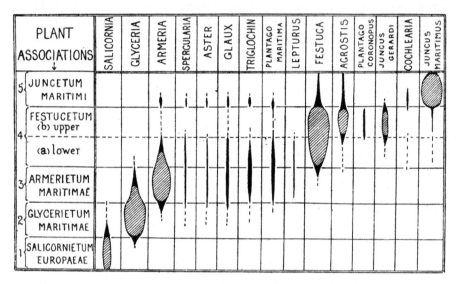

FIG. 153. ZONATION OF ANGIOSPERMIC COMMUNITIES AT YNYSLAS
SALT MARSH, DOVEY ESTUARY (CARDIGANSHIRE)

The ranges of the principal species in the zoned communities are shown. Dominance or subdominance is indicated by diagonal shading. From Yapp, 1917.

Z. nana (Chapman, 1934); *Ruppia rostellata* may also be present (Philip, 1936).

(1) **Salicornietum herbaceae** (Samphire or Glasswort Marsh) (Pls. 137–8, phots. 344–6). In the great majority of salt marshes that have been studied this is the first (earliest) community of the salt marsh proper. It is dominated by and (apart from algae) often exclusively composed of herbaceous annual species of *Salicornia*, which may be included in the aggregate *S. herbacea*: segregate or closely allied species such as *S. dolichostachya, S. stricta, S. ramosissima*, may also be present in some quantity. The colonisation is at first by isolated plants, which are very liable to be uprooted by wave action on the mobile mud (Wiehe, 1935) especially within range of daily tides (Fig. 156). At a higher level the numbers increase, but the Salicornietum remains open with bare mud between the plants[1] still

[1] Pl. 137, phot. 344–5.

PLATE 137

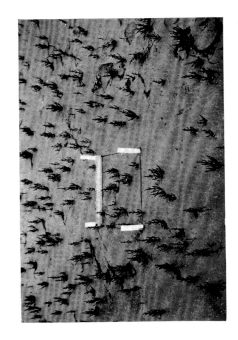

Phot. 344. Very sparse Salicornietum averaging one plant per sq. ft. subject to daily covering by the tide. The pegs enclose a square foot. Dovey estuary. *P. O. Wiehe* (1935).

Phot. 345. Denser though still sparse Salicornietum (av. 24 plants per sq. ft.) just above the range of daily tides. *P. O. Wiehe* (1935).

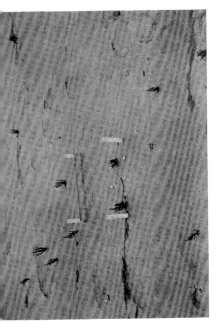

Phot. 342. Bare cracked mud of the Severn estuary with severely eroded cliff whose level top bears old salt marsh grassland. One plant of *Aster tripolium* has colonised the mud. *C. G. P. Laidlaw.*

Phot. 343. Looking down the Severn estuary. Deeply cut channel with slopes of bare unstable mud: eroded cliff on the right above. *C. G. P. Laidlaw.*

Bare Mud and Salicornietum

PLATE 138

Phot. 346. Salicornietum with primary depression pan. Dovey salt marshes. *R. H. Yapp* (1917).

Phot. 347. *Spartina townsendii* invading sparse Salicornietum. *Fucus* sp. in background and extreme foreground. Hayling Island, Hants. *S. Mangham.*

SALICORNIETUM AND *SPARTINA*

occupied by the algae of the lower zone. The first phanerogam to become associated with *Salicornia* is usually *Glyceria maritima*, the most character-istic of the salt-marsh grasses and commonly occurring through a consider-able range of level. In some marshes *Glyceria* may colonise the mud *pari passu* with *Salicornia* (Marsh, 1915), or even act as a pioneer (Heslop

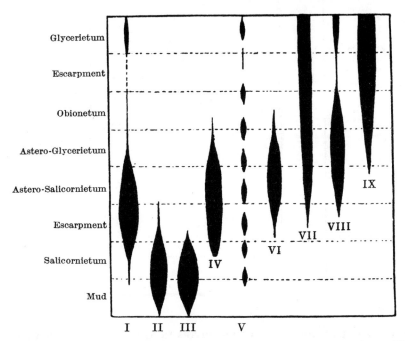

FIG. 154. ZONATION OF ALGAL COMMUNITIES IN THE MUDDY SALT MARSH AT CANVEY ISLAND IN RELATION TO THE COMMUNITIES OF ANGIOSPERMS

I. General community of Green Algae, of great range but most developed in the lower zones. II. Marginal community of Diatoms. III. Marginal community of Blue-green Algae. These two have a maximum development in the unstable mud of the channel and in the adjacent Salicornietum. IV. *Ulothrix flacca* community—transitory, pre-vernal, covering at that time all the lower zones of the marsh. V. *Enteromorpha minima*—*Rhizoclonium* community—epiphytic, throughout the marsh, only observed at Canvey. VI. *Anabaena torulosa* community—middle zones of the marsh. VII. Filamentous Diatom community on mud, middle and upper zones (not seasonal). VIII. Autumn community of Blue-green Algae. IX. *Phormidium autumnale* community, middle and upper zones. Most of the algal communities are markedly seasonal. From Carter, 1933, p. 390.

Harrison, 1918; Morss, 1927), but usually there is a pure Salicornietum preceding the advent of the grass. Other species which may settle in the pioneer Salicornietum are *Aster tripolium* and *Suaeda maritima*, the latter especially but not exclusively on sandy substrata. The dwarf *Fucus caespitosus*, embedded in the soil, is characteristic of the Salicornietum at Scolt Head. The other algae are broadly the same as those of the lower zone. Fungi are few and far between (see p. 833).

At Scolt Head the lower fringe of the Salicornietum is exposed to the air only at low tide except for a very short period in March, but the upper extremity is exposed for long periods, amounting in all to about seven times the periods of submergence (Chapman, 1934).

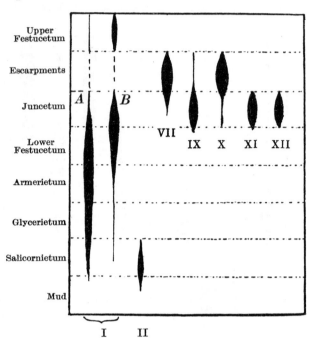

FIG. 155. ZONATION OF ALGAL COMMUNITIES IN THE SANDY SALT MARSH AT YNYSLAS (DOVEY ESTUARY) IN RELATION TO THE COMMUNITIES OF ANGIOSPERMS

Compare Figs. 153, 154.

I. General community of Green Algae, of greatest range. The green components (A) are developed most strongly in the lower and middle, the blue-green components (B) in the middle and upper zones. II. Marginal community of Diatoms, only in unstable mud and the Salicornietum. (Communities III–VI inclusive of the Canvey Marsh are absent.) VII. Filamentous Diatom community, confined to escarpments and Juncetum, owing to the closed vegetation of most of the other zones. (Community VIII of Canvey Marsh is absent.) IX. *Phormidium autumnale* community, Juncetum (mainly) and escarpments only. X. *Rivularia-Phaeococcus* community, almost confined to escarpments. XI, *Pelvetia canaliculata* (ecad *muscoides*) community, and XII, *Catenella opuntia-Bostrychia scorpioides* community, are confined to the Juncetum. The last three communities are absent at Canvey, except X, which exists on the escarpments to a slight extent. From Carter, 1933, p. 394.

Wiehe (1935) has published an interesting study of the effects of the tides on Salicornietum. He finds that a "threshold" time of 2 or 3 days nudisturbed by tides is necessary for the establishment of a dense population of *Salicornia*. Daily tides drag the seedlings from their anchorage in the mud (Fig. 156).

(1*a*) **Spartinetum townsendii** (Pl. 138, phot. 347; Pl. 139, phots. 348–9).

On deep mobile mud, often too mobile for the successful establishment of the Salicornietum, though approximately on the same level, the tall strongly growing perennial grass *Spartina townsendii* (Fig. 157), a hybrid of *S. alterniflora* (an American species) with the European *S. stricta*, forms almost pure communities on the south coast of England, especially in Southampton

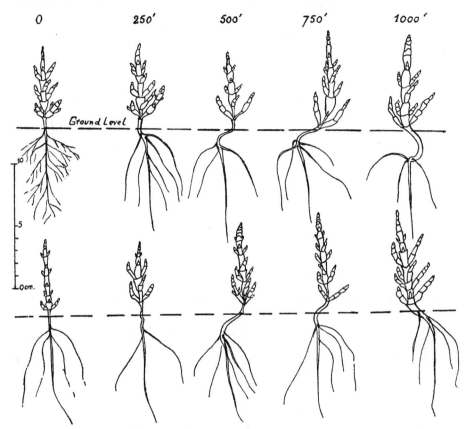

FIG. 156. *SALICORNIA HERBACEA* IN THE SALICORNIETUM OF THE YNYSLAS SALT MARSH (DOVEY ESTUARY)

The plants show increasing distortion of the axis in passing from left to right of the figure. Those at 500 ft. and more from the zero point are subject to the effect of daily neap tides, which drag the seedlings from the erect position and frequently dislodge them altogether. From Wiehe, 1935.

Water and Poole Harbour, as also on the opposite coast of France. This hybrid was first reported from Southampton Water in 1870, and since then has extended its area very rapidly. It has also been planted in several other places on our coasts, and in some of these is well established and spreading. Pl. 139 illustrates the remarkable transformation that can occur on stretches of tidal mud as the result of the activity of the rice-grass.

    S. townsendii is a much larger plant than the native *S. stricta* which plays a subordinate part in some of our salt marshes. The stout stem bears stiff erect leaves: from the bases of the stems stolons radiate in all directions, binding the soft mud; and feeding roots, mostly horizontal in direction,

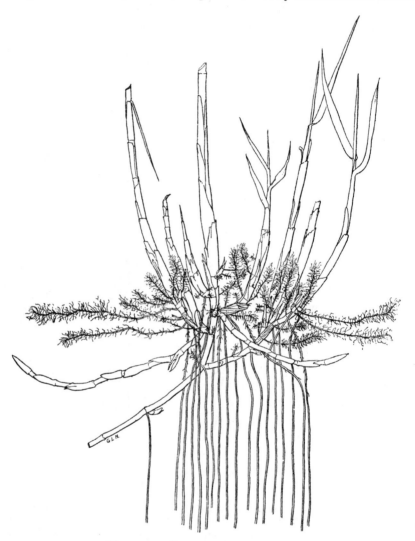

FIG. 157. *SPARTINA TOWNSENDII*

Showing bases of stout ascending aerial shoots, horizontal stolons and feeding roots, and deeply penetrating vertical anchoring roots. From Oliver, 1926.

ramify through the surface layers of the mud, while stouter anchoring roots extend vertically downwards (Fig. 157). The leaves offer broad surfaces to the silt-bearing tidal water, and their points catch and hold fragments of seaweed and other flotsam. The thick forest of stems and leaves breaks up the tidal eddies, thus preventing the removal of mud which has once

PLATE 139

Phot. 348. View across Holes Bay, Poole Harbour, Dorset, showing *Spartina townsendii* colonising the soft mud. June 1911. *R. V. Sherring.*

Phot. 349. View from approximately the same spot showing dense Spartinetum developed in 13 years. June 1924. *F. W. Oliver.*

SPARTINETUM TOWNSENDII

PLATE 140

Phot. 350. *Salicornia herbacea* (foreground), *Aster tripolium* and Glycerietum maritimae near Gedney Drove End, The Wash. *R. H. Yapp.*

Phot. 351. *Salicornia dolichostachya*, on bare mud, and dense Glycerietum developed between 1911 and 1921. Berrow mud flats, Somerset. *H. S. Thompson* (1921).

Phot. 352. Another view of the Glycerietum, same locality. Luxuriant flowering tuft of *Aster tripolium* (centre) on edge of channel and *Triglochin maritimum* (left). *H. S. Thompson* (1921).

GLYCERIETUM MARITIMAE

settled on the marsh. In this way large areas of mobile mud are fixed, the surface level is raised, and "reclamation" eventually made possible. At the same time the tidal currents and channels are profoundly altered and the whole aspect of the intertidal zone completely changed. No other species of salt-marsh plant, in north-western Europe at least, has anything like so rapid and so great an influence in gaining land from the sea.

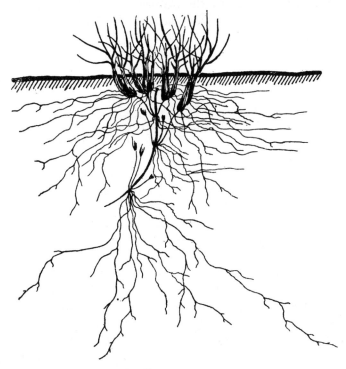

FIG. 158. *GLYCERIA MARITIMA*

Tiller formation giving tufted habit under grazing (tillers separated for drawing).
¾ natural size. From Yapp, 1917.

The Spartinetum townsendii is a remarkably pure community, containing only occasional examples of species flourishing at about the same level, such as *Salicornia herbacea* and *Aster tripolium*. The Salicornietum is thus very much restricted in areas in which the rice grass has successfully established itself.

(2*a*) **Glycerietum maritimae.**[1] The upper edge of the Salicornietum is invaded, and *Salicornia* eventually replaced as a dominant, either by *Glyceria maritima* or *Aster tripolium*. *Glyceria maritima* (Fig. 158) is a vigorous perennial grass which quickly extends horizontally and begins to form a turf. After it is once established *Glyceria* rapidly collects silt and often forms low flat hummocks, thus restricting the space available for the

[1] Pl. 140.

single plants of *Salicornia*. The plant is generally shallow rooted (2–4 in. = 10 cm.), but under exceptionally favourable conditions (good aeration) roots may penetrate to a much greater depth, reaching 2 ft. (60 cm.) or even more. Sometimes a community is formed in which *Glyceria* and *Salicornia herbacea* are co-dominant, with a little *Suaeda maritima* and other species of herbaceous *Salicornia*, such as *S. disarticulata* and *S. ramosissima* (Heslop Harrison, 1918). In the Severn estuary (Priestley, 1918) the following species occur in this zone besides the dominant:

Salicornia herbacea	Spergularia marginata
Suaeda maritima	Triglochin maritimum
Plantago maritima	Glaux maritima

The last four species usually occur at higher levels of the marsh, and *Glaux* flowers very little in this habitat.

In the estuary of the River Nith (Dumfriesshire) on the north side of the Solway Firth (Morss, 1927), where *Salicornia* plays quite a minor role, *Glyceria maritima* is the pioneer on all the areas studied, and is predominant on most parts of the marshes. *Cochlearia officinalis* occurs freely in the middle Glycerietum of the Nith estuary and also on the soft mud of the sides of creeks. It affects comparatively open soil and may also occur (at a much higher level) in the Juncetum maritimi, where the shade of the tall sea-rush keeps the surface of the soil bare.

In the Bouche d'Erquy, a sandy Breton salt marsh, studied by Professor F. W. Oliver and his colleagues and students in the early years of the century, most of the area of the marsh is occupied by Glycerietum, composed of alternating low flat hummocks or tracts of somewhat higher lying flat ground, and shallow channels or lower lying flats, the grass being regularly associated with *Suaeda maritima* on the former and with *Salicornia herbacea* in the latter. All this area, uncovered during the whole of the neap tide cycle, is used as sheep pasture.

(2*b*) **Asteretum tripolii.** In some marshes, in the lower zones of which Glycerietum is absent or plays a relatively insignificant part, the Salicornietum is invaded instead by *Aster tripolium*. Thus at Scolt Head (Chapman, 1934) the new dominant displaces *Salicornia*, which remains in the barer areas between. The following species are present:

Aster tripolium var.		Pelvetia canaliculata
discoideus	d	forma libera
Salicornia herbacea	ld	Bostrychia scorpioides
Spartina stricta	ld	
Suaeda maritima var.		
flexilis	la	

Along the banks of the creeks, farther up in the marsh but at about the same horizontal level, an Asteretum is developed in which *Suaeda maritima* var. *flexilis* is replaced by *Salicornia perennis*, and *Pelvetia* by *Fucus caespitosus* and *F. volubilis*. The soft mud deposited on the banks of creeks cut by the tide in consolidated marshes is a favourite habitat of *Aster*.

At Canvey Island in the Thames estuary (Carter, 1933) *Aster* also plays a prominent part at this level, invading the Salicornietum along with *Glyceria* and remaining abundant after *Glyceria* has established dominance, but it is not itself dominant. A number of green and blue-green algae are prominent in this zone (see Fig. 154). *Ulothrix flacca* is dominant in February, giving place in March to *Enteromorpha prolifera*, which covers the *Ulothrix* like a blanket during the spring and summer and decreases in August, when a mixture of dark blue, olive green or brown Cyanophyceae appear, including several species of *Oscillatoria* and *Phormidium*. *Microleus chthonoplastes* forms local streaks on the mud, and circular colonies of *Anabaena* and *Nodularia* occur near the moist margins of the zone. Diatoms are frequently present among the Cyanophyceae. This phase lasts till October when the mud becomes increasingly bare.

The marshes up to this point are included in Chapman's Group Z (see Fig. 152), belonging to the lower marsh.

(3) **Limonietum vulgaris** (Sea-lavender marsh). The common sea lavender, *Limonium vulgare* (formerly known as *Statice limonium*) invades the upper edge of the Glycerietum in many salt marshes, and becomes co-dominant. Flowering throughout the later summer its lavender-purple blossoms make magnificent sheets of colour. This is a very typical community of what may be called the middle marsh and often occupies wide areas. It forms a transition to the communities of the upper marsh (Group X).

In the Tees marshes the following species occur (Heslop Harrison, 1918):

*Limonium vulgare and *Glyceria maritima   co-d

*Aster tripolium	a	*Suaeda maritima	o
Plantago maritima	a	*Salicornia herbacea (agg.)	r
Spergularia salina	f	Limonium humile	vr
*S. marginata	f	*Triglochin maritimum	vr
*Armeria maritima	f		

The eight species marked with an asterisk also compose the flora of the Limonietum at Holme in Norfolk (Marsh, 1915).

In some salt marshes the sea lavender appears to be quite absent, e.g. in the Dovey and Nith marshes. Yapp (1917) and Morss (1927) suggest that this may be due to heavy grazing.

(4) **Armerietum maritimae** (Thrift marsh). The thrift or sea pink, *Armeria maritima* (Fig. 159), often dominates a community at a slightly higher level than the sea lavender. Here we have typically a close turf composed of a number of species in which the compact rosettes of *Armeria*, together with the tufts of *Glyceria*, form the largest part. *Armeria* cannot tolerate such high salt concentrations as *Salicornia* and *Glyceria*. It roots more deeply (to 23 in. in the Nith estuary) and grows more vigorously with

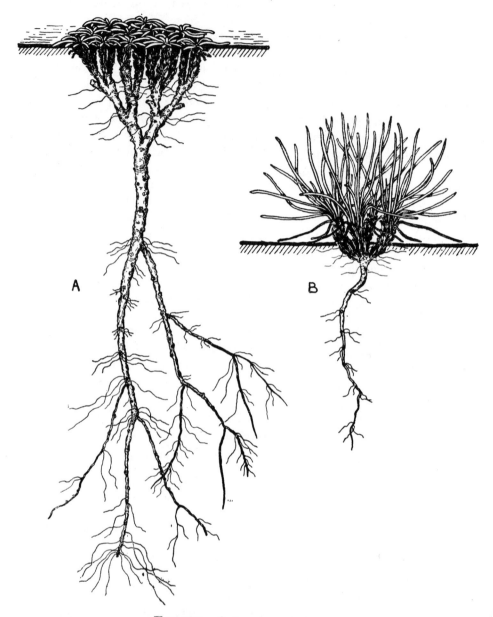

FIG. 159. *ARMERIA MARITIMA*

A. From the Armerietum—rosette habit under grazing.

B. From the Juncetum—more diffuse habit under partial shade and protection from grazing. About ¾ natural size. From Yapp, 1917.

decreasing salt, only giving way before the still less halophytic competitors of the upper marsh. The thrift stands grazing well and its compact rosettes (Fig. 159 A) often form a considerable part of the turf of sea meadows. The thrift has a much longer flowering season than the sea lavender, beginning in April and being at its best in May and June, when the massed flowers show sheets of rose-pink, comparable in beauty with the purple stretches of sea lavender in the following months.

In the Tees marshes a rather open Armerietum has the following species:

Armeria maritima and Glyceria maritima   co-d

Plantago maritima var.		Suaeda maritima	o
latifolia	f	Aster tripolium	o
Spergularia spp.	f	Limonium vulgare	o
Salicornia herbacea	lf	Triglochin maritimum	r

In another part of these marshes the dark green turf of the Armerietum is much closer and the annuals are practically absent. Here *Festuca rubra*, *Glaux maritima* and *Artemisia maritima* occur, though none is more than rare or occasional.

In the Dovey marshes on the coast of Wales (Yapp, 1917) the Armerietum (Pl. 141, phot. 354; Pl. 143, phot. 359; Pl. 144, phot. 364) contains the following species:

Armeria maritima   d
Glyceria maritima   co-dominant in the lower part of the zone

Triglochin maritimum	Aster tripolium
Plantago maritima	Spergularia sp.
Glaux maritima	Pholiurus (Lepturus) filiformis

Green algae. Three green algae compete for space in the Dovey Armerietum—*Rhizoclonium*, *Vaucheria* and *Enteromorpha percursa*, the first-named dominant during winter and spring, *Vaucheria* and Cyanophyceae prominent in the late summer when the marsh is exceedingly dry and *Rhizoclonium* and *Enteromorpha* are not visible. *Vaucheria* probably tolerates drought better than other green forms because its filaments penetrate the soil to some depth (Carter, 1933, p. 388).

Soil fungi. The soil fungi of the Dovey salt marshes were investigated by Elliott (1930). This is one of the very few natural plant formations whose soil fungi have been studied.

In the Salicornietum fungi were few and far between. The species found there all occurred also in the Glycerietum and Armerietum (from which most of the samples examined were taken) except *Chaetomium spirale* and *Macrosporium commune*. The soil is a badly aerated stiff tenacious clay (*p*H 8). Generally speaking the same species were common to the two, and those found at a depth of $3\frac{1}{2}$ in. (c. 8·75 cm.) were for the most part the same as at $1\frac{1}{2}$ in. (c. 3·75 cm.). Direct microscopic examination of the soil yielded little information. Conidia and fragments of hyphae were rarely met with and then only in association with organic matter such as dead or dying roots, stems and leaves. In pure culture, however, forty-eight species were isolated, but the fungi were not abundant, or at least not very active, since many fewer colonies were developed in culture from a suspension of a gram of waterlogged salt-marsh soil than from a gram of cultivated soil. Samples taken in June (soil temperature at $3\frac{1}{2}$ in. 19·5° C.) gave the largest crops.

Of the whole number of species recorded, most of which are known above ground as saprophytes, twenty-seven were Fungi Imperfecti, fourteen Ascomycetes and seven Mucorales. Besides these several septate sterile mycelia were found, and one, showing clamp connexions, was presumably a Basidiomycete. No evidence of active parasitism was found.

The commonest species were *Torula allii*, *Penicillium hyphomycetis* and *Fusarium oxysporium* var. *resupinatum*. The first two were not previously recorded as soil fungi. Almost equally common were *Trichoderma lignorum*, *T. könig*, *Hormodendron cladosporoides*, *Mucor circinelloides*, and *Periconia felina*. Several of the species isolated grew in culture by sending out a few long branches, fruiting as they grew, followed by others taking the same course and coiling round the pioneer branches so that rope-like strands were formed, from which similar strands branched out later. In this way the fungus quickly covered a large area of substratum.

Both *Glyceria maritima* and *Armeria maritima* are mycorrhizal plants (though *Glyceria* at least can be cultivated in the absence of mycorrhiza), the endophytic fungus extending throughout the plant body of the former and probably also of the latter. It is likely that some of the mycorrhizal fungi are the same as those recorded from the soil, and one was seen in the root of *Glyceria* which appeared identical with *Monilia pruinosa* occurring in the soil. *Stachylidium cyclosporum* and *Cladosporium herbarum* appeared in cultures from the rootlets of *Glyceria* and these may also perhaps be mycorrhizal. *Glyceria* has a compact and densely lignified root stele which resists the decay that destroys the cortex, but the roots of *Armeria* have little lignified tissue and possess large air spaces. At a depth of a foot the taproots of *Armeria* are little more than hollow tubes of cork, the whole of the interior having decayed from the attacks of fungi and bacteria. This superior resistance of *Glyceria* to decay helps to explain its much greater efficiency in binding the tidal silt. Where tidal erosion has taken place the soil of the Glycerietum shows a tangle of roots and rhizomes which is absent in the Armerietum.

It is to be noted that the salt marsh dominated by these middle communities has long been used as sheep pasture and that the excrement is no doubt an important factor in determining the fungal vegetation. It should also be noted that the culture media employed (bread-, potato- and raisin-agar) may have had a selective effect on the fungi.

"**General salt-marsh community.**" In *Types of British Vegetation*,

PLATE 141

Phot. 353. Early stages of primary salt marsh formation. Thin Glycerietum with hummocks of *Armeria* (which traps silt more quickly). Winding channels beginning to be established. Dovey estuary. *R. H. Yapp* (1917).

Phot. 354. Later stage: Glycerietum denser, *Armeria* hummocks more numerous and larger. Primary depression pan in foreground, narrow channel crossing the picture from left to right. *R. H. Yapp* (1917).

GLYCERIETUM AND ARMERIETUM

PLATE 142

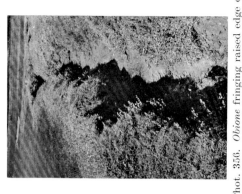

Phot. 356. *Obione* fringing raised edge of deeply cut channel. S. Hastings.

Phot. 355. General salt marsh community: *Glyceria maritima*, *Obione portulacoides* and *Limonium vulgare* (left), *Limonium reticulatum* (right), *Armeria maritima* (flower heads on extreme left), *Plantago maritima*, *Spergularia marginata* (centre). Holme-next-the-sea, Norfolk.

R. H. Compton

Phot. 358. Old salt marsh, largely covered with Obionetum and with

Phot. 357. Obionetum succeeding Glycerietum. Near Gedney Drove End.

written when but few British salt marshes had been carefully surveyed, a "general salt-marsh association",[1] in which no species was generally dominant, was described as occupying the middle zones of the formation; but most subsequent workers have recognised specific dominants of all the zones. Chapman (1934, p. 111), however, describes a "general salt-marsh association" at Scolt Head island with the six species listed below "more or less equally important":

Limonium vulgare	Spergularia salina (media)
Armeria maritima	Triglochin maritimum
Glyceria maritima	Obione portulacoides

He thinks that the co-dominance of these plants in the general salt marsh is due to the separation in space of their root systems, which reach different depths (Figs. 160, 161), and to the difference in time of their flowering seasons. It is clear that the "general salt-marsh" community corresponds in a general way with the Limonietum and the Armerietum, but that in many if not most marshes the sea lavender and thrift (in conjunction with *Glyceria*) are definitely dominant in zones of slightly different level. Pl. 142, phot. 355, shows this community at Holme, Norfolk.

Salisbury (in Oliver and Salisbury, 1913) gives the following representative list for the "saltings" (general salt-marsh community) on the south side of "Blakeney Harbour" (estuary of the River Glaven):

Armeria maritima	a	Spergularia salina	f
Limonium vulgare	a	Salicornia disarticulata	o
Glyceria maritima	a	Suaeda maritima	r
Salicornia ramosissima	a	Cochlearia officinalis	r
Glaux maritima	f		

(4 a) **Obionetum.** *Obione (Atriplex) portulacoides* is a mealy greyish white undershrub which plays an important part in many salt marshes, especially on the eastern coast of England.[1] The common large bushy form of the plant (*O. portulacoides* (L.) Gaertn. var. *latifolia* (Gussone) Chapman) has a wide vertical range on the salt marshes of north Norfolk (6·66–10·17 ft. above O.D. at Scolt Head, Chapman, 1937); but at the lower levels it is especially dominant on the banks of the tidal channels which become established quite early in the development of the marsh. Here the *Obione* occupies a relatively well-drained soil, and it is probably this factor which mainly determines its habitat. The water channels deepen considerably as the marsh develops and eventually form deep creeks up which the high tides run to gain access to the upper levels of the marsh. When the creek is full the silt-bearing water pours over the sides on to the surrounding flats. A portion of the silt is strained off by the fringing vegetation, consisting largely of *Obione*, and thus raises the level of the bank above that of the adjacent marsh flats.[2] In this way the bank becomes exceptionally well drained and *Obione* increases in luxuriance, forming a conspicuous greyish

[1] Pl. 142. [2] Phot. 356.

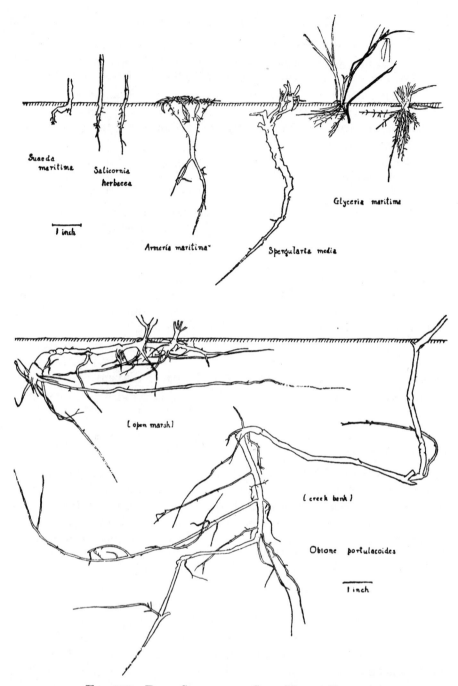

FIG. 160. ROOT SYSTEMS OF SALT MARSH PLANTS

Suaeda maritima and *Salicornia herbacea* (shallow rooted): *Glyceria maritima*, *Armeria maritima*, *Spergularia media* (medium depths), *Obione portulacoides* (different depths of penetration). *Obione* shows by far the deepest roots in the well aerated habitat of the creek bank. Plover Marsh, Scolt Head (Norfolk). From Chapman, 1934.

white band by which the courses of the creeks can be traced from a distance.[1]
Where the plant is growing with maximum luxuriance there is little or no
room for other species, and from the creek banks it spreads over the drier
marshes, forming a dense carpet one to two feet in depth and almost
completely obliterating their former vegetation.[2] The almost ubiquitous

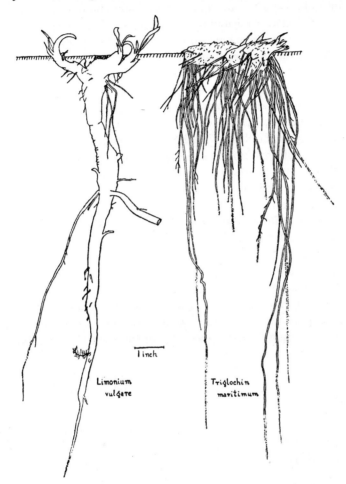

FIG. 161. ROOT SYSTEMS OF *LIMONIUM VULGARE* AND *TRIGLOCHIN MARITIMUM*

Plover Marsh, Scolt Head (Norfolk). From Chapman, 1934.

Glyceria maritima is said by Salisbury (Oliver and Salisbury, 1913) to be
the only species which grows successfully in this thick Obionetum.

Obione also becomes dominant on areas where blown sand has covered
the salt marsh ("sandy Obionetum" of Chapman, 1934). Here it grows in
slightly raised tussocks owing to its action in arresting and binding the

[1] Pl. 142, phot. 356,   [2] Pl. 142, phot. 357–8.

838   *Salt-marsh Formation*

sand. The small prostrate form (*O. portulacoides* var. *parvifolia* Rouy) with very much smaller leaves is locally abundant on sandy salt marshes in Norfolk (Chapman, 1937). The two varieties maintain their character when transplanted into each other's habitat.

(4*b*) **Suaedetum fruticosae.** *Suaeda fruticosa*, a Mediterranean species, which occurs (only locally but very abundantly where it does occur) on the south and east coasts of England, is another plant with a considerable vertical range. It forms dark green bushes 2 or 3 ft. high, and, like *Obione*, seems to depend on the drainage factor. It is specially associated with shingle beaches, where these overlie salt marsh, and the belts of dark green *Suaeda* mark the boundary between shingle and marsh (see Chapter XLII).

Associated with *Suaeda fruticosa* on the north Norfolk coast are two other Mediterranean species, *Limonium reticulatum* and *Frankenia laevis*.[1] They are also abundant in the "lows" or hollows of the sand-dune system (often floored on this coast with mixed sand and shingle brought in by exceptionally high tides) and on flat tracts of redistributed sandy shingle lying between the dunes and the salt marshes. They occur too in some of the upper marsh communities such as Chapman's "sandy Obionetum" at Scolt Head. *Limonium binervosum* (*bellidifolium*) is also found on the shingly sand of the lows and flats, but it is more abundant at a somewhat higher level on the sides of the old lateral shingle spits or "hooks" separating the marshes of the Marams at Blakeney, in a zone even more rarely visited by the high spring tides (see Chapter XLII, p. 889).

(5) **Festucetum rubrae.** Where the soil is very sandy (Dovey, Severn, Holme) this community occupies a zone above the "general salt marsh". At Holme *Festuca* invades the upper edge of the Armerietum (Marsh, 1915). The following is a list of species:

Festuca rubra	d	Spergularia marginata	o–f
Juncus gerardi	ld	Armeria maritima	o
Plantago coronopus	f	Pholiurus (Lepturus)	
P. maritima	f	filiformis	o
Glaux maritima	f–a	Triglochin maritimum	o
Agropyron pungens	o		

The first three of these species and *Agropyron pungens* are confined to the upper zones of the salt marsh and *Glaux* is characteristic of the higher levels, though it occurs in some marshes lower down. *Agrostis stolonifera* is locally dominant or co-dominant at the higher levels of the Festucetum in the Dovey marshes, and *Hordeum nodosum* occurs in this community in the Severn estuary. In the close turf of the upper Festucetum of the Dovey

Mosses and Algae   marshes two mosses are recorded—*Pottia heimii* and *Amblystegium serpens* var. *salinum*.

In the lower part of the Festucetum of the Dovey estuary *Rhizoclonium* is more successful in the summer, *Vaucheria*, accompanied by a number of

[1] Pl. 143, phot. 360.

blue-green algae and diatoms, in the wetter winter months. In the late summer, below the green algae, there is often a tough stratum of Cyanophyceae, and the drier the conditions the more prominent these become in relation to the green forms. The algal flora of the soil of the higher Festucetum most nearly approaches that of non-saline soils. In the summer Cyanophyceae predominate and are more abundant than in the lower zones, *Rivularia* and *Nostoc* being the most important, the latter in the wetter months (Carter, 1933).

(6) **Juncetum maritimi.** Where the upper edge of salt marsh is relatively undisturbed it is commonly occupied by a community in which the sea-rush (*Juncus maritimus*) is dominant.[1] This grows much taller than any of the accompanying species, and its dominance tends to destroy the close turf of the sea meadow, providing more open soil and maintaining moist air between its shoots. This often leads to the reappearance of more

Moister conditions

moisture-loving plants from the lower zones, such, for example, as *Cochlearia officinalis*, *Plantago maritima*, *Aster tripolium* and even *Salicornia* spp., and to an increase in the number and variety of the algae.

At Blakeney the Juncetum maritimi contains:

Armeria maritima	a	Limonium vulgare	f
Glyceria maritima	a	Plantago maritima	f
Glaux maritima	f	Spergularia salina	f
Salicornia spp.	f		

In the Dovey marshes, where it follows the Festucetum:

Festuca rubra	a	Glaux maritima	f
Agrostis stolonifera	a	Plantago maritima	f
Armeria maritima	f	Spergularia salina	o
Cochlearia officinalis	f	Aster tripolium	o

The algal vegetation of the Dovey Juncetum is more stable than in the lower zones of the marsh (Carter, 1932) and includes *Vaucheria*

Algae

and *Rhizoclonium* among the green algae, *Lyngbya aestuarii*, *Lyngbya* spp. and *Microcoleus chthonoplastes* among the blue-green: of brown seaweeds there are *Sphacelaria* and *Pelvetia canaliculata*; and of red *Catenella opuntia* and *Bostrychia scorpioides*. The last three large algae are very local in occurrence and prefer stretches of bare mud between the *Juncus* plants.

Transition to land vegetation. The upper edge of the salt marsh, only touched by the highest spring tides, is invaded by non-halophytic species of considerable variety—pasture plants on the drier soils, fresh-water marsh plants where the soil is wet but the water supply is mostly fresh. Among the latter *Oenanthe lachenalii* and *Samolus valerandi* are characteristic. *Phragmites communis* may also occur at this level when the salt content is low, as well as other species of *Juncus*, *Phalaris arundinacea*, *Ranunculus*

[1] Pl. 143, phot. 359.

flammula, etc. and submaritimes such as *Scirpus maritimus* and *S. taber-naemontani* (cf. Chapter XLIII).

Salt-marsh pasture. The middle and upper levels of salt marsh are very generally used as sheep pasture and are sometimes grazed by cattle. Where the sandy Glycerietum ((2a), p. 829), associated with *Suaeda maritima* on the drier higher lying areas, and with *Salicornia herbacea* on the wetter slightly lower lying flats, is extensive and is not covered by the neap tides, so that it is exposed for nearly a fortnight at a time as in the Bouche d'Erquy (p. 830), this is the main area on which the sheep depend. In other less sandy marshes, as at Blakeney and Scolt Head, *Glyceria maritima* becomes important and often co-dominant in later stages of the succession, as in the Limonietum and the Armerietum; and here the so-called "sea meadow" is developed, interrupted only by the Obionetum, on the upper levels of the marsh. In very sandy marshes *Glyceria*, as we have seen, is replaced by *Festuca rubra* in this upper zone. The short close turf of the sea meadow, whether Glycerietum, Armerietum or Festucetum, is determined largely by the grazing.

Drainage channels. The surface of a primary salt marsh is never quite level, because of the unequal growth and unequal powers of trapping silt of the different marsh-building species. Thus there is a local formation of hummocks by the plants which hold silt in greatest quantity (Pl. 141, phot. 353) and the flow of the tide tends to take winding courses round the largest hummocks. On the one hand the hummocks extend horizontally as well as vertically and often coalesce into ridges and continuous raised areas, while their growth in height slows down: this process tends to the development of flat uniform stretches of marsh carpet. On the other hand the channels followed by the tide between the hummocks are at first shallow and shifting, but they are gradually deepened by the flowing water. A high spring tide rushes up the channels with considerable force, especially if backed by a strong wind, but it is mainly at the ebb of such a tide that erosion of the channels occurs. The growth of the marginal plants increases their power of trapping silt and thus the banks of a channel rise while its bed is deepened by the scour of the water. In this way regular drainage systems showing both erosion and deposition are established in the older marsh, and these are very similar to river systems with main streams and tributaries, though the action of the water is intermittent and both up and down the channels. The greater the tidal range and the vertical height of the marsh the stronger the scour and the deeper the channels are cut. The edges of the deep channels in mature marsh are raised above the general level owing to the trapping of silt by the larger marsh plants, such as *Obione*, which line the crests of the banks (Pl. 142, phot. 356).

Owing to various causes the flow of water up and down a particular channel may become insufficient to keep its bed scoured and clear of

PLATE 143

Phot. 359. Juncetum maritimi abutting on sward of *Festuca rubra* with *Armeria maritima* in flower. Dovey salt marshes. *R. H. Yapp* (1917).

Phot. 360. Society of *Frankenia laevis* in mature salt marsh. Blakeney Point. *J. Massart.*

UPPER SALT MARSH

PLATE 144

Phot. 363. Old pan drained into channel and invaded by *Glyceria* at upper end. *R. H. Yapp* (1917).

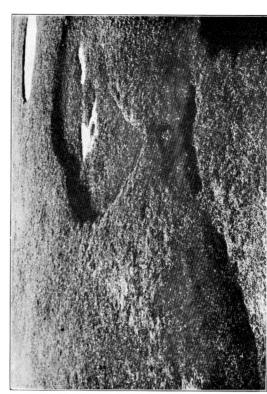

Phot. 365. Formation of subterranean channel by bridging. Drained and vegetated pan behind. *R. H. Yapp* (1917).

Phot. 362. Formation of pans from channel. *R. H. Yapp* (1917).

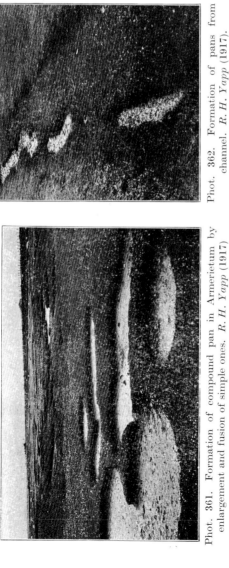

Phot. 361. Formation of compound pan in Armerietum by enlargement and fusion of simple ones. *R. H. Yapp* (1917)

Phot. 364. General view of channels and pans in salt marsh turf. Dovey estuary. *R. H. Yapp* (1917)

vegetation. The channel is then invaded by the marsh plants, and sometimes bridged across the top, for example by stolons of *Glyceria maritima*.
Thus parts of such a dormant or semi-dormant channel may "grow up"
or be "roofed in" by silt-trapping marsh vegetation. The roofing in of a
channel which still conducts a certain amount of water may result in its
becoming "subterranean" for parts of its course, opening to the surface
at intervals (Pl. 144, phot. 365; cf. the similar phenomena in raised bog,
Chapter xxxv).

Formation of "pans". If a group of hummocks formed in a primary
marsh coalesce so as to surround an area which then forms a relative
depression (Pl. 141, phot. 354) the water is hindered from draining away
at the ebb, and a "primary depression pan" is formed in which water
remains after the ebb. This water may slowly percolate away if the substratum is sufficiently permeable, or it may remain for a long time, from
one spring tide cycle to the next, and slowly evaporate, the water level
rising of course if there is heavy rainfall. Such pans are normally bare of
the ordinary salt marsh vegetation, probably because of their stagnancy
and the great variations of salt content due to alternating evaporation
and precipitation. The sequence of conditions in stagnant pans and their
algal vegetation have not, however, been adequately investigated.

The pans formed in the earlier phases of salt marsh development, e.g. in
the Salicornietum (Pl. 138, phot. 346 and Pl. 141, phot. 353) are often
transitory, but some of them persist into the later stages, and these are
generally remarkably stable and permanent, e.g. in the *Armeria* sward.
Pans may also be formed by the blocking up of a channel at intervals,
when the scour has become insufficient to keep the channel open, and
vegetation has extended across it. "Channel pans" may be recognised by
the fact that they occur in series, marking the course of the former
channel (Pl. 144, phots. 362, 364).

If a pan is secondarily drained into a drainage channel, vegetation may
invade its surface (Pl. 144, phot. 363), and ultimately it may be obliterated.
Very often however this invasion is partial and smaller "residual pans"
are formed on the floor of the original one (Pl. 144, phot. 365). Erosion of
the sides of a pan may occur as a result of swirling water entering the pan
at a high spring tide, and in this way the turf intervening between neighbouring pans may be cut through and compound pans formed (Pl. 144,
phot. 361).

For a detailed description and discussion of drainage channels and pans
in the salt marshes of the Dovey estuary the paper by Yapp, Johns and
Jones (1917) should be consulted. On some salt marshes pans may be
formed in other ways than those described.

Algal communities of the salt marsh. Chapman (1938) recognises a number
of algal communities of the salt marsh distinct from those dominated by
flowering plants. These, he says, are not in general correlated with marsh level,

and the factors which determine their distribution are still largely unknown. In view of the important part played by algae in the salt-marsh formation it seems worth while to reproduce Chapman's list here:

I. General "association" of Chlorophyceae:

    (a) Low sandy community.

    (b) Sandy community.

    (c) Muddy community.

II. Marginal Diatom community.

III. Marginal community of Cyanophyceae.

IV. Vernal *Ulothrix* community.

V. *Enteromorpha minima* community.

VI. Community of gelatinous Cyanophyceae.

VII. Community of filamentous Diatoms.

VIII. Autumn community of Cyanophyceae.

IX. *Phormidium autumnale* community.

X. *Rivularia-Phaeococcus* community.

XI. Dwarf *Pelvetia* community.

XII. *Catenella-Bostrychia* community.

XIII. *Pelvetia limicola* community.

XIV. *Enteromorpha clathrata* community.

XV. *Fucus limicola* community.

XVI. Pan community.

Some of these were previously described by Carter (1933). Cf. Figs. 154, 155. For details the reader is referred to Chapman's paper.

REFERENCES

CARTER, NELLIE. A comparative study of the alga flora of two salt marshes. Part I. *J. Ecol.* 20, 341–70. 1932. Part II. 21, 128–208. Part III. 385–403. 1933.

CHAPMAN, V. J. "Ecology" in Steers, *Scolt Head Island*. Cambridge, 1934.

CHAPMAN, V. J. A note upon *Obione portulacoides* (L.) Gaertn. *Ann. Bot., Lond.*, N.S. 1. 1937.

CHAPMAN, V. J. Studies in Salt Marsh Ecology. *J. Ecol.* 26, 144–79. 1938.

ELLIOTT, JESSIE S. B. The soil fungi of the Dovey salt marshes. *Ann. Appl. Biol.* 17. 1930.

HESLOP HARRISON, J. W. A survey of the lower Tees marshes and of the reclaimed areas adjoining them. *Trans. Nat. Hist. Soc. Northumb.*, N.S. 5, pt. I. 1918.

MARSH, A. S. The maritime ecology of Holme-next-the-Sea, Norfolk. *J. Ecol.* 3, 65–93. 1915.

MORSS, W. L. The plant colonisation of Merse lands in the estuary of the River Nith. *J. Ecol.* 15, 310–43. 1927.

OLIVER, F. W. Blakeney Point Reports, 1913–29. *Trans. Norfolk Norw. Nat. Soc.*

OLIVER, F. W. *Spartina* problems. *Ann. Appl. Biol.* 7. 1920.

OLIVER, F. W. *Spartina townsendii*, its mode of establishment, economic uses and taxonomic status. *J. Ecol.* **13**, 74–91. 1925.

OLIVER, F. W. and SALISBURY, E. J. The Topography and Vegetation of the National Trust Reserve known as Blakeney Point, Norfolk. *Trans. Norfolk Norw. Nat. Soc.* **9**, pt. 4. 1913.

PHILIP, G. An enalid plant association in the Humber Estuary. *J. Ecol.* **24**, 205–19. 1936.

PRIESTLEY, J. H. The pelophilous formation of the left bank of the Severn Estuary. *Bristol Nat. Proc.*, 4th Series, **3**, pt. 1. 1911.

RICHARDS, F. J. The salt marshes of the Dovey Estuary. IV. The rates of vertical accretion, horizontal extension, and scarp erosion. *Ann. Bot., Lond.*, **48**, 1934.

WIEHE, P. O. A quantitative study of the influence of tide upon populations of *Salicornia europea* [*herbacea*]. *J. Ecol.* **23**, 323–32. 1935.

YAPP, R. H., JOHNS, D. and JONES, O. T. The salt marshes of the Dovey Estuary. Part II. The salt marshes (Yapp and Johns). *J. Ecol.* **5**, 65–103. 1917.

Chapter XLI

THE FORESHORE COMMUNITIES.
COASTAL SAND-DUNE VEGETATION

(2) THE FORESHORE COMMUNITIES

On open sea shores undisturbed by man, and where the substratum is not composed of rock, a scattered vegetation of flowering plants is often met with along the zone barely reached by the highest spring tides. The existence of these littoral communities depends on the maintenance for a time of conditions that are not too constantly and violently disturbed. The substratum must be reasonably stable, not subject to erosion by wind or waves and not overwhelmed by sand or shingle. Such conditions are realised on the upper limit of a fringing shingle beach (p. 868) where shingle is sparsely scattered over a clayey or loamy substratum; and again at the seaward foot of sand dunes which are not growing very actively nor being eroded by wind or the sea. These habitats are of course *liable* to considerable disturbance, e.g. by specially high tides, and correspondingly the foreshore communities are essentially "migratory".

Tidal drift. The soil of the foreshore communities is fed by sea drift (organic debris largely composed of dead seaweeds) thrown up by exceptionally high tides, and this humus supply is important for the luxuriant growth of the plants. The soil also contains a considerable amount of sea salt, and the plants may fairly be reckoned as halophytes, some being characteristically succulent.

Characteristics. The foreshore communities are very open, consisting of more or less widely and irregularly spaced individuals, because the conditions are not stable enough for long periods to permit of the development of closed vegetation, and for the same reason they are inconstant. They are composed of scattered annuals, nearly all belonging to the families Chenopodiaceae, Cruciferae or Polygonaceae, with a few perennials (*Beta, Crambe, Mertensia*), which settle in this zone when and where they can, some favouring sandy and some muddy shores, disappearing and reappearing as the conditions fluctuate, and marking a true "tension line" between tidal and non-tidal habitats. The characteristic species are few, but they are sometimes reinforced, especially on shingly and muddy coasts, by annual weeds which can tolerate a certain amount of salt in the soil; and where the conditions are sufficiently stable, by perennials like *Beta maritima* and *Crambe maritima*.

Flora. The following are the most characteristic species of this community:

Atriplex glabriuscula	f	A. patula	o
A. hastata	f	Beta maritima	f
A. littoralis	f	Cochlearia spp.	o–f

Mertensia maritima	r	*Sandy shores*	
Polygonum littorale	o	Cakile maritima[1]	f
P. raii	o	Crambe maritima	r
Raphanus maritimus	o	Salsola kali[2]	f

This foreshore vegetation has never been given serious study, so that it is impossible to say more about the exact conditions of its existence, nor about the particular requirements of the individual species. On sandy shores where dunes are formed it immediately precedes, and is sometimes reckoned as the first stage of the dune succession (psammosere).

(3) THE SAND-DUNE FORMATION

Conditions of existence. The accumulations of blown sand known as sand dunes are not confined to sea coasts. On the contrary, by far the greatest dune areas of the world are the so-called "continental" sand dunes of desert and arid or semi-arid regions generally, while inland sandy regions in a relatively though not excessively dry climate may show smaller local dune areas. Coastal dunes are also formed on the shores of large bodies of fresh water such as the Great Lakes of North America. Wherever plants can establish themselves in mobile dune areas their life forms have a general resemblance, for in order to survive they must be able to cope with the sand which is continually blowing over them.

In this country sand dunes are almost confined to the sea coast where collectively they cover a considerable area, second only to that occupied by the salt marshes; and it is mainly because of this that their plant covering is included in maritime vegetation, for the plants which grow on sand dunes are not halophytes and will not, for the most part, endure immersion in salt water.

The supply of sand for the formation of coastal dunes comes from shoals formed offshore on a flat coast and exposed at low tide, and from gently sloping sandy beaches over which the tide advances and recedes for a long distance. At low tide great areas of sand are thus laid bare, and as it dries the grains are driven landwards by onshore winds. At any small obstacle the stream of sand grains is checked and accumulated round it, both on the windward and still more on the leeward side,[3] for there the air is relatively quiet, and the lighter grains are carried by eddies up to the top of the miniature dune, and then slide down the longer and gentler leeward slope. On a fixed obstacle, such as a large stone or a lump of stranded seaweed, the sandhill can grow no higher than the top of the obstacle. Thus a brushwood fence planted across a sand-laden windway will accumulate sand up to the

Effect of growing plants height of the fence. But a growing plant can push its shoots upwards and sideways as it becomes covered with sand (though different species vary greatly in this power), and the embryo dune thus grows in height and width until it is arrested from other causes.

[1] Pl. 145, phots. 366–8.   [2] Pl. 146, phot. 371.
[3] Pl. 148, phot. 379; Pl. 149, phot. 381.

It is in this way that large and lofty systems of dunes are formed by plants which have exceptional powers of pushing up their shoots and of continually forming fresh roots in the moist sand just below the surface. Species which have only limited powers of such adjustment form dunes of

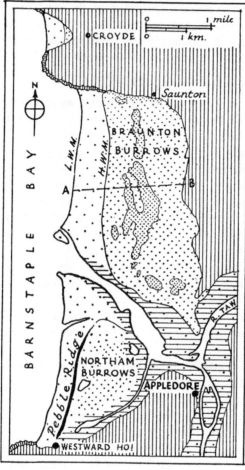

FIG. 162. SKETCH MAP AND SECTION OF BRAUNTON BURROWS (NORTH DEVON)

Sand-covered area dotted, height of dunes indicated by increasing closeness of dots. Horizontal lines are added to show the mud (mixed with sand) bordering the estuaries and bearing saltmarsh in places. Cultivated land represented by vertical ruling.

1. Flat sand covered at high tide. 2. Foredunes. 3. Very mobile dunes rising to 50 ft. 4. Brackish "slacks". 5. Mobile dunes rising to 100 ft. 6. Second line of slacks. 7 and 8. Low stable sandhills.

Horizontal scale of section twice that of map: vertical scale five times horizontal scale. After Watson, 1918.

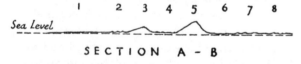

SECTION A - B

little height. Sand-dune complexes are formed in bays facing prevailing onshore winds (Fig. 162) and in other situations where blowing sand is trapped.

Parallel dune ranges. If a new series of dunes arises on the seaward side of a range already formed the latter is more or less protected from the wind, the supply of blown sand is cut off so that it ceases to grow in height,

PLATE 145

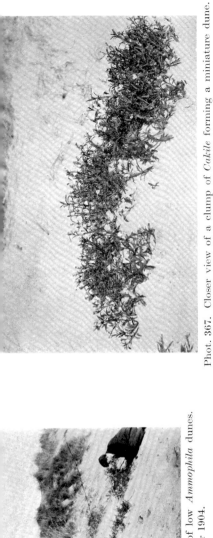

Phot. 366. *Cakile maritima* in flower in front of low *Ammophila* dunes. Near Rye, Sussex. November 1904.

Phot. 367. Closer view of a clump of *Cakile* forming a miniature dune.

Phot. 369. *Honckenya peploides* (with isolated *Ammophila* shoots forming miniature dunes in front of the main ridge). Holme-next-the-Sea. *R. H. Compton.*

Phot. 368. Zonation on seaward face of dunes near Hunstanton, Norfolk. From front to back: (1) sand mixed with shingle, highest limit of ordinary spring tides; (2) foreshore community of *Cakile maritima*, *Salsola kali*, *Honckenya peploides*; (3) foredune, Agropyretum juncei; (4) Ammophiletum with *Eryngium maritimum*. *S. Mangham.*

FORESHORE AND FOREDUNES

PLATE 146

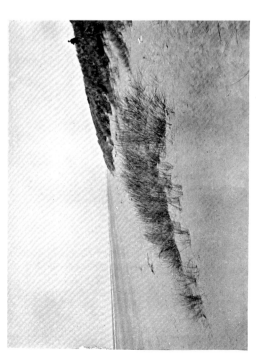

Phot. 370. Tidal drift and low dunes dominated by *Agropyron junceum*. Wind-eroded *Ammophila* dunes behind. Southport, Lancs. *W. Ball.*

Phot. 371. Foreshore community of *Salsola kali*. Sparse Agropyretum in front. *Ammophila* dunes behind. Paul Graebner and F. W. Oliver. Near Southport, Lancs. *Mrs Cowles* (1911).

Phot. 372. Another part of the shore near Southport. *Ammophila* is here the first colonist. *Mrs Cowles.*

Phot. 373. Foredunes formed by *Ammophila arenaria* on the left and *Elymus arenarius* on the right. Hemsby, Norfolk.

and thus a series of parallel dune ranges is often formed.[1] The protected older dunes gradually become stabilised and covered with a continuous carpet of vegetation, interrupted only where strong winds gain access and form the so-called "blow-outs" (Pl. 153).

Sand-dune vegetation shows a regular successional series of communities (psammosere) beginning with those forming and inhabiting the embryo dunes on the sea shore and ending with the stabilised vegetation on the landward side.

MOBILE DUNE COMMUNITIES

(*a*) **Sea couchgrass consocies** (Agropyretum juncei). This community, which is not represented in all dune systems, is dominated by *Agropyron junceum*, a plant of somewhat similar habit to that of the common couchgrass or "twitch" (*Agropyron repens*) which is such a pestilent weed of clayey arable land. It can withstand short immersions in seawater and is thus able to grow within reach of the high spring tides; and its extensively creeping rhizomes ramify through the sand and send up new aerial shoots through the fresh sand blown over the plant.

Foredunes. The sea couchgrass is peculiarly well fitted to act as a pioneer in accumulating sand. Its underground runners penetrate deeply into the substratum, even shingle, and are thus protected from exposure by erosion, while the leaves are often prostrate and thus tend to prevent the removal by wind of accumulated sand. In this way *Agropyron* hummocks rapidly spread and in a few years a single seedling may form a dune 15–20 ft. across and 3 or 4 ft. high (Oliver, Blakeney Point Report, 1929). *Agropyron* cannot however grow indefinitely in height, and the dunes it forms, often called *foredunes*, are thus comparatively low (Pl. 145, phot. 368).

Associated species. Very commonly associated with the sea couchgrass is the succulent sea sandwort *Honckenya* (*Arenaria*) *peploides*, which itself may produce miniature dunes (Pl. 145, phot. 369); and since the low dunes are formed by these two species on the sandy foreshore just along the limit of high spring tides, representatives of the sandy facies of the foreshore community, such as *Cakile maritima* and *Salsola kali* (Pl. 145, phot. 366–8), as well as the littoral species of *Atriplex*, frequently occur between the miniature sand hills. *Cakile* and *Salsola*, although annuals, may form tiny dunes during their season of growth, and their persistent dead bodies hold the sand throughout the winter, thus assisting the establishment of *Ammophila* (see below) at the slightly raised level.

The first five species to appear on the foreshore of the "Far Point" at Blakeney, when the shingle bank had attained the level of the highest spring tides were *Salsola*, *Cakile*, *Agropyron*, *Honckenya* and *Ammophila*. Occasional plants of species more characteristic of the Ammophiletum, such as the horned poppy (*Glaucium flavum*), the sea holly (*Eryngium mariti-*

---

[1] Cf. Pl. 133, phot. 334.

mum) and the common sea spurge (*Euphorbia paralias*) may also be found on the larger and older *Agropyron* dunes. A rayed variety of the common groundsel (*Senecio vulgaris* var. *radiatus*) is frequently met with in the Agropyretum of the Lancashire dunes.

(*b*) **Marram grass consocies** (Ammophiletum arenariae).[1] This is the principal consocies of mobile sand dunes, and is almost the sole agent in the building of the main dune ranges. Seedlings of *Ammophila* establish themselves on sand or sandy shingle just above the reach of high tides.[2] Unlike *Agropyron junceum* the marram grass cannot endure immersion in salt water and is therefore unable to colonise sand reached by the sea. Furthermore, its underground runners spread by preference in pure sand rather than shingle, and in the young stage its leaves stand up like a shaving brush, so that the plant is more liable than *Agropyron* to destruction by erosion (Oliver, Blakeney Point Report, 1929).[3] On the other hand, once established in deep sand, it has far more vigorous powers of vertical and lateral extension and is able to grow up through many metres of sand provided this is not deposited too quickly. When a high dune has been partly removed by wind, as often happens, the exposed face of sand down to the base level, representing a vertical section of the dune, is seen to be completely penetrated by the rhizomes and roots of *Ammophila*, mostly dead, but those nearer the surface still living. The capacity to form fresh adventitious roots as the shoots grow up, at higher and higher levels in the moist layer of sand which is constantly maintained a little below the dried surface, is the power which enables any plant to dominate and increase the height of a dune. On the great sand dunes of the southern shore of Lake Michigan a certain number of trees and shrubs possess this power, while others do not, and it is the former alone whose living tops maintain themselves on the surface of dunes more than 100 ft. (*c*. 30 m.) high. Those which cannot root in this way are killed when they are covered by sand.

Limitations and powers of Ammophila

On many sandy shores the Agropyretum is absent, and here *Ammophila* forms the foredunes as well as the main ranges. Another grass which plays a similar role, but by no means so universally, is the sea lyme grass (*Elymus arenarius*). The two are seen forming foredunes side by side in Pl. 146, phot. 373. In some sandy bays of the west coast of Ireland, as at Dog's Bay in Connemara, *Ammophila* is absent, and here the dunes are all low and are entirely dominated by *Agropyron* (Pl. 154, phot. 397).

The speed with which new dunes are formed by *Agropyron* and *Ammophila* under favourable conditions is well illustrated by the development of the "Far Point" beyond Blakeney headland (cf. Fig. 170, p. 875). Based on the continuation of the main shingle spit westwards and the subsequent formation of numerous "hooks", sand dunes have developed and become

---

[1] Pls. 146–9.　　　　　　　　　　　　　　　[2] Pl. 148, phot. 377.
[3] The different habit of the two grasses is well seen in Pl. 147, phot. 375.

PLATE 147

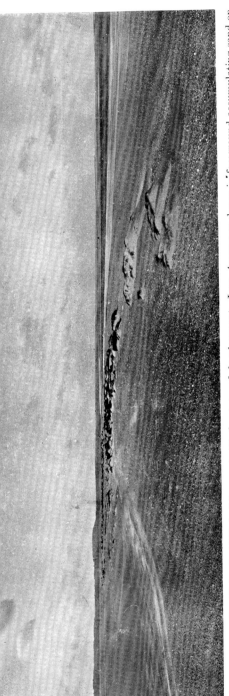

Phot. 374. The Far Point in 1927, looking west. Early stage of development. Low *Agropyron* dunes in foreground accumulating sand on the shingle. *Ammophila* and *Agropyron* complexes in the distance and middle distance. *F. W. Oliver.*

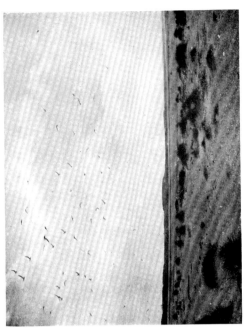

Phot. 376. Terns rising from their nests in foredunes. *F. W. Oliver* (1927).

Phot. 375. *Agropyron* and *Ammophila* recently covered by tide in foreground. Single *Ammophila* tufts in distance to the left. *F. W. Oliver* (1927)

FOREDUNES ON BLAKENEY FAR POINT

PLATE 148

Phot. 378. Blakeney Far Point in 1935, looking east from near extreme end. New scattered dunes on the sandy shingle in front. Consolidated dunes behind on the right.

Phot. 380. The Far Point seen from the dunes of the Headland, Blakeney Point. Consolidated Ammophiletum to the left, scattered foredunes to the right. 1935.

Phot. 377. Colonisation of sandy shingle by *Ammophila* seedlings. Blakeney Norfolk. *F. W. Oliver* (1910).

Phot. 379. Active accumulation of sand by *Ammophila*, especially on the lee (right hand) side of each tuft, main dune ridge behind. Blakeney Point. *R. H. Compton* (1910).

consolidated on the sandy shingle of the spit and hooks in a very few years. In 1928 there were only a few scattered *Agropyron* and *Ammophila* dunes, whose development had been slow during the preceding years.[1] In 1928 (Oliver *in litt.*) *Ammophila* seeded very heavily and the seeds germinated everywhere, forming numerous "lawns" of seedlings in the following year. Since then development and consolidation have been extremely rapid, so that now (1937) there are many acres of continuous consolidated dunes[2] whose crests are 2·5 m. above the beach, in place of the few scattered pioneer dunes of 1927–8. New hook formation and new colonisation by the dune pioneers is still proceeding (Fig. 170, 1937).

Features of Ammophiletum. The marram grass community is never closed on the surface, the dominant forming separate tufts or clumps with dry loose sand between.[3] These interspaces are colonised by scattered individuals of a few other non-maritime species, such as ragwort (*Senecio jacobaea*),[4] the hawkweed *Hieracium umbellatum*, and the two thistles *Cirsium lanceolatum* and *C. arvense*. The sand fescue (*Festuca rubra* var. *arenaria*) is the first grass to associate itself with *Ammophila*, and other maritime species which are frequent on the edges of the Ammophiletum, though sometimes found on older dunes, are the two sea spurges, *Euphorbia paralias* (of fairly general occurrence in England except on the north-east coast) with the rarer *E. portlandica* (only on south-western and western coasts), the sea holly (*Eryngium maritimum*) and the sea convolvulus (*Calystegia soldanella*). The rare *Brassica monensis* occurs in this community in the Isle of Man (Moore, 1931).

While the marram grass is practically the sole agent in forming and holding together the structure of the main dunes it does little to consolidate the surface sand (Salisbury, 1913). Where it is present alone, or almost alone, the wind still freely removes the loose sand from between the tufts of the grass, so that unless the rate of supply of sand is at least equal to the rate of removal the dune will be eroded. Consolidation of the surface is often effected by the sand fescue (*Festuca rubra* var. *arenaria*). In relatively sheltered places the sand sedge (*Carex arenaria*) is also effective in binding the superficial sand with its long rhizomes, which grow just below the surface in remarkably straight lines whose course is marked by a series of aerial shoots bearing rosettes of leaves above and bunches of roots below. The sea spurges and the sea convolvulus also possess creeping stems below the surface of the sand and carry out a similar function, but on a much less extensive scale.

The physiognomy of the main mobile dune complex, with its steeply rolling, highly irregular crests and valleys, is very characteristic. Stretches of perfectly bare sand on the slopes and in the hollows alternate with clumps and tufts of marram, which occupy most of the crests. This is due

[1] Cf. Fig. 170, 1921. [2] Seen in the distance as a dark band in Pl. 148, phot. 380.
[3] Foreground of phot. 380, Pl. 148. [4] Pl. 154, phot. 394.

to the complex distribution of wind currents and eddies, as the streams of quickly moving air are deflected and broken up by the slopes and crests. Throughout this maze of elevations and depressions the same types of surface form are constantly repeated, following the laws of motion of the loose wind-driven sand, partially fixed here and there by the marram grass.

Mobile dunes are often called "white" or "yellow" dunes according to the prevailing colour of the fresh sand, in contrast to fixed "grey" dunes dominated by the lichens *Cladonia* and *Cetraria*.

Flora. The following list of species is given by Watson (1918) for the very mobile Ammophiletum of the west coasts of England:

Ammophila arenaria	d		
Anagallis arvensis	o	Bryum argenteum	r
Cynoglossum officinale	r	Potentilla anserina	r
Erodium cicutarium	r	Sedum acre	o
Euphorbia paralias	o	Senecio jacobaea	o
Leontodon hispidus	r	Teucrium scorodonia	r
Nepeta hederacea	r	Viola canina	r

For the Ammophiletum of the headland at Blakeney Point, Norfolk, the following species are recorded by Salisbury (Oliver and Salisbury, 1913).

Ammophila arenaria   d
Festuca rubra var. arenaria   sd

Abundant:

Cerastium semidecandrum	Sedum acre
C. tetrandrum	Senecio jacobaea
Cirsium arvense	Stellaria pallida (boraeana)
Erophila verna (agg.)	
Myosotis collina	Tortula ruraliformis
Phleum arenarium	

Frequent:

Anagallis arvensis	Erodium cicutarium
Cirsium lanceolatum	Galium verum

Rare:

Agropyron junceum	Silene maritima

Very rare:

Corynephorus canescens	Hieracium pilosella
Elymus arenarius	Solanum dulcamara

From the north coast of the Isle of Man, where the supply of sand is very meagre, so that the dunes do not exceed a few feet in height, Moore (1931, p. 121) records the following species from the Ammophiletum:

Ammophila arenaria	d		
Atriplex sp.	o	Hypochaeris radicata	o
Brassica monensis	r	Leontodon nudicaulis	o
Cakile maritima	o	Matricaria inodora	r
Calystegia soldanella	o	Ononis repens	f
Eryngium maritimum	la	Senecio jacobaea	f
Festuca rubra var. arenaria	f		

PLATE 149

Phot. 381. Wind-blown *Ammophila* accumulating sand, Blakeney Far Point. *F. W. Oliver* (1927).

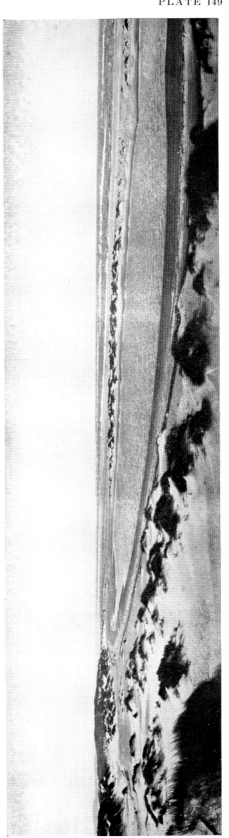

Phot. 382. Great Sandy Low, Blakeney Point, holding water left by the last high spring tide. In the foreground and between the low and the sea are young *Ammophila* dune complexes. These have since formed consolidated dune ridges. Old consolidated dune ridge on the left behind. *F. W. Oliver* (1910).

ACTIVE AMMOPHILETUM

PLATE 150

Phot. 383. Moss carpet of fixed calcareous dune. *Tortula ruraliformis, Brachythecium albicans, Camptothecium lutescens, Hypnum cupressiforme, H. chrysophyllum: Ammophila arenaria, Eryngium maritimum, Leontodon nudicaulis, Centaurium littorale, Asperula cynanchica*, etc. Castle Gregory, Co. Kerry. *R. J. L.*

Phot. 384. Moss carpet of fixed calcareous dune near Rosapenna, Co. Donegal. *Campto-thecium lutescens* and *Cylindrothecium concinnum* the most abundant species; also a rich phanerogamic flora (not shown) with *Ammophila* relics. *R. J. L.*

FIXED DUNE WITH MOSS CARPET

Variability of flora. A comparison of these three lists illustrates the considerable variation of the accessory flora of "white" or "yellow" dunes in different localities. This variability is doubtless largely due to accidental seeding (at Blakeney partly from the arable mainland), and most of the species are extremely inconstant since they have little or no means of coping with superincumbent sand. Their successful germination and establishment depend no doubt on periods of damp weather when the surface of the sand is temporarily stable. From some areas of Ammophiletum other vascular plants are entirely absent.

In the very wet climate of the west of Ireland it would seem that almost any creeping plant which roots at the nodes can fix sand on a limited scale. *Ranunculus repens*, *Cirsium arvense* and *Potentilla anserina* especially have been found playing this role on the smaller dunes of Connemara, Mayo and Donegal.

Role of mosses in fixation.[1] Owing to their very limited rooting systems neither mosses nor lichens can establish themselves on very mobile soils such as loose incoherent sand. They are therefore absent from the typical Ammophiletum in the earlier stages of its development. But when a ridge of dunes bearing Ammophiletum finds some shelter from wind and from blowing sand by the growth of a new range on its seaward side, certain species of these groups appear. Thus Richards (1929), at Blakeney Point, found four species in this situation: *Tortula ruraliformis*, confined to sand dunes, and characteristic of places on the dunes where a moderate amount of fresh sand is being received; *Brachythecium albicans*, also chiefly on these comparatively young dunes, but persisting through the whole succession in fair abundance; *Ceratodon purpureus*; and *Bryum* sp., perhaps *B. capillare*. All four species form separate patches growing between the dense tufts of *Ammophila*. No lichens had yet appeared.

On the west coasts of England (Watson, 1918), besides the four species of moss occurring at Blakeney, the sand dune form of *Camptothecium lutescens* is abundant on young dunes as well as three species of *Bryum* (o). These indeed, are recorded by Watson from the Agropyretum juncei, which more typically has no bryophytes at all; but if the development of the foredunes is considerable and sheltered areas are produced the conditions would be much the same as in similar situations on the main dunes. The occurrence of the calcicolous *Camptothecium* and the greater variety of the moss flora are doubtless due to the fact that in some of the areas described the sand is largely composed of comminuted seashells. On the highly calcareous dunes of the west of Ireland, also, *Camptothecium* is very abundant and *Hypnum cupressiforme* and *H. chrysophyllum* are frequent (Pl. 150).

Nature of dune sand. Owing to the varying nature of the sand which composes them dunes differ very much in their original content of calcium

---

[1] Pl. 150; Pl. 153, phot. 393; Pl. 154, phot. 394.

carbonate. The two extreme cases are those composed almost entirely of minute shells or comminuted fragments of larger shells, nullipores and corallines, and those which are composed almost entirely of quartz grains, more or less coloured by iron salts. Typically these form "white" and "yellow" dunes respectively. Every grade of mixture of quartz

White and yellow dunes grains and calcareous shells or fragments may occur. The effect is seen particularly in the flora of the fixed dunes—where the sand contains much lime the vegetation is rich in calcicolous species.

FIXED DUNE COMMUNITIES

As the surface of sand becomes gradually consolidated and less mobile by the action of the various plants which settle down in the relative protection afforded by the marram grass, and on the landward side by the dune crests themselves, many other species colonise the soil and eventually establish a continuous carpet of vegetation. Thus the mobile dunes gradually pass, with protection from wind and fresh blowing sand, into fixed dunes.

Variability of vegetation. The plants which actually make up fixed dune vegetation are extremely varied, and form, according to geographical situation, climatic region, and various physiographic, edaphic, and particularly biotic factors, a considerable series of different communities. Thus sand dunes in most forest climates eventually develop forest; in western Europe, including parts of the British Isles, often *Calluna*-heath; where they are used for grazing or as rabbit warrens, grassland, often bearing local scrub.

In the British Isles natural forest communities do not occur on fixed dunes, though plantations of various trees in suitable situations, especially pines, and particularly the Austrian pine, *P. nigra* var. *austriaca*, are often quite successful. The failure to develop forest may be partly due to violent coastal winds, but is mainly because of the paucity of suitable seed parents in the neighbourhood, alluded to in earlier chapters in connexion with the widespread lack of spontaneous forest establishment, and this is especially marked near the coast. The development of Callunetum on dunes has been recorded in eastern Scotland (W. G. Smith, 1903), on Walney Island off the coast of north Lancashire (Pearsall, 1934), and on the south coast of England, in Dorset (Good, 1935),[1] but the stages have not been closely studied.

Most of the British fixed dunes are occupied by some form of grass or mixed grass and herb vegetation, with scrub frequently well developed here and there. Such dunes have either been left as "wasteland", almost entirely deserted, or are more commonly used as rabbit warren or for golf links.

The consolidated sand of the low dunes of the Ayreland of Bride at the

[1] See Pl. 126, phot. 309.

northern end of the Isle of Man bears a turf in which the marram grass is still abundant, and the majority of species of the Ammophiletum are still present, but have been joined by numerous others (Moore, 1931, p. 123):

Agrostis spp.	f	Hypochaeris radicata	o
Ammophila arenaria	a	Jasione montana	o
Brassica monensis	r	Lotus corniculatus	f
Cakile maritima	r	Ononis repens	f
Carduus tenuiflorus	o	Orchis maculata	r
Carex arenaria	la	Rosa spinosissima	l
Cerastium semidedandrum	o	Sedum anglicum	a
Erodium cicutarium	o	S. acre	o
E. maritimum	o	Senecio jacobaea	o
Eryngium maritimum	o	Silene maritima	o
Festuca rubra var. arenaria	a	Thymus serpyllum	f
Galium verum var.		Viola canina var.	
maritimum	f	ericetorum	o
Hieracium pilosella	o		

The consolidation is mostly brought about by *Carex, Festuca, Thymus* and *Lotus.*

From the dunes at the north end of Walney Island, Pearsall (1934) records the following species from the fixed dune community:

Ammophila arenaria	a	Lotus corniculatus	f, la
Calystegia soldanella	f	Ononis repens	la
Festuca rubra var. arenaria	a	Rosa spinosissima	la
Galium verum	f, la	Salix repens (hollows)	la
Geranium sanguineum	f, la	Thalictrum dunense	la

The *p*H of the soil is between 6 and 6·5.

Mosses and lichens. In the formation of closed communities on the sand surfaces, lichens and mosses usually play the most important part, but not before the loose sand surface has been to some extent stabilised by the higher plants or protected by the formation of new dunes on the seaward side. Among lichens, species of *Peltigera* and *Cladonia* are prominent, and it is the abundance of the latter genus which has given the name of *grey dunes* to certain fixed dunes, as contrasted with the mobile **Grey dunes** white or yellow dunes. Of mosses *Tortula ruraliformis* is the most important pioneer, often forming luxuriant carpets on protected patches of bare sand in the Ammophiletum itself (Salisbury in Oliver and Salisbury, 1913).

Watson (1918) gives the following list of mosses and lichens, for the less mobile dunes, still dominated by *Ammophila*, of the west coasts of England:

Mosses

Tortula ruraliformis	d	Bryum argenteum	f
T. ruralis	o	B. caespiticium	o
Camptothecium lutescens		B. inclinatum	f
(sand dune form)	a	B. pendulum	f
Barbula convoluta	f	Trichostomum crispulum	f
Brachythecium albicans	a	T. flavovirens	f
Ceratodon purpureus	f		

Lichens

Peltigera canina	a	P. spuria	o
P. rufescens	a	Collema crispum	o

For "somewhat stable" and "almost fixed" sandhills he gives the following lists:

Mosses

Tortula ruraliformis	d	Camptothecium lutescens	a–d
Acaulon muticum	o	Dicranella heteromalla	o
Amblystegium serpens	o	Dicranum scoparium	o
Barbula convoluta	a	Ditrichum flexicaule	o
B. fallax	a	Encalypta rhabdocarpa	r
var. brevifolia	a	E. streptocarpa	r
B. gracilis	o	Eurhynchium confertum	o
B. hornschuchiana	o	E. megapolitanum	r
B. rubella	a	Hylocomium splendens	o
B. unguiculata	o	var. gracilius	o
B. vincalis	o	H. squarrosum	f
Brachythecium albicans	a	Hypnum cupressiforme	a
B. glareosum	o	var. ericetorum	a
B. rutabulum	f	var. tectorum	o
Bryum argenteum	f	Pleuridium alternifolium	o
B. atropurpureum	r	Swartzia inclinata	r
B. caespiticium	o	Thuidium abietinum	r
B. capillare	o	T. filiberti	r
B. donianum	r	Tortula ruralis	r
B. inclinatum	f	T. subulata	o
B. pendulum	a	Trichostomum fragile	o
B. roseum	o	T. mutabile	o
Ceratodon conicus	o	var. cophocarpum	o
C. purpureus	a	var. littorale	o
Climacium dendroides			
f. depauperata	o		

Liverworts

Frullania dilatata	r	Scapania aspera var.	
Lophocolea bidentata	o	inermis	o
L. cuspidata	f	S. aequiloba	–
Ptilidium ciliare	r		

Lichens

Bacidia muscorum	o	L. scotinum var. sinuatum	a
Biatorina caeruleonigricans	o	L. tenuissimum	r
Bilimbia sabuletorum	o	Peltigera canina	a
Cladonia furcata	a–d	P. horizontalis	r
C. fimbriata	o	P. rufescens	a
C. pungens	f	P. spuria	o
f. foliosa	f	Ramalina farinacea	o
Collema ceranoides	o	Rhizocarpon petraeum	
C. cheileum	r	(stones)	r
C. crispum	f	Squamaria crassa	r
C. pulposum	o	Urceolaria bryophila	f
Evernia prunastri	o	f. lichenicola	f
Leptogium lacerum	a	U. scruposa	f
pulvinatum	o	Usnea hirta	o

Phases of succession. The succession from yellow to grey dunes at Blakeney Point is marked, according to Richards (1929) by the enlargement and fusion of the separate patches of the four mosses mentioned (p. 851) as colonising protected areas of the Ammophiletum (*Tortula ruraliformis, Brachythecium albicans, Ceratodon purpureus* and *Bryum* spp.) into wide carpets, and by the appearance of tufts of *Hypnum cupressiforme* var. *tectorum.* At the same time the following lichens appear:

Cladonia fimbriata	Parmelia caperata
C. furcata	P. physodes
C. pyxidata	Peltigera polydactyla
Evernia prunastri	P. rufescens
var. stictoceras	

Nearly all these lichens are represented by young thalli growing in moss tufts, and "it is probably safe to say that no significant colonisation of the dunes by lichens could take place without the aid of mosses".

On the typical grey dunes the mosses are surpassed both in number of species and in abundance by the lichens. The most abundant moss is *Hypnum cupressiforme* var. *tectorum,* forming wide, golden-brown patches which tend to break up in the middle after reaching a certain size, probably due to the scratching of rabbits. *Brachythecium albicans* and especially *Tortula ruraliformis* are still abundant, and *Ceratodon* remains fairly frequent. Some of the lichens have now become very abundant, especially *Cladonia furcata*; two new species, *Cladonia foliacea* and *Cetraria aculeata,* appear and become conspicuous, and later *Cladonia sylvatica* joins them. In some places, indeed, mosses are almost entirely ousted by lichens, among which the large circular thalli of *Parmelia physodes* and *P. saxatilis* are conspicuous.

Flora of typical grey dune. The following list (Richards, 1929) illustrates the cryptogamic flora of a particular well-established grey dune at Blakeney Point:

Bryophytes

Brachythecium albicans	r	Dicranum scoparium var.	
Bryum sp.	o	orthophyllum	la
Ceratodon purpureus	lf	Hypnum cupressiforme var.	
Cephaloziella starkii	vr	tectorum	o–f
(round the roots of an		Tortula ruraliformis	la
Ammophila tuft)			

Lichens

Cladonia fimbriata	vr	Lecidea uliginosa	r
C. floerkeana	la	Parmelia physodes	lf
C. foliacea	la	P. saxatilis	r
C. furcata	a–va	Peltigera canina	+
C. pyxidata	o–a	P. polydactyla	+
C. sylvatica	la–o	P. rufescens	o
Evernia prunastri var.		Usnea florida var. hirta	vr
stictoceras	r		

Vascular plants. Unfortunately there is no record of the phanerogams occurring on the particular areas studied by Richards, but one dune which bore nearly all the above-mentioned bryophytes and lichens was occupied by Caricetum arenariae.

When the surface of the soil has been stabilised and enriched with humus by the growth of the bryophytes and lichens the tufts of *Ammophila* become sparser and less luxuriant, with many dead leaves, they rarely or never flower, and are sometimes obviously moribund. The contrast between the vigorous green marram grass of the mobile dunes and the impoverished, sparsely scattered, dingy tufts of the fixed dunes at once strikes the observer.[1] Though it may persist for a long time in this condition the plant has obviously become a dying relict. Various explanations of this loss of dominance and vigour have been proposed, of which the two most plausible are increasingly severe competition for available water, and the effect of lack of oxygen, or of increased carbon dioxide or some other toxic product of the new vegetation carpet. But the problem has not received serious experimental investigation.

Degeneration of Ammophila

The vascular plant covering of fixed dunes is very varied, and shows a mixture of maritime with non-maritime species. On dunes whose soil is not very calcareous the latter are preponderantly "grass-heath" species, but mixed with these are a good many others of various categories, including weeds of light arable soil which are able to find a foothold in the still open communities.

The following general list is taken from Salisbury's lists (Oliver and Salisbury, 1913) for two areas of fixed dune, "the Long Hills" and "the Hood", at Blakeney Point:

*Agrostis maritima	l	Melandrium album	vr
Aira praecox	f–a	Myosotis collina	f
*Ammophila arenaria	a	M. versicolor	r
Anagallis arvensis	f	Phleum arenarium	f
*Armeria maritima	lf	*Plantago coronopus	lf
Bromus hordeaceus (mollis)	r	†Polypodium vulgare	l
*Carex arenaria	ld	Rumex acetosella	f
*Catapodium (Desmazeria)		*Sagina maritima	o
loliaceum	vr	Sedum acre	f
Cerastium semidecandrum	f–a	Senecio jacobaea	a
C. tetrandrum	f–a	S. sylvaticus	r
C. vulgatum	vr	Sherardia arvensis	vr
Cirsium arvense	f	*Silene maritima	la
C. lanceolatum	r	Solanum nigrum	o
*Corynephorus canescens	ld	Stellaria pallida	f–a
Erodium cicutarium	f	Taraxacum erythro-	
Filago minima	lr	spermum	o–f
Galium verum	la	Urtica dioica	vr
Geranium molle	f	Valerianella olitoria	f
Hypochaeris glabra	f	Veronica arvensis	o–f
H. radicata	r	V. officinalis	lf
Luzula campestris	la		

* Maritime species. † On humus at the base of *Ammophila* tufts.

One plant of *Athyrium filix-femina* was found in a disused rabbit hole.

[1] Pl. 150.

The Blakeney Point dunes are of rather special interest. Though of small extent and not particularly rich in species, they have been comparatively well studied, the relative age of the different ridges is more or less accurately known, and the whole area is separated from the nearest land vegetation by more than a mile (about 2 km.) in a direct line across the estuary of the River Glaven.

"The Hood" at Blakeney Point. The oldest dune extant (estimated at about 300 years) is the small area known as "the Hood" separated from the rest of the system by about a quarter of a mile (400 m.). It is a curved patch of old dune sand, not more than 200 m. across in greatest diameter, considerably worn down by wind, and almost completely surrounded by shingle. The sand bears the majority of the species given in the list on p. 850 (Salisbury, 1913), but is specially notable for the local dominance of *Corynephorus canescens*, a rare and local species of our east coast dunes. This occurs abundantly in sheltered places on the edge of the old dune ridge. *Carex arenaria* is found in similar situations. *Ammophila* is still present in scattered tufts—the typical impoverished state on fixed dunes. In 1927 (Richards, 1929) most of the vegetation consisted of a very close sward of tall *Carex arenaria*, and the cryptogamic plants were completely different from those of the typical grey dunes described on pp. 853–5, being much more like those of an inland grass heath. The lichens had completely lost their dominance and were relatively scarce, while the moss vegetation had become rich and abundant (14 species in all), a number occurring which are not found elsewhere on the Point, including tall "hypnoid" mosses such as *Brachythecium purum*, *Hylocomium triquetrum* and *Hypnum cupressiforme* var. *ericetorum*, well adapted for living in closed communities of higher plants. The Hood is also almost the only place where liverworts occur on the Point.

The Hood, therefore, illustrates very well the evolution of fixed dune vegetation towards grass heath, though the dominant phanerogams are still entirely "maritime" species.

Calcareous dunes. The Blakeney dunes contain very little lime, but the sand of many dune systems, as already mentioned (p. 852), is largely composed of minute shells and comminuted fragments of larger shells and calcareous algae, and a large number of calcicolous species occur. Thus at Dog's Bay, on the south coast of Connemara (Co. Galway)[1] the sand consists almost entirely of minute shells. *Ammophila* is entirely absent and the low dunes of the centre of the spit fully exposed to the wind are dominated by *Agropyron junceum* and heavily eroded. The soil of the fixed dune community has 75 per cent of calcium carbonate, pH 8·1. The vegetation contained the following species:

Festuca rubra var. *arenaria*   d

Achillea millefolium	o f, la	Luzula campestris	o
Agrostis sp.	o	Plantago lanceolata	f
Anthyllis vulneraria	o–f	Polygala vulgaris	o
Asperula cynanchica	a	Prunella vulgaris	o
Bellis perennis	f	Ranunculus bulbosus	o
*Carex arenaria (relict)	la	Rumex acetosa	o
Cerastium vulgatum	o	Sedum acre	o
Chrysanthemum		Trifolium dubium	o
leucanthemum	o	T. pratense	o
Hypochaeris radicata	f	T. repens	f–a
Lotus corniculatus	f		

[1] Pl. 154, phot. 397.

In an older fixed and pastured community on dune sand accumulated between rocks there were in addition:

Campanula rotundifolia	Linum catharticum
Carex diversicolor (flacca)	Leontodon nudicaulis
Euphrasia sp.	*Plantago coronopus
Galium verum	*P. maritima
Hieracium pilosella	Sagina nodosa
Koeleria cristata	Spiranthes spiralis

* Maritime species.

From the same locality Praeger records further:

Anacamptis (Orchis) pyramidalis	Centaurea scabiosa
Arabis brownii (ciliata)	Euphrasia salisburgensis
Blackstonia perfoliata	Sesleria caerulea
Carlina vulgaris	

Out of approximately forty species in each of these lists, from Blakeney Point and Dog's Bay respectively, it will be observed that there are only seven species common to the two. The Dog's Bay list contains several markedly calcicolous plants, and if we except the very few maritime species, might have been taken from a non-maritime highly calcareous grassland.

Rosapenna　　On the extensive fixed dunes at Rosapenna in Co. Donegal, several hundred metres from the sea, the soil of which showed a pH of 8·3 and a carbonate content of 48·6 per cent, the mosses are on the whole dominant over much of the area, especially in the hollows.[1] The following species occur in this mixed moss community:

Brachythecium albicans	lf	Ditrichum flexicaule	o
Camptothecium lutescens	a, ld	Hypnum chrysophyllum	r
Ceratodon purpureus	o	H. cupressiforme	f
Cylindrothecium concinnum	va	Tortula ruraliformis	f, ld

with the lichens *Peltigera canina* and *P. rufescens* (lf).

Among the cryptogamic vegetation there are scattered flowering plants which sometimes close to form a continuous carpet:

Ammophila arenaria (scattered, relict)	Fragaria vesca
Arenaria serpyllifolia	Galium verum
Bellis perennis	Gentiana campestris
Cardamine hirsuta	Geranium molle
Centaurium umbellatum	Hieracium pilosella
Cerastium vulgatum	Holcus lanatus
Cirsium arvense	Leontodon nudicaulis
C. lanceolatum	Linum catharticum
Crepis capillaris	Lotus corniculatus
Daucus carota	Luzula campestris
Erodium cicutarium	Poa compressa
Euphrasia sp.	Potentilla anserina
Festuca rubra var. arenaria	Prunella vulgaris
	Ranunculus repens

[1] Pl. 150, phot. 384.

Senecio jacobaea
Sedum anglicum
Sonchus asper
Taraxacum officinale

Thymus serpyllum
Trifolium repens
Veronica chamaedrys
Viola curtisii

This list again has only seven or eight species in common with each of the other two, and there is no species which is characteristically maritime except *Ammophila*, though *Daucus* and *Erodium* may be considered submaritime.

Dune grassland. When rabbits are abundant on a fixed dune area, or when it is used as sheep or cattle pasture, woody plants have little chance of establishing themselves except sporadically.

Under pasturage by sheep or cattle a continuous turf develops, made up of a mixed assemblage of grasses, herbs and mosses, but usually reduced in numbers and modified in composition as a result of systematic grazing. At Castle Gregory on the north coast of the Dingle peninsula in Co. Kerry the following species occurred in a moist flat low-lying cattle pasture, evidently derived from a "slack", on calcareous dune sand:

Castle Gregory

Festuca rubra, thinly dominant

Agrostis stolonifera	a	Leontodon nudicaulis	f
Bellis perennis	f	Linum catharticum	a
Carex arenaria	a	Lotus corniculatus	f–la
C. diversicolor (flacca)	a	Potentilla anserina	o
Euphrasia sp.	a	Prunella vulgaris	a
Hydrocotyle vulgaris	la	Ranunculus repens	o
Hypochaeris radicata	a	Salix repens	f–la
Juncus gerardi	f	*Samolus valerandi	o
J. maritimus	ld	*Trifolium repens	f

Camptothecium lutescens   a

* In Juncetum.

Here submaritime, marsh and calcicolous elements are all in evidence. At Tramore in Co. Donegal a sloping cattle pasture on dune sand showed practically no maritime elements:

Tramore

Festuca rubra		d	
Achillea millefolium	f–a	Prunella vulgaris	f–a
Bellis perennis	a	Ranunculus repens	o–f
Carex arenaria	f–a	Senecio jacobaea	r–o
C. diversicolor (flacca)	la	Taraxacum erythro-	
Cerastium vulgatum	f, la	spermum	o
Galium verum	va	Trifolium repens	a
Lotus corniculatus	f	Veronica chamaedrys	la
Tortula ruraliformis	a	Camptothecium lutescens	f
Plantago lanceolata	va		

Pteridium aquilinum often invades non-calcareous dune grassland, just as it does other grassland on light soils, and closed Pteridieta may become established.

Acidic dunes. Walney Island. Pearsall (1934) describes a grassland community occupying the more rounded and lower dunes on Walney Island behind the first fixed dune community mentioned on p. 853. This community contains the following species:

Most constant species:

Aira praecox	Luzula campestris
Anthoxanthum odoratum	Polygala serpyllifolia
Carex arenaria	Potentilla erecta
Festuca rubra (most abundant)	Salix repens (la, in hollows)

Widely distributed:

Campanula rotundifolia	Holcus lanatus
Galium saxatile	Viola canina
Brachythecium purum	Polytrichum juniperinum
Dicranum scoparium	

The pH of the soil is between 4·8 and 5·5.

At points farthest from the sea this dune grassland is gradually passing into heath, though the process is not complete. The constant species are now:

Agrostis tenuis	Erica cinerea
Aira praecox	Rumex acetosella
Calluna vulgaris	
Carex arenaria	Polytrichum juniperinum

Frequent, but more local:

Sedum anglicum	Dicranum scoparium
Sieglingia decumbens	

And in the hollows:

Nardus stricta	Juncus squarrosus

The pH is 4·2 at the surface and 4·6–4·8 at a depth of 3 in. Leaching has evidently been considerable, judging from the very marked reaction with alcoholic ammonium thiocyanate.

Dune heath. Ayreland of Bride. Heath forms a zone within the low dunes on the Ayreland of Bride in the Isle of Man (Moore, 1931), and though it is difficult to demonstrate actual primary succession from one to the other this has probably occurred, judging from the occasional *Ammophila* present, which can scarcely be interpreted otherwise than as a persistent relict from a dune stage. The heath has the following composition, the numbers indicating the individuals on a line transect about 190 yards long:

Agrostis spp.	2	Festuca ovina	19
Ammophila arenaria	8	F. rubra	2
Calluna vulgaris	10	Hieracium pilosella	1
Carex arenaria	12		
Deschampsia flexuosa	1		
Erica cinerea	20	Dicranum scoparium	1
Erodium cicutarium	1	Polytrichum juniperinum	3
		P. strictum	1

Jasione montana	1	Ulex gallii	10
Lotus corniculatus	3	Viola canina var.	
Polygala vulgaris	1	ericetorum	1
Rosa spinosissima	3		
Sedum anglicum	6	Cladonia cervicornis	1
Silene maritima	, 1	Peltigera sp.	1
Thymus serpyllum	1	Lycoperdon caelatum	1

The sample of heath on which this transect was taken is obviously incompletely developed, since the sand dune elements are still considerable and the heaths have no overwhelming dominance over the grasses and herbs. Pl. 124, phot. 303 shows a later stage of development.

Subseres initiated by rabbit scratching show a stage of colonisation by *Cladonia cervicornis* and *Parmelia physodes* (which are said **Subseres** to furnish humus by scorching during summer), then by *Polytrichum juniperinum*, followed by *P. strictum*, *Dicranum scoparium* var. *spadiceum* and *Hypnum cupressiforme*. Soil in which sand is predominant (the substratum of the whole area is largely composed of pebbles) is rapidly colonised by the suckers of *Rosa spinosissima*. More pebbly soil is colonised by *Sedum anglicum*, which increases rapidly both by seed and vegetatively, and then by *Thymus serpyllum* and *Jasione montana*. Other species follow, including the grasses *Festuca* spp., *Deschampsia* and *Agrostis*, which form a turf containing such herbs as *Viola canina*, *Lotus corniculatus*, *Erodium cicutarium*, *Ononis repens*, and *Galium verum*. Finally *Erica cinerea* and *Calluna vulgaris* enter and become dominant. Where the acidity is marked (*p*H 5·8) the heaths may, however, directly colonise a lichen and moss carpet.

South Haven peninsula (Dorset). Here Good (1935, pp. 387–9) has traced a complete psammosere passing from the Ammophiletum of the dunes to typical Callunetum (Pl. 126, phot. 309). After the appearance of a number of grasses and herbs among the *Ammophila* the vegetation is invaded by *Erica cinerea* and later by *Calluna* which ultimately becomes completely dominant, the marram grass progressively disappearing. In some places *Ulex europaeus* replaces the heaths.

Vegetation of "slacks"

On the west coasts the damp or wet hollows left between the dune ridges, where the ground water reaches or approaches the surface of the sand, are known as "slacks". The following general account is summarised from Watson (1918):

Flora of permanent or semi-permanent pools. These may contain:

Amblystegium filicinum var. whiteheadii	H. intermedium
	H. revolvens
Hypnum aduncum, with numerous varieties	H. scorpioides
	H. wilsoni and var. hamatum
H. cuspidatum	
H. elodes	Potamogeton perfoliatus
H. giganteum	Zannichellia palustris

Around the margin the vegetation does not differ from that of an inland pool:

Eleocharis palustris	Hydrocotyle vulgaris
Epipactis palustris	Iris pseudacorus
Glyceria maxima (aquatica)	Juncus effusus
Holcus lanatus	Phalaris arundinacea
Hypnum aduncum and varieties	Pellia fabbroniana

Dune marshes.[1] These again show a vegetation which contains only a slight representation of maritime or submaritime species:

Anagallis tenella	sd	Juncus maritimus	l
Carex spp.	a	Littorella uniflora	ld
Galium palustre	a	Lycopus europaeus	a
Glaux maritima	o	Mentha aquatica	a
Hydrocotyle vulgaris	sd	Orchis incarnata	o
Iris pseudacorus	o	Parnassia palustris	lf
Juncus acutus	l	Sagina nodosa	f
J. bufonius	o	Samolus valerandi	f
J. effusus	f	Scirpus holoschoenus	l
J. inflexus	o	Teucrium scordium	l

Together with a great variety of Hypna (Harpidia):

Hypnum aduncum, and several varieties	a	Hypnum polygamum	a
		var. stagnatum	f
H. cuspidatum	f	H. scorpioides	o
H. giganteum	o	H. sendtneri	–
H. intermedium	lf	H. wilsoni	o
H. lycopodioides	f	var. hamatum	lf

And other bryophytes and algae:

Amblystegium filicinum	f	Pellia fabbroniana	f
var. whiteheadii	la	Riccia crystallina	o
Bryum neodamense	r		
B. pseudotriquetrum	f	Mougeotia sp.	a
		Tribonema bombycina	a
Cephalozia bicuspidata	o	Vaucheria sessilis	a
Moerckia flotowiana	f	Various Cyanophyceae	a

In moist places of the "slacks" of Braunton Burrows on the north coast of Devon (4 and 6 in Fig. 162, p. 846) occur patches of algae and liverworts:

Arthopyrenia areniseda	Riccia crystallina
Collema glaucescens	Vaucheria dichotoma
C. pulposum	V. sessilis

Good (1935, pp. 390 ff.) describes the vegetation of various damp places and marshes on the South Haven peninsula.

Salix repens dunes. On many of our western dunes, as on those of the north coasts of France and Belgium, the damp soil of the slacks is colonised by the creeping willow (*Salix repens*).[2] If quantities of sand are not constantly blown on to the Salicetum, the latter maintains itself at or near the

---

[1] Pl. 151, phot. 387. [2] Pl. 151, phot. 385.

PLATE 151

Phot. 385. Salicetum repentis occupying a depression ("slack") in the dune complex at Southport, Lancs. *W. Ball.*

Phot. 386. Edge of "slack" with low dune dominated by *Salix repens*. Eroded high dunes behind. *Mrs Clements, I.P.E.* 1911.

Phot. 387. Vegetation of a wet "slack"—*Parnassia palustris, Anagallis tenella, Centaurium littorale*, etc. *J. Massart.*

"SLACKS" OR DUNE VALLEYS

PLATE 152

Phot. 388. Scrub of *Hippophaë rhamnoides* (sea buckthorn) on back of main dune ridge at Hemsby, Norfolk. *Rubus rusticanus* is abundantly associated and *Ammophila* still conspicuous in the gaps.

Phot. 389. Detail of dune scrub: *Hippophaë* (right), *Rubus rusticanus* (left), *Polypodium vulgare* (centre), *Lonicera periclymenum*, *Festuca rubra*, *Holcus lanatus* and *Ammophila* (relict). Hemsby.

Phot. 390. Detail of dune scrub. *Hippophaë*, *Ammophila*, *Calystegia soldanella*. Hemsby. *J. Massart.*

Phot. 391. Detail of dune scrub. *Ulex europaeus*, *Pteridium*, *Polypodium vulgare*. Hemsby. *J. Massart.*

DUNE SCRUB (HIPPOPHAËTUM)

original level, the surface soil remaining more or less wet and then supporting a variety of marsh and damp grassland species. These include *Selaginella selaginoides* (Anglesey), and a long list of bryophytes (see Watson, 1918,
p. 142) with eleven species of *Bryum*, five of *Hypnum*, and five of *Aneura*,
as well as other liverworts, such as *Petalophyllum ralfsii* and *Moerckia
flotowiana*, characteristic of this habitat.

When enough sand is supplied to form low dunes or hummocks through
which the shoots of *Salix* grow up, carrying the surface well above the
ground water, a drier community is established,[1] including, on the Lancashire
dunes, such plants as *Pyrola rotundifolia* and *Monotropa hypopitys* which
live in the humus accumulated from the decay of the willow leaves.

If more sand is constantly supplied from the neighbouring *Ammophila*
dunes the shoots of the willow continue to grow up through it, producing
new roots in the damp layer just below the surface, and thus form mobile
dunes of considerable height capped by the living shoots of the willow.
These larger *Salix* dunes, in all stages of development and wind erosion,
are a conspicuous feature of some of our west coast sandhills, as they are of
the Channel coasts of Belgium and north-eastern France. Their flora hardly
differs from that of the ordinary mobile *Ammophila* dunes.

At Sandscale opposite Walney Island (Pearsall, 1934) the older dunes
become almost completely covered by *Salix repens* developed from the
carpet which has established itself in the slacks. Here the ground water
level is high, held up by an underlying stratum of clay. The larger slacks
may show all stages from bare sand to a continuous carpet of *Salix repens*
or occasionally *Potentilla anserina*. The following species occur in the earlier
stages of colonisation:

Carex arenaria	Listera ovata
C. oederi	Orchis maculata
Centaurium umbellatum	O. purpurella
Epipactis palustris	Parnassia palustris
Equisetum variegatum	Potentilla anserina
Galium verum	Salix repens
Gentiana amarella (agg.)	Samolus valerandi
Juncus articulatus var. litoralis	

At a later stage *Salix repens* becomes dominant and begins to collect
sand, forming low dunes on which a drier and somewhat acid facies develops, with:

Carlina vulgaris	Pyrola minor
Carex arenaria	Rhacomitrium lanuginosum
Festuca rubra var. arenaria	Tortula sp.

These dunes resemble the Southport dunes in a general way, and are
much less acid than those on Walney Island. The pH value of the soil of
the advanced *Salix* dunes is about 6 (lowest recorded 5·4). This may per-

[1] Pl. 151, phot. 386.

haps be accounted for by the height of the water table or the protection from leaching afforded by the covering of *Salix*. Pearsall says there are indications that woodland might develop here, starting from the damper places.

Effects of disturbance. The surface and vegetation of fixed dune areas are generally subject to various kinds of disturbance. Sand is often blown on to it from the mobile dunes nearer the sea, and in this way smaller secondary dunes arise, sometimes colonised by *Ammophila* or by *Salix repens*; or, if the sand is comparatively slight in quantity, by some of the pioneer mosses, such as *Tortula ruraliformis*, previously mentioned. A very important factor is disturbance by rabbits. Most dune areas are infested by these animals, which find the easily excavated sand peculiarly suitable for making burrows, and some such areas are preserved as rabbit warrens. Because they do not eat the bryophytes their abundance may lead to the dominance of these plants. (Cf. Chapter xxvii, p. 543 and Pl. 102, phot. 248.) The rabbits not only browse on the plants, but also, by their continual scratching in particular spots, destroy the carpet of vegetation, break up the stabilised surface of the sand and give the wind a purchase. In this way "blow-outs"[1] are started, and these, when at all extensive, throw back the course of succession to an earlier stage. "Blow-outs" are also (and more generally) initiated by specially violent winds, particularly from an unusual direction. Thus, in addition to the normal course of the main successional sequence, every stage of slow sand deposition, of erosion and rejuvenation, may be met with on a fixed dune area.

Ultimate vegetation. The ultimate vegetation of fixed dunes is various, as we have already seen. Three main types may be distinguished, Callunetum, scrub, and grassland. Where Callunetum develops the heather colonises the stabilised surface of the sand and heath conditions are established.

Dune scrub may consist of any of the common spinous shrubs which can colonise light sandy soils, e.g. gorse (*Ulex europaeus* and in the west *U. gallii*), bramble (*Rubus fruticosus* agg.), blackthorn (*Prunus spinosa*), or species of rose. The burnet rose (*Rosa spinosissima*) is often abundant in highly calcareous dune areas. Elder (*Sambucus nigra*), that ubiquitous "weed shrub", which is markedly resistant to rabbit attack, may also be present in quantity. The most characteristic dune shrub, confined in Britain to this habitat, is the sea buckthorn (*Hippophaë rhamnoides*), **Hippo-phaëtum** but this is very local in occurrence, and confined, as a native plant, to some of the east coast dunes,[2] with *Polypodium vulgare* conspicuous beneath it (Pl. 152, phots. 389, 391). Elsewhere it is frequently planted. The species does not in fact occur in maritime habitats over most of its range. In central Europe, though not in the British Isles, it inhabits shingle banks by the sides of rivers and similar situations. The establishment, history and fate of dune scrub has not been closely studied.

[1] Pl. 153.   [2] Pl. 152.

PLATE 153

Phot. 392. An old blow-out in half fixed *Ammophila* dune near Southport, Lancs, partly revegetated. *Mrs Cowles, I.P.E.* 1911.

Phot. 393. An old blow-out in old fixed dune near Carrigart, Co. Donegal, partly revegetated. Moss carpet in foreground and on the left. *R. J. L.*

"Blow-outs"

PLATE 154

Phot. 395. View including highest point (c. 150 ft.) in the great dune complex of Tramore, Co. Donegal. Planted *Ammophila* in foreground. *R. J. L.*

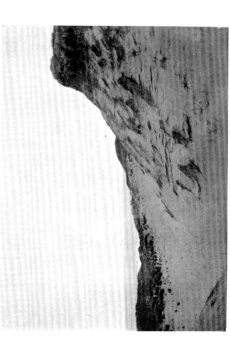

Phot. 397. Spit of mobile, highly calcareous dunes, worn down by wind erosion, between two bays. *Agropyron junceum* dominant, *Ammophila* absent. Fixed dune grassland in foreground and distance. Dog's Bay, near Roundstone, Connemara.

Phot. 394. Back of half fixed dunes at Blakeney Point. *Ammophila* and *Senecio jacobaea.* Complete fixation of the sand is effected by the mosses (*Tortula ruraliformis* dominant) in the foreground. *R. H. Compton.*

Phot. 396. Wind erosion of high dunes at Tramore. Regeneration on the left. Tory Island on the horizon.

Dune soils. The various composition of dune sand has been mentioned on pp. 851–2. The pH values fall with the development and progress of fixed dunes owing to progressive leaching. The initial alkalinity is due primarily to sea salt, and the sand of the foredunes is always on the alkaline side of neutrality. But dune sand largely composed of shells shows much higher pH values and contrasts with the relatively neutral dunes and even more strongly with those which develop high acidities.

Salisbury (1922) investigated some of the chemical characters of the dune sand at Blakeney Point in the different stages of succession. The youngest dunes had an appreciable but low carbonate content, generally under 1·0 per cent. (The highest value in the area, 4·2 per cent, was obtained from the drift zone of a lateral shingle hook.) This gradually decreased in passing to the older sandhills, many samples from which yielded no carbonates at all. The average pH values and humus contents (loss on ignition) obtained by Salisbury are shown in Table XXVI:

Table XXVI

	pH	Humus (loss on ignition) per cent
Youngest dunes	7·1 (max. 7·4)	0·36
Main ridge	7·03	0·5
Older ridges	6·9	{ 0·52 { 0·86
Oldest ridges	{ 6·8 (min. 5·5) { 6·24	{ 1·15 (max. 6·34) { 2·7

These figures clearly show both progressive leaching and progressive humus accumulation with time. The drift of the two processes is graphically shown for the Blakeney Point and Southport dunes in Figs. 163, 164 (Salisbury, 1925).

Quite comparable data were obtained by Moore (1931) from the sand of the very pebbly soil in different zones of the Ayreland of Bride in the Isle of Man. "Coarse sand" enormously preponderated over the other fractions separated by mechanical analysis. Table XXVII shows some selected figures:

Table XXVII

	pH	Humus (loss on ignition) per cent	Carbonates per cent
Embryo dune	7·6	0·86	2·73
Seaward face of dune	6·9	Nil	3·32
Consolidated dune	7·0	1·24	0·75
Grass heath	6·4	5·54	Nil
Heath	5·8	10·86	Nil

The sand of dunes consisting mainly or almost entirely of minute shells and fragments of larger shells, or of nullipores and coralline algae, show very much higher carbonate contents and pH values. Table XXVIII gives

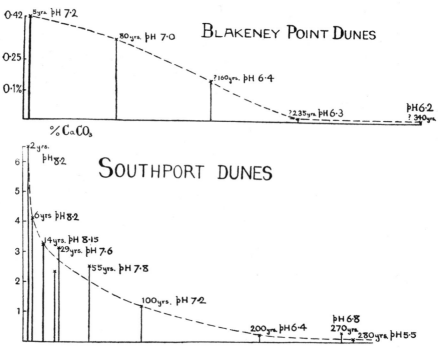

FIG. 163. RATES OF LEACHING AND DECREASE OF pH VALUES WITH AGE OF SAND DUNE SOILS

Note the much greater initial lime content and higher pH values of the Southport dunes compared with those of Blakeney Point. In both cases the pH value is reduced to the neighbourhood of 6 in about 200 to 300 years. From Salisbury, 1925.

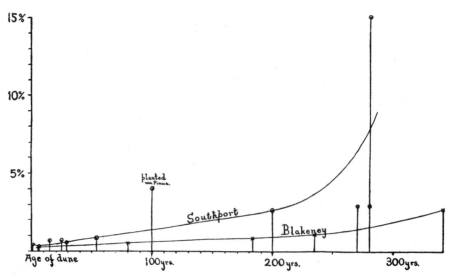

FIG. 164. INCREASE IN ORGANIC CONTENT WITH INCREASE OF AGE OF DUNE SOILS

Note the higher content and greater rate of increase at Southport. From Salisbury, 1925.

data from single samples of partly fixed dunes on the west coast of Ireland.

Table XXVIII

	pH	Carbonates per cent
Castle Gregory (Co. Kerry)	7·3	17·9
Tramore (Dunfanaghy, Co. Donegal)	8·3	25·4
Rosapenna (Co. Donegal)	8·3	48·6
Dog's Bay (Roundstone, Co. Galway)	8·1	75·0

The highly calcareous nature of these dunes is reflected in the vegetation (see pp. 857–9).

The foredunes and mobile dunes at Walney Island (Pearsall, 1934) show pH values of 7·5 and 7, falling to 6–6·5 on the fixed dunes (p. 853), 4·8–5·5 in the dune grassland, and 4·2–4·8 as heath develops (p. 860).

REFERENCES

CAREY, A. E. and OLIVER, F. W. *Tidal Lands: a study of shore problems*. London, Blackie and Sons, 1918.

GOOD, RONALD. Contributions towards a survey of the plants and animals of the South Haven peninsula, Studland Heath, Dorset. II. General ecology of the flowering plants and ferns. *J. Ecol.* **23**, 361–405. 1935.

McLEAN, R. C. The ecology of the maritime lichens at Blakeney Point, Norfolk. *J. Ecol.* **3**, 129–48. 1915.

MOORE, E. J. The ecology of the Ayreland of Bride, Isle of Man. *J. Ecol.* **19**, 115–36. 1931.

OLIVER, F. W. Blakeney Point Reports, *passim*, 1913–29. *Trans. Norfolk Norw. Nat. Soc.*

OLIVER, F. W. and SALISBURY, E. J. Topography and Vegetation of Blakeney Point, Norfolk. *Trans. Norfolk Norw. Nat. Soc.* **9**, 1913 (also issued separately).

PEARSALL, W. H. North Lancashire sand dunes. *Naturalist*, pp. 201–5. 1934.

RICHARDS, P. W. Notes on the ecology of the bryophytes and lichens at Blakeney Point, Norfolk. *J. Ecol.* **17**, 127–40. 1929.

SALISBURY, E. J. The soils of Blakeney Point: a study of soil reaction and succession in relation to plant-covering. *Ann. Bot.* **36**, 391–431. 1922.

SALISBURY, E. J. Note on the edaphic succession in some dune soils with special reference to the time factor. *J. Ecol.* **13**, 322–8. 1925.

Types of British Vegetation, pp. 339–52. Cambridge, 1911.

WATSON, W. Cryptogamic vegetation of the sand dunes of the west coast of England. *J. Ecol.* **6**, 126–43. 1918.

Chapter XLII

(4) SHINGLE BEACHES AND THEIR VEGETATION

Nature of shingle. Shingle beaches, which fringe many miles of our southern and eastern coasts, form another well-marked maritime habitat, though their total area in the British Isles is not nearly so great nor is their distribution so wide as that of the sand dunes. They are formed of water-worn and more or less rounded pebbles of very various size, derived by wave erosion from hard rocks or from flints originally embedded in chalk, driven along the coast by shore currents and eventually accumulated in banks or flat expanses on low-lying shores.

Below the limits reached by ordinary high tides they are destitute of plant life, for the pounding of the shingle in a rough sea, apart from the force of the waves themselves, makes existence impossible within the zone of surf. But above this zone the substratum becomes accessible to invasion.

Sand and shingle. On most beaches sand and shingle are mixed in various proportions, and where sand forms any large part of the substratum the vegetation above high water of spring tides is essentially arenicolous. For example, the low dunes of the Ayreland of Bride on the north coast of the Isle of Man, described in the last chapter, are formed from scanty blowing sand, the substratum of beach, dunes and heath—based on old "raised beach"—being composed of pebbles. At Blakeney Point and Scolt Head, on the Norfolk coast, more considerable dunes are formed above a substratum of shingle, which is often in evidence in the "lows" or valleys between the dune ridges. The presence of shingle tends of course to stabilise the sand, but the vegetation depends on the latter.

Beaches composed almost entirely of pebbles show, however, quite distinctive features as a plant habitat, and these have been investigated mainly by Oliver and his fellow-workers (1911–29, see References, pp. 894–5).

Types of shingle beach. Oliver (1912) classifies coastal shingle beaches as follows:

(1) *Fringing shingle beaches* are the simplest and commonest type (Pl. 156, phot. 399; Pl. 158, phot. 406). The shingle, driven by a shore current parallel with the coast, forms a strip in contact with the land along the top of the beach. *Examples*: most of the shingle beaches along the Sussex coastal plain.

When the current leaves the shore other types are produced.

(2) *The shingle spit* (Fig. 165). When the coast line changes its direction while the shore current continues in a straight line the latter necessarily leaves the shore, and the shingle it carries forms a bank or causeway diverging from the land and often extending for several miles. The apex is

liable to sudden deflection landward as a "hook". *Examples*: Hurst Castle and Calshot spits, Orfordness, Blakeney bank (Pl. 133, phot. 335; Pl. 134; Pl. 136, phot. 341 and Figs. 168–70).

(3) *The shingle bar* (Fig. 166) is formed if a shingle spit once more reaches the land, cutting off a closed lagoon between spit and land. *Examples*: Chesil Bank (Pl. 155, phot. 398), which may however have originated as a fringing beach, the lagoon behind (Fleet) being later excavated by erosion: Looe Bar (Cornwall).

(4) The *apposition shingle beach* (Fig. 167) is formed when new shingle is deposited on the flank of an earlier beach, where it accumulates till lifted

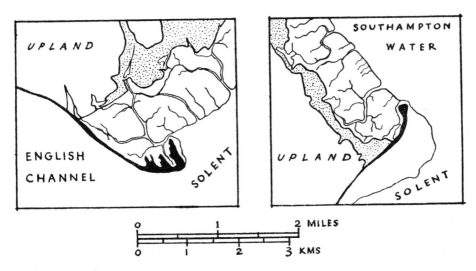

FIG. 165. TWO SHINGLE SPITS (HURST CASTLE AND CALSHOT SPITS),
ON THE SOLENT, HAMPSHIRE

Note the continuation into the curved spit of the line of fringing shingle beach (black). The spits protect areas of soft mud, exposed at low tide and traversed by drainage channels. On this mud salt marsh is developed. The dotted areas on the landward side are mature saltings and partly reclaimed.

above tidal limits by an exceptionally high tide caused by a gale. If the process is repeated a succession of closely approximated, more or less parallel banks are formed, producing a very extensive area of shingle (Pl. 161). *Example*: Dungeness.

The fringing and apposition types of shingle beach have not been investigated in detail. The former is always relatively narrow, and the strip colonisable by vegetation usually extremely narrow. On the latter, owing to their great extent, the most advanced (non-maritime) vegetation occurs.

Most of the British work on shingle vegetation has been done on the spits and on the famous "bar" of Chesil Bank.

Morphology and development of the shingle spit. A longshore current carrying shingle tends to continue in a straight line when the coast line, along which the shingle has been deposited as a fringing beach up to that point, turns landward. The continuation of the fringing beach then forms a spit, diverging at an angle from the inward curve of the coast and protecting the waters of the bay or estuary in this inward curve. The shingle

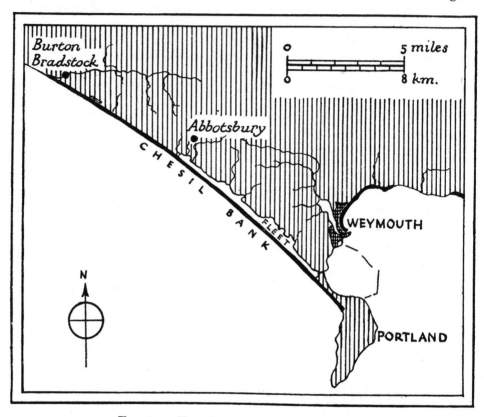

FIG. 166. THE CHESIL BANK SHINGLE BAR

The western end of the Chesil Bank and the beaches east of Weymouth are of the fringing type. Compare Phot. 398 (Pl. 155) which is taken from the air about two miles south-east of Abbotsbury looking south-eastward along the Bank towards Portland.

spit itself is usually gently curved inwards, its convex side facing the open sea (Figs. 165, 168), and the apex generally shows a marked landward deflection known as a *hook*. Besides the terminal hook there are commonly several others ("laterals") running landwards from the sub-terminal portion of the main spit, and these may form quite complex systems, as at Blakeney Point and Scolt Head (Fig. 168). Each of these sub-terminal or "lateral" hooks was at one time the apical hook, deflected landwards when it was the growing apex of the main spit. Thus the spit grows in a manner

PLATE 155

Phot. 398. The south-eastern half of the Chesil Beach shingle bar seen from the air. The lagoon called the Fleet separates the bar from the mainland to the left; Portland Harbour, with shipping, beyond. The distance from the front of the picture to the harbour is more than 6 miles (nearly 10 km.). The high ground of Portland (left centre distance) is another 2 miles along the beach. "Deltas" project into the Fleet, forming a sinuous line, and the dark edging is the belt of *Suaeda fruticosa*. Photograph from *The Times*.

The "camm" with its "ravines" and "buttresses" is plainly seen in the foreground along the inner side of the beach.

CHESIL BEACH FROM THE AIR

PLATE 156

Phot. 399. Sheep feeding on *Silene maritima* at the end of a long drought (September 1911). Near Burton Bradstock. The beach is here a fringing beach. *F. W. Oliver.*

Phot. 400. Chesil Beach seen across the Fleet, showing the "camm" with its ravines and buttresses, the fringing belt of *Suaeda fruticosa* on the edge of the water, and the "back" of the beach above. About 2 miles east of Abbotsbury. *F. W. Oliver*, 1911.

CHESIL BEACH

analogous to that of a sympodial rhizome, the lateral hooks corresponding with successive apices.

The actual cause of hook development, i.e. of the landward deflection of the apex of the spit, appears to be the rapidly increasing force of the incoming tidal current, greatly augmented by the waves formed by a gale blowing in the same direction, which fills the bay or estuary inside the spit, as this lengthens and encloses a greater area of tidal ground. The volume of water required to fill this space in a given time obviously increases as the square of the length of the spit if this makes an angle of 45° with the coast line. Spits rarely in fact make such a wide angle with the shore line, and the space to be filled is progressively diminished by the silting up of the bay or estuary, but the scour of the tide is certainly greatly increased by the widening of the mouth as the spit diverges farther and farther from the

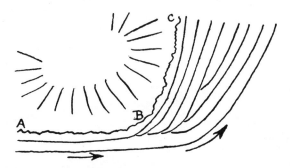

FIG. 167. DIAGRAM OF APPOSITION SHINGLE BEACH

The current here followed the shore line round the corner at B so that successive strips
of shingle were added outside the stretch B–C. From Oliver, 1912.

mainland. To this increased force of the tidal current, backed by wave action on the flood when a strong wind is blowing, is also due the frequent deflection of the hooks backwards (i.e. towards the proximal end of the main spit) seen especially in the Blakeney Point system (Figs. 168, 169, 170). Oliver actually observed the formation of an L-shaped end to a sub-terminal hook during a gale in 1911, the hook having recently been exposed to the tidal scour by the wasting of the terminal dune-covered head of the main bank, which had formerly protected the sub-terminals (Oliver, 1913).

The concentration of hooks in the sub-terminal region of the spit is probably due to the increased proportional strength of the tidal current entering the bay or estuary with growth in length of the main spit, as compared with the shore current responsible for lengthening the main spit. Thus the whole spit system passes through a period of youth, when the shore current is alone active, resulting in the continuous lengthening of the main spit, to a period of maturity or hook formation when the incoming tidal current becomes strong enough, backed by a gale, to deflect the tip. Since this deflecting action is intermittent, depending on the occurrence of

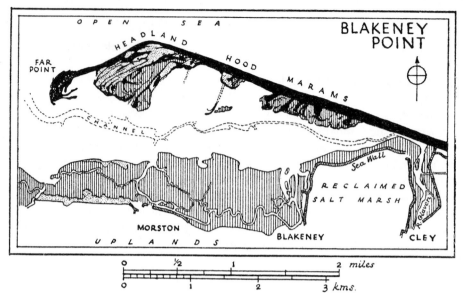

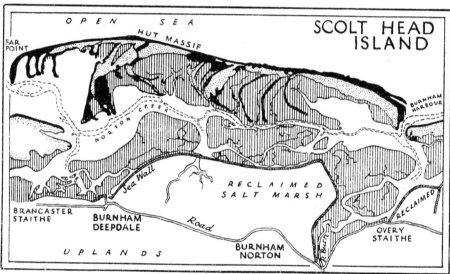

FIG. 168. BLAKENEY POINT (1923–24) AND SCOLT HEAD ISLAND (1925)

Two shingle spit systems on the north coast of Norfolk with associated sand dunes and salt marshes. Shingle *black*, sand dunes *dotted*, salt marshes *vertically ruled*. Sea, coastal and estuarine sand and mud, reclaimed marshes and uplands *white*. The general physiographic correspondence of the two systems is obvious at a glance.

The shore current runs from east to west, and the growing tips of the banks are consequently at their western extremities. Note that the banks are curved inwards towards their distal ends. The "hooks" or lateral banks connected with the inner side of the main banks form complex irregular groups, each group corresponding to a period of active hook formation (prevalence of storms?) with intervening periods of continuous growth of the main bank. Most of the hooks are more or less curved, with the convexity facing west or north-west. Each hook was originally an apex of the main bank, though its shape has often been much modified subsequently. The "Far Points" are the most recent additions, each forming a main group of hooks.

The sand dunes overlie the broader expanses of shingle: the salt marshes are formed mainly in the concavities of the hooks, where protection is greatest.

gales, there is now an alternation of hook formation and growth of the main spit.

It is necessary to understand the developmental morphology of shingle spits because the different parts of the system bear different vegetation.

Mobility of shingle. Except in the apposition type, which will be referred to later, and the narrow landward edge of the fringing type, the shingle of a sea beach never reaches a state of complete repose. The mobility of the stones is however very different in the different regions of a beach, and on decrease in mobility depends the possibility of the existence of vegetation.

There are three factors responsible for the movement of shingle. The first is *wave impact* on the seaward face of the beach, which keeps the shingle constantly on the move with every incoming tide, most violently of course when an onshore gale is blowing. This throws up shingle on to the "storm shelf" (Figs. 172, 174) and even over the crest of the beach when there is a high spring tide combined with an onshore gale. The crest and back of the Blakeney bank are awash under such conditions, so that fans of shingle spread over the salt marsh behind (Pl. 157, phot. 401) and the whole bank gradually travels landwards. Where different stretches of a shingle bank are subject to unequal incidence of wave action or are unequally resistant to it a "sagging" of the bank landwards can be observed. The landward travel of the main bank will necessarily encroach on the basal parts of the lateral hooks and thus abbreviate their length (Fig. 171, p. 876). This may affect the vegetation of that part of the main bank which has incorporated the base of the hook.

The second factor is *percolation* of water through the beach. As a result of its open structure shingle is readily traversed by water. At Blakeney at high tide, especially if a heavy sea is running and its level is well above that of the salt marshes ("saltings") on the inner side of the bank, sea water normally traverses the bank, breaking out in numerous springs just above the level of the saltings. A certain amount of displacement of shingle then occurs, but it is relatively insignificant because of the small height of the bank—only 6 or 7 ft. (*c.* 2 m.) above ordinary high spring tide mark.

"Camm" formation. On the Chesil Bank, however, where the crest rises 25 ft. (*c.* 8 m.) above high water mark and the slope often approaches the critical angle of rest of the shingle, the results of this landward percolation are important. Seen from across the Fleet (the lagoon separating the shingle beach from the land) the bank rises steeply above the low flat terrace which fringes the Fleet, and is seamed by a series of ravines ("cans") 10 or 12 ft. deep, which with the intervening buttresses form a conspicuous "camm" (Pl. 156, phot. 400 and Fig. 173). The back and sides of a "can" are inclined at the angle of repose (about 34°), and the floor is generally scored by a shallow gully connecting with a detrital fan of shingle projecting into the Fleet some 8 or 10 ft. beyond the edge of the terrace (Pl. 158, phot. 404).

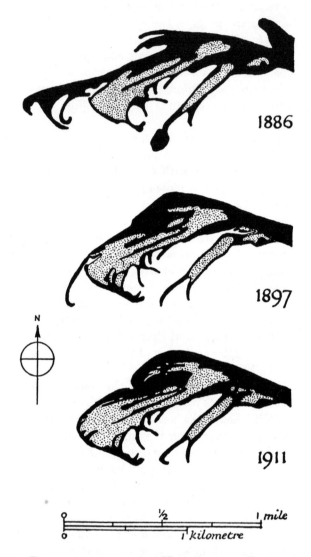

Fig. 169. Development of the Headland of Blakeney Point,
(Norfolk), 1886–1911

Shingle *black*, sand dunes *dotted* (salt marshes not shown).

The hooks forming a "Far Point" complex in 1886 were obliterated by 1897 and a new single hook formed which had itself disappeared by 1911. Note also the progressive development of new dunes on the northern side of the Headland.

From Oliver and Salisbury, 1913. (The first two maps are from 6 in. O.S. maps, the third is based on a survey by the Botanical Department, University College, London.)

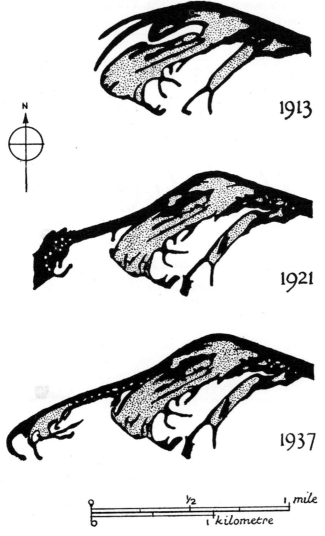

FIG. 170. DEVELOPMENT OF THE FAR POINT AT BLAKENEY
1913–1937 (continuation of Fig. 169)

By 1913 the growth of the main spit was resumed, and by 1921 it had extended half a mile, the apex broadening into an expanse of shingle (the Far Point) bearing scattered embryo dunes. Between 1927 and 1935 these consolidated into considerable dune areas and several new hooks were formed, the latest (westernmost) in 1936–37. Note the modifications of several old hooks. The charts of 1913 and 1921 (based on air photographs) are from *Blakeney Point Reports* (1917–19 and 1920–23); the Far Point in the 1937 chart is from a plane table survey by the Botanical Dept., University College, London, by kind permission.

The slopes of the buttresses are inclined at a lower angle than those of the "cans" and merge without material change of angle into the gentle slope at the back of the beach crest (see Fig. 173 and Fig. 179, p. 886).

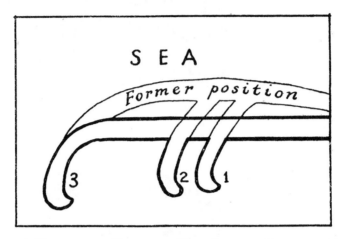

FIG. 171. DIAGRAM OF PART OF A MAIN SHINGLE SPIT WITH HOOKS

1 and 2 are "hooks or former apices of the main bank, 3 is the present apex—all deflected inwards. The thin lines indicate the former position of the main bank, which has been driven inwards by cumulative wave impact and has embedded the bases of the hooks. From Oliver, 1912.

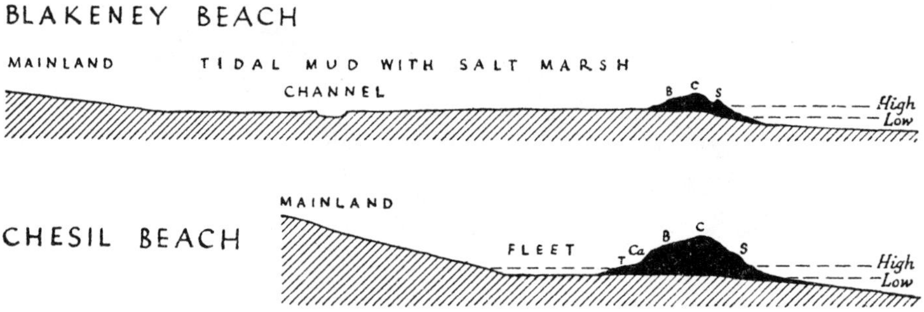

BLAKENEY BEACH

FIG. 172. BLAKENEY AND CHESIL SHINGLE BANKS

Profiles showing the relation of the shingle (*black*) to the marine terrace. S=Storm shelf, C=crest, B=back, Ca=camm, T=terrace.

Vertical scale about four times the horizontal, but drawing not accurately to scale. After Oliver, 1912.

These ravines ("cans") are formed by the sea water running through from the front of the beach and gushing out at the back. At the points of exit shingle is displaced, dislodging the stones above, which slide down their "slope of repose". Thus the "cans" are cut deeper into the bank. The amount of shingle shifted in ravine formation is large, and the ravine-buttress or camm structure of the bank above the terrace is important in determining the vegetation. The general result of the process is clearly the transfer of shingle from the higher parts of the bank to the terrace below, to which it is added, while the stones which are carried farthest form the fans which project into the Fleet. Active "can" formation and cutting back seems to be an intermittent phenomenon, some "cans" apparently remaining quiescent for long periods. The whole process is a minor feature

FIG. 173. PART OF THE "CAMM" ON THE INNER FACE OF CHESIL BANK

View taken from the terrace about 20 ft. from the edge of the Fleet looking up a gully or "can" with bare floor. The terrace bears scattered plants and a bush of *Suaeda fruticosa*. On the flat top of the nearest "buttress" and on the "back" of the bank (at the top of the picture) are plants of *Silene maritima*. Reproduction (by kind permission) of a sketch by Prof. T. G. Hill. From Oliver, 1912.

of the landward creep of the whole bank, primarily caused by wave impact, and made up of the piling of fresh shingle from the sea on the storm shelf and crest and its gradual transference down the landward face.

A third factor affecting the mobility of a shingle bank is *undercutting* by tidal or other currents, e.g. the current of a river impinging on the inner side of a bank. This of course results in wasting the bank and may retard its general inward creep.

Structure of a shingle spit or bar. The primary division of a shingle spit is into *main bank* and *laterals* (*hooks*), of which the former is, broadly speaking, mobile, while the latter are immobile.

On the main bank of a spit like that at Blakeney or a bar like Chesil Beach we can distinguish the following topographical features (Fig. 172, 174). On the seaward face, which is normally the steeper, there is a series of "steps" corresponding with different high tide marks, culminating in the *storm shelf* (S), not reached by ordinary spring tides. Above this is the *crest* (C) or summit of the bank, built up only by quite exceptional tides when these coincide with onshore gales. The lower steps, where the mobility of the shingle is at a maximum, are destitute of vegetation. On the storm

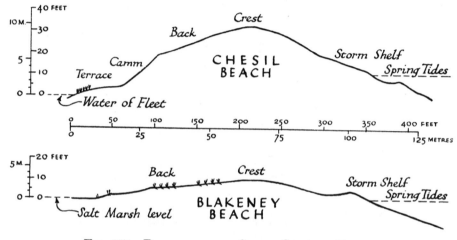

FIG. 174. BLAKENEY AND CHESIL SHINGLE BANKS

Profiles drawn to the same scale. Note the three zones of *Suaeda fruticosa* on the back of the Blakeney beach. After Oliver and Salisbury, 1913.

shelf the vegetation, when present at all, is very sparse indeed, consisting mainly of prostrate species of *Atriplex* and sometimes *Beta maritima* (cf. the foreshore community p. 844). The crest of the bank, like the storm shelf, bears very little vegetation, though there may be a few poor specimens of plants derived from the back of the bank. Both crest and storm shelf are obviously very much exposed to wind and spray, making plant establishment difficult, and the species which do establish themselves are occasionally drenched with sea water. On a low bank like that at Blakeney the *back* (B) of the bank slopes gently to the marshes. On the lofty Chesil Bank it slopes more steeply, up to 10–15° on the general face (B), while on the sides of the ravines (Ca) the angle is the critical angle of rest of the shingle (about 34°). Below is the flat terrace (T) receiving shingle from the bank, most actively from the "cans", and encroaching on the Fleet within. There is no terrace on the landward side of the Blakeney bank, the back slope passing straight on to the marshes, over which spread fans of shingle. The back of the bank is by far the most important plant habitat and bears the vegetation characteristic of mobile shingle (Pl. 157, phots. 401–2; Pl. 158, phot. 405; Pls. 159–60).

The laterals of a shingle spit show later stages of succession and possess a much more numerous and less characteristic flora.

Shingle and sand. The shingle of our coastal beaches is sometimes almost pure but very often mixed with sand. Sand and pebbles are both constantly thrown up by the sea and both are carried up on to shingle beaches, to which wind-borne sand may also be contributed; but much of this is again washed down by retreating waves and on the ebb of the tide. Because of the very different carrying power required to move sand and stones the tides and waves tend to sort them out one from the other, so that comparatively pure strata of each are often met with. Thus the storm shelf normally consists of pure shingle. On the other hand, in boring through an old shingle beach, strata representing very various degrees of mixture of the two constituents are met with (Hill and Hanley, 1914).

Water content. The water content of shingle beaches is considerable. "In the hottest summers and on the hottest days water may be found within a few inches of the surface of a shingle bank." This water is commonly "fresh", wells sunk in shingle yielding potable water, and on the inner slope of the main bank at Blakeney Point a series of samples at a depth of 5 cm. yielded from 0·07 to 0·44 per cent of chlorides (sea-water 3·19 per cent). At a depth of from 5–9 ft. from the surface of the shingle, on a level with the standing water table at the surface of the adjacent salt marsh, the percentage of chlorides varied quite similarly, from 0·08 to 0·45, much higher values being obtained only near the edge of the marsh, from which there was probably some lateral percolation of salt water. This fresh water is no doubt derived largely from rain, but there is evidence that dew is an important contributor to it, and "internal condensation", i.e. dew formation on the surfaces of the stones in the interior of the beach, may also be a considerable factor. "Diffusion of salt from the sea to this fresh water must be so slow that the additions of fresh water are sufficient to prevent a relative increase in the salinity" (Hill and Hanley, 1914). Shingle plants do not apparently ever suffer from drought.

Humus. The humus of young shingle beaches, like that of the youngest dunes, is entirely formed from the disintegration of tidal drift, and the supply is always sufficient to provide adequate food for the shingle plants which establish themselves. The considerable amounts present in such beaches can easily be tested by delving among the pebbles with the hands, which quickly become soiled by the dark humous material in the interstices between the stones. Salisbury (1922) found the loss on ignition of the fine material from the youngest lateral at Blakeney Point to be 0·486 per cent with a pH of 7·6. On the old stabilised laterals which are covered with vegetation, the humus content was 3·19 per cent and the pH 6·9. The high "elbows" of some intermediate laterals show more extreme figures (humus 22·45 per cent, pH 6·38) and bear a much larger population of non-maritime plants.

FLORA AND VEGETATION

There are no species dominants of shingle beaches in the sense in which *Ammophila arenaria* and *Agropyron junceum* are dominant on sand dunes, and the vegetation is very poor floristically, except on certain old stabilised beaches where the flora is mainly composed of inland species, just as old stabilised dunes support inland heath vegetation. A few species are, however, very characteristic of shingle, and of these several occur also on sand dunes.

Suaeda fruticosa. One species, the shrubby sea-blite, *Suaeda fruticosa*, occupies a special position.[1] It is a perennial sometimes reaching 4 ft. (1·2 m.) in height with a woody stem sometimes 2 in. (5 cm.) in diameter and numerous more or less upright branches arising close together near the ground. The sub-cylindrical, dark green, purplish or crimson leaves are evergreen in the south but fall in late autumn in the more exposed situations on the Norfolk coast. In habitat, over most of its range, in anatomy, and presumably in physiology, the species is a halophyte, but in Britain, where it is extremely local in occurrence, it grows abundantly and luxuriantly only in connexion with shingle beaches, and mainly with the shingle spits of the north coast of Norfolk and the Chesil beach in Dorset. According to Salisbury (1922) *Suaeda fruticosa* grows in ordinary garden soil with extreme luxuriance.

Distribution at Blakeney Point. At Blakeney Point *Suaeda fruticosa* is almost restricted to the main shingle beach and to the numerous laterals, though it occurs also near the edges of neighbouring salt marsh. On old laterals it forms a dense zone on either flank[2] and is continuous over the crests of the lower laterals, which are almost always reached or covered by the higher spring tides. It also occurs on the landward slope of the main beach,[3] but only where this is protected from wave action by the lateral beaches, which act as groynes. From the more exposed stretches of the inner edge of the main beach it is absent (Pl. 157, phot. 403).

Along the inner margin of the main beach, where this abuts on well developed salt marsh the shingle advances intermittently on the marsh, driven down by the highest tides flowing over the bank when backed by an onshore gale, and forms local fans of stones spreading over the marsh (Pl. 158, phot. 404 and Fig. 175). The spring tides flow from the estuary over the marshes (i.e. in the opposite direction), and may rise several feet up the landward side of the bank, leaving a drift line of debris, but the onset of the tide here is always quiet and there is no wave action.

---

[1] See also Chapter XL, p. 838.   [2] Pl. 146, phot. 341.   [3] Pl. 157, phot. 401.

Dissemination and establishment. In a good year this drift is laden with ripe seeds of *Suaeda*, a handful containing several hundred seeds, and it is in this zone that germination occurs in great numbers, the young plants rapidly developing long tap-roots which reach down to the moister layers of the shingle. Their density often amounts to 20–30 per sq. ft. (*c*. 900 sq.

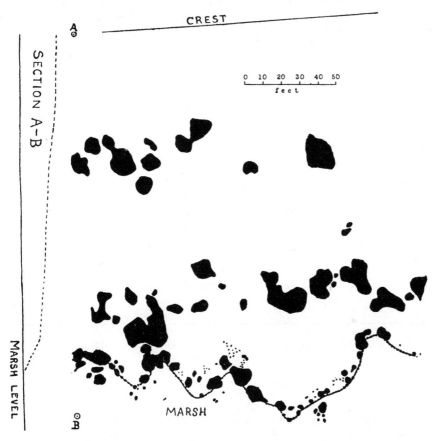

FIG. 175. DISTRIBUTION OF *SUAEDA FRUTICOSA* IN THREE ZONES ON THE BACK OF BLAKENEY MAIN SHINGLE BANK

Fans of shingle projecting on to the marsh. Dotted areas crowded with *Suaeda* seedlings. Vertical scale of section is twice the horizontal scale. From Oliver and Salisbury, 1913.

cm.), the maximum density being reached in the bays between the shingle fans where the thickest drift is deposited (Fig. 175). The conditions for germination and growth are extremely favourable since the aspect is full south and there is complete protection from north winds as well as from wave action. The shoots of *Suaeda* attain a height of 2–3 in. in the first year, 6 in. in the second, while in five years a bushy plant is developed 18 in. (45 cm.) in height and an inch in stem diameter.

Behind this littoral zone there are two other zones of *Suaeda* bushes higher up the bank, one just above the shingle fans and the third near the crest of the beach (Figs. 175, 178),[1] and here also seedlings occur and establish themselves, though in much smaller numbers, probably because of the sparsity of tidal drift at the higher levels. The higher lying seedlings, especially those near the crest, show less luxuriant growth since they are much more exposed to wind. The existence of the three zones is attributed by Oliver and Salisbury (1913*b*) to the occurrence of great storms at intervals of several years, throwing quantities of shingle over the crest of

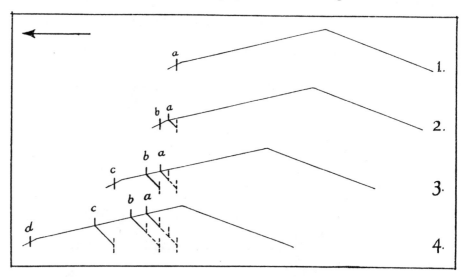

FIG. 176. ORIGIN OF THREE *SUAEDA* ZONES IN RESPONSE TO INWARD
TRAVEL OF BEACH

The plants originate as seedlings on the edge of the marsh (*a* in 1, *b* in 2, *c* in 3, *d* in 4) and grow up through the shingle which travels from the crest down the back of the beach, and covers the bushes. Thus the plants "climb" the back of the beach as the crest shifts its position inwards. The interrupted lines indicate the dead proximal portions of *Suaeda* shoots buried in the shingle. From Oliver and Salisbury, 1913*b*.

the beach and down on to the marsh behind. During the intervals between these the shingle fans fringing the marsh are quiescent and for a number of years opportunities occur for the establishment of fresh *Suaeda* seedlings. The two upper zones of *Suaeda* bushes are thus interpreted as the records of former periods of quiescence of the inner fringe of the beach (Fig. 176). The bushes of these zones have grown up through the shingle with which
Effects of shingling over   they have been partially covered during storms that have forced the whole bank landwards, and this is shown by their habit. When shingle has flowed over a plant the lower part of the shoot system is bedded in the shingle and tilted forwards, the buried base

[1] Pls. 134 and 157, phot. 401.

showing repeated branching and tufts of roots arising at irregular intervals. In the middle and upper zones, where each individual bush covers a large area and is more exposed to shingle flow the original main axis of the plant is rarely recognisable, its place being taken by an underground system of prostrate shoots thickly beset with roots, and mainly directed more or less horizontally towards the downward slope of the bank. From these arise crowded leafy shoots which together form the existing sub-aerial portion

FIG. 177. BRANCH OF *SUAEDA FRUTICOSA*

The plant has been beaten down and covered with shingle, and has sent up new vertical shoots from its distal portion. One-third natural size. From Oliver and Salisbury, 1913*b* (drawing by Sarah M. Baker).

of the bush (Fig. 177). The horizontally running underground shoots with their numerous roots form a dense mat in the surface layers of shingle, checking the travel of stones down the landward slope of the beach. The older parts of these buried shoots die off and then rapidly disintegrate. Three or four feet from the insertion of the green aerial shoots the "rhizome" is usually dead and cannot be traced for more than another foot or so to the mouldering end. This rapid disintegration is probably important as a contribution to the food supply of the shingle vegetation.

The prostrate habit of the underground stems of *Suaeda fruticosa* seems to be imposed by its conditions of life: they do not *grow* horizontally. On the dormant shingle of the laterals the habit of the plant is erect.

According to the observations of Oliver and Salisbury (1913*b*) the more *Suaeda fruticosa* is "shingled over" the more vigorous its vegetative response, and this applies also to other characteristic and abundant shingle plants, such as *Silene maritima* and *Honckenya (Arenaria) peploides*.

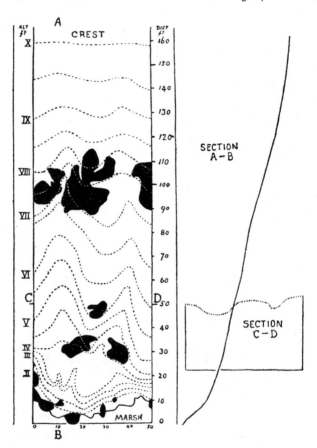

FIG. 178. CONTOURED STRIP OF BLAKENEY MAIN BANK FROM CREST TO MARSH

Two gullies are shown with *Suaeda* bushes of the three zones mainly on the ridges between. Vertical scale of section five times the horizontal scale. From Oliver and Salisbury, 1913*b*.

After a winter in which the shingle has been much disturbed there is not only an unusual vegetative display but also an extraordinary profusion of flowers on these two species.

Shallow gullies run in the shingle from near the crest of the beach to the shingle fans on the edge of the marsh. These represent the lines of travel of the shingle, and the clumps of *Suaeda* in the two upper zones lie, on the whole, between them (Fig. 178). This suggests that the plants effect a real

PLATE 157

ot. 401. The back of the beach at the Marams from one of the mature salt marshes, showing the three zones
Suaeda fruticosa, the lowest (youngest) on the edge of the marsh, the highest (oldest) just short of the crest. Cf.
s. 175, 178. *F. W. Oliver.*

ot. 402. The crest of the beach looking west. Saltings to the left, open sea to the right. Clumps of *Suaeda
fruticosa*. The tall upright plants are *Rumex crispus* var. *trigranulatus* and *Glaucium flavum*. *F. W. Oliver*, 1908.

ot. 403. Main beach from salt marsh just west of the Marams. Here the inner side of the beach is devoid of
aeda fruticosa owing to the tidal scour from the west (left). On the right is the last (youngest) lateral of the Marams,
ed with a dark belt of *Suaeda* which also occurs at its junction with the main bank. *F. W. Oliver*, 1927.

BLAKENEY BEACH

PLATE 158

Phot. 404. Looking north-west along the Fleet: "deltas" of shingle project from the terrace into the Fleet, with the belt of *Suaeda fruticosa* behind and the "back" of the beach above. *F. W. Oliver* (1911).

Phot. 405. The back of the beach with patches of *Silene maritima*. Belt of *Suaeda fruticosa* on the edge of the Fleet. *F. W. Oliver* (1911)

Phot. 406. The Chesil Bank west of Abbotsbury; here a comparatively narrow fringing beach, poorly vegetated. Belt of *Tamarix anglica* on the right (inner edge of beach). *F. W. Oliver* (1911).

CHESIL BEACH

arrest of shingle travel, a suggestion confirmed by the heaping up of stones behind individual bushes or clumps. The gullies tend to be permanent when once formed, carrying fresh shingle driven down from the crest by a new series of storms.

Distribution on Chesil beach. Besides the north Norfolk beaches the other British locality where *Suaeda fruticosa* occurs abundantly and grows luxuriantly is the great Chesil shingle bank on the coast of Dorset, an account of whose structure has been already given (pp. 873, 876–8). Here *Suaeda* lines the low terrace, which abuts on the Fleet for a great part of its length (Pl. 156, phot. 400, and Pl. 158). It will be recalled that the shingle of this terrace is fed from the main mass of the bank through a series of steep-sided ravines caused by the active percolation of water through the beach from the sea at high spring tides, not, like the shingle fans spreading over the salt marsh at Blakeney, by the travel of shingle over the general surface of the bank. Fig. 179 is a chart of one of these ravines, locally known as "cans".

Correspondingly the Chesil *Suaeda* shows no power of climbing the bank, but merely sends finger-like extensions of its zone on to the floors of some of the gullies, which lie scarcely higher than the surface of the terrace itself. This incapacity to spread up the bank may probably be accounted for by the excessive mobility of the sides of the "cans" and the stability of the intervening buttresses. The stability of the Chesil shingle, except in the gullies, is evidenced by the abundance and variety of the lichens which cover its stones. The *Suaeda* bushes abutting on the stable shingle are the least healthy and some are moribund. Furthermore, the arrival of tidal drift bringing humus and seeds is much scantier than at Blakeney, for most of the Fleet is little affected by the tides owing to its very narrow and protected entrance and its great length. Altogether the conditions for active invasion and spread of the *Suaeda* are much less favourable than at Blakeney.

Suaeda fruticosa, then, though primarily a halophyte, finds, in this **Preference for shingle habitats** country, its best conditions for growth on shingle, whose soil may show quite a low chloride content; and it is actually benefited by partial covering with travelling stones. It seems probable that the freer aeration of its root system is one factor that leads to its preference for this habitat, and the parallelism with the behaviour of *Ammophila* (see p. 856) is noteworthy. *Obione (Atriplex) portulacoides* seems to share this preference for well aerated soil, though to a less marked degree.

Other flowering plants. Salisbury (in Oliver and Salisbury, 1913a, p. 25) enumerates 60 species of flowering plants as occurring on the main shingle bank at Blakeney. Of these 20 are marked as "very rare" and 16 more as "rare". Six at least of the rare or very rare species are clearly casuals on shingle, while others are occasional colonists from the adjoining salt marsh or sand dune formations.

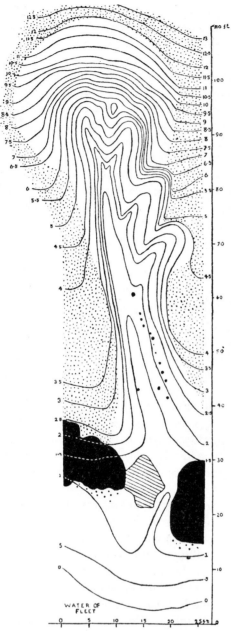

FIG. 179. CONTOURED CHART OF RAVINE ("CAN") IN THE CAMM OF THE CHESIL BANK

Stable shingle dotted, mobile shingle white. *Suaeda fruticosa* black. *Obione portulacoides* diagonally ruled. From Oliver and Salisbury, 1913*b*.

The following twelve species are distinguished by Salisbury as "the more important":

Agropyron junceum
A. pungens
Festuca rubra
Glaucium flavum
Honckenya peploides
Rumex crispus var. trigranulatus
Sedum acre
Senecio vulgaris (forma)
Silene maritima
Sonchus arvensis var. angustifolius
S. oleraceus
Suaeda fruticosa

Suaeda fruticosa has already been dealt with at some length. Next in order of importance come *Honckenya peploides*[1] and *Silene maritima*,[2] which are abundant on all the less barren portions of the bank. The former is a characteristic sand plant of the foredunes (p. 847). It often occurs also on secondarily bared areas where the loose sand is not too deep, and is notable for its power of burrowing into underlying shingle. Its presence on shingle is almost certainly determined by the existence of considerable sand among the stones. It approaches the sea more closely than the other shingle plants (compare its position on foredunes), sometimes occurring on the seaward side of the crest of the bank. Though somewhat succulent, and easily enduring occasional immersion by spring tides, it is not quite a complete halophyte, since prolonged immersion in salt water turns its aerial shoots yellow. *Silene maritima*, while not infrequent on partly fixed dunes, is

[1] Pl. 159, phot. 408.
[2] Pl. 156, phot. 399; Pl. 158, phot. 405; Pl. 160, phot. 409; Pl. 161, phot. 413.

PLATE 159

Phot. 407. *Suaeda fruticosa* (rabbit eaten) in a "low" of sandy shingle which is covered at the highest spring tides. The bushes accumulate a little blown sand. *Mrs Cowles* (1911).

Phot. 408. *Honckenya peploides*, with one plant of *Rumex crispus* var. *trigranulatus*, on the back of Blakeney main bank. The crest of the bank, with the sea beyond, is seen in the distance. *Mrs Cowles* (1911).

SHINGLE VEGETATION AT BLAKENEY POINT

even more characteristic of shingle and these two species often form extensive carpets on the banks. Their creeping stems are able to endure covering by stones as by sand, and with their deep roots and abundant aerial shoots they certainly contribute considerably to the stabilisation of the beach.

The horned poppy (*Glaucium flavum*) and a variety of the curly dock (*Rumex crispus* var. *trigranulatus*)[1] are the next most abundant species. Individuals of both are conspicuous by their erect habit and are numerous on all the less sterile portions of the bank, i.e. along the stretches where there is abundant tidal drift which brings the seeds and also the humus materials. *Glaucium* is a perennial in more protected situations on the landward side, but on the crest a biennial, not surviving its second (flowering) year. *Sedum acre*, the yellow stonecrop, is also a common plant on the same drift-laden stretches of the beach and it also behaves as a biennial, the flowering plants of one summer being almost entirely replaced by seedlings in the next, so that the greatest abundance of the species shifts from place to place.

Agropyron pungens is a characteristic plant of the edge of the zone above the marsh just touched by the high spring tides, while *A. junceum* is a pioneer psammophyte which can extend its runners in shingle (see pp. 847–8).

Other common species are the red fescue, the two sow-thistles and the common groundsel. *Festuca rubra* usually occurs between the ranks of the *Suaeda* bushes, on the sheltered side, and thus (with *Sonchus oleraceus* and *S. arvensis* var. *angustifolius*) is characteristic of the most stable shingle. (Cf. p. 889 and Pl. 136, phot. 341.) The ephemeral *Senecio vulgaris* is fairly common throughout the length of the bank.

In troughs and shallow depressions ("lows") on the landward side of the beach, of greater surface stability and with finer shingle, the grasses *Poa annua*, *Aira praecox*, *Catapodium* (*Desmazeria*) *loliaceum* and *Pholiurus* (*Lepturus*) *filiformis* occur, the first named common and characteristic, *Catapodium* local and *Pholiurus* rather rare.

Halophytes. Apart from *Suaeda fruticosa* the following halophytes occur on the main shingle bank at Blakeney Point:

Artemisia maritima	Limonium vulgare
Aster tripolium	Obione portulacoides
Glyceria maritima	Plantago maritima

Glyceria maritima and some *Obione* occur away from the salt marshes, but in regions where drift has accumulated, carried up by storms. The remaining four species and *Obione* are found in the shingle opposite the fringing marshes, mostly on the edge of the latter where the marsh soil has

[1] Pl. 157, phot. 402; Pl. 160, phot. 409; Pl. 161, phot. 413.

recently been covered with shingle, but isolated plants occur well up on the bank. *Artemisia maritima* was found in more than a dozen separate stations, sometimes in large patches, and mainly on the landward edge of and quite close to the lowest line of *Suaeda* bushes. It occupies a corresponding position on the lateral spits, and must thus be regarded, together with most of the other shingle halophytes, as surviving from the uppermost edge of salt marsh now covered by shingle. The plants of *Aster, Limonium* and *Plantago* close to the inner edge of the beach still have their roots in the marsh soil below.

Psammophytes. Besides *Agropyron junceum, Honckenya peploides* and *Silene maritima*, referred to above, the following sand dune plants occur as rarities on the shingle:

Carex arenaria	Eryngium maritimum
Calystegia soldanella	Senecio jacobaea
Elymus arenarius	Stellaria pallida (boraeana)

Ammophila arenaria is a not infrequent component of the shingle vegetation, and may be a relict of old dunes, now vanished, on the older part of the bank.

The sea-holly (*Eryngium maritimum*), typically a sand dune plant, occurs at Blakeney only on the shingle.

Sandy or gravelly heath plants are rare on the shingle, but *Lotus corniculatus, Festuca ovina, Holcus lanatus* and perhaps *Arrhenatherum elatius* may be reckoned as very sporadic colonists belonging to this category.

Habit of shingle plants. All the shingle plants, apart from *Suaeda*, are very low growing, except *Glaucium, Sonchus* and *Rumex*. The two former pass most of their life in the state of rosettes, only sending up their flowering shoots in the milder season. *Rumex* protects its young leaves with the old withered foliage, and all the flowering spikes may be killed by the salt-laden spray driven before a violent wind.

Vegetation of lateral banks

While the main bank as a whole shows the earlier and characteristic stages of shingle vegetation, which is here "open", the plants for the most part being scattered without very definite arrangement, the stabilised laterals exhibit the later stages of succession leading to non-maritime vegetation. The stages of succession can on the whole be very accurately traced from the younger to the older banks, on which the same zoning is constantly repeated.

Each bank presents three main topographical habitats: the sloping *flanks*, steeper on the side turned towards the apex of the main bank, i.e. originally the seaward side; the flattened or slightly convex *crest*; and the *high elbow*, or crest of the sharp bend of the L-shaped termination. The zonation from below upwards is as follows, the first three zones occupying the flank of the bank.

(1) The *Suaeda fruticosa* zone with dense bushes about 2 ft. (60 cm.) high and of varying width.[1] On its outer (lower) side it abuts on the zone of *Artemisia maritima*, here taken as the uppermost salt marsh zone. *Obione portulacoides* is common in the *Suaeda* zone, especially where the bank abuts on marshes in which it is dominant. *Aster tripolium* (scattered) and *Glyceria maritima* (sometimes frequent) are also present. *Cochlearia anglica* is restricted to this zone, but is rather rare and on the older laterals only.

Suaeda bushes in decreasing number occur in the higher zones up to the crest, but they are very scattered and of poor growth above the limit of drift at the top of zone 2.

(2) The *Festuca rubra* zone, of variable width and sometimes broken, but consisting of an almost pure sward of the grass.[1] At the upper edge of this there is often a subsidiary zone of *Agropyron pungens* (also var. *aristatum*), with which is associated *Cochlearia danica*, and more rarely *Atriplex littoralis*. Here is the upper limit of the drift.

(3) The *Limonium binervosum* zone, with much less abundant *Frankenia laevis*. The substratum is sandy mud covered with bare shingle. Numerous scattered plants of *Obione*, prostrate in habit and only 2 or 3 in. high, *Armeria maritima*, and stunted *Glyceria maritima* and *Plantago coronopus* are also present. This zone occupies the whole top of the fifth lateral, which is broad and flattened and devoid of any crest.

(4) The *crest*, which is of greater width than any of the other zones, is occupied by a sward dominated by *Armeria maritima*, *Silene maritima* and *Agrostis stolonifera*. The following associated species occur:

Arenaria serpyllifolia	a	*Pholiurus filiformis	o
Artemisia maritima	vr	Plantago coronopus	a
Catapodium (Desmazeria)		*Sagina maritima	o
loliaceum	l	Sedum acre	a
*Cerastium tetrandrum	f	*Trifolium arvense	f
Honckenya peploides	r	T. procumbens	o
Leontodon autumnalis	vr	*T. striatum	l
Limonium binervosum (relict)	f		

* Mainly on the tracks along the crests, and probably determined by the relative bareness of the banks and the low growth of the vegetation.

(5) The *high elbow* bears several "crest" species, and many others which do not occur elsewhere on the shingle banks. The following is a complete list:

Aira praecox	f	Bromus hordeaceus (mollis)	f
Arenaria serpyllifolia	o	Cirsium lanceolatum	vr
Armeria maritima	a	Erodium cicutarium	o
Bellis perennis	vr	Festuca ovina	o

[1] Pl. 136, phot. 341.

"Festuca maritima"	vr	Rumex acetosella	a
Galium verum	f	R. crispus var. trigranu-	
Geranium molle	r	latus	r
Hordeum murinum	r	Sedum anglicum	r
Koeleria cristata	vr	Senecio jacobaea	r
Lotus corniculatus	a	S. sylvaticus	vr
Plantago lanceolata	a	Silene maritima	f
Poa pratensis	a	Vicia angustifolia	o

It will be noted that the most abundant plants, except the thrift, are non-maritime species occurring on gravelly heaths, while *Galium*, *Aira*, *Festuca ovina*, *Koeleria*, *Sedum* and *Senecio sylvaticus* also belong to light inland soils. *Bellis*, *Cirsium* and *Hordeum* are only found close to the watch house and their introduction is doubtless due to human agency.

This "high elbow" vegetation clearly represents the most advanced stage of succession in the whole Blakeney Point system.

The foregoing lists are taken from the more or less stable zonation of the mature banks. The younger banks are mostly dune-covered, but one shows a younger and still open phase of the crest community, with *Honckenya peploides* and *Silene maritima*, recalling the open vegetation of the main bank, but with *Glaucium* absent, *Rumex trigranulatus* only occasional, and the addition of *Armeria maritima* and such species of the crest or high elbow of the more stable banks as *Lotus corniculatus*, *Plantago coronopus* and *Rumex acetosella*, as well as *Filago minima*.

Vegetation of Chesil beach. This has not been investigated in detail, but owing to the absence of laterals and their more varied flora as seen at Blakeney, and probably also owing to the poverty of drift on the inner margin bordering the Fleet, it is not so rich in flowering plants. The occurrence of *Suaeda fruticosa* on the Chesil terrace has already been described (p. 885). The main vegetation of the crest is dominated by *Lathyrus maritimus*, a species which does not occur at Blakeney, but which according to Oliver (1912) is at its best on mobile shingle. The stable buttresses between the percolation gullies are occupied by *Silene maritima*, which also dominates the stable shingle of the flat crest of the beach at its western end. Remarkable evidence of the good and permanent water content of shingle beaches is cited by Oliver (1912) à *propos* of this plant. At the end of the long drought of 1911, when the pasture on the neighbouring downs was so parched that sheep could find no food there, they wandered down on to the shingle bank to feed on the shoots of the sea campion, which were perfectly green and fresh (Pl. 156, phot. 399). Near the same spot *Polygonum amphibium* was found growing luxuriantly on the shingle. None of the shingle beach vegetation observed in 1911 suffered from the drought.

Other characteristic shingle species. Another species plentiful on the stable buttresses of the Chesil Bank is *Geranium purpureum* (Pl. 160, phot. 410), a very characteristic plant of stable shingle which is common on

PLATE 160

Phot. 409. *Rumex crispus* var. *trigranulatus* and *Silene maritima* (in flower) on the back of Blakeney main bank.

Phot. 410. *Geranium purpureum* Vill. in flower and fruit on shingle at the back of the present tidal beach, near Rye. This is a characteristic species of the south-eastern shingle beaches. November 1904.

Phot. 411. Freshwater marsh with *Triglochin palustre* in a "low" of an apposition beach near Rye, Sussex. Zone of *Glaucium flavum*, etc. on the slope of the "full" to the left.

SHINGLE BEACH VEGETATION

PLATE 161

Phot. 413. Shingle "fulls" with *Rumex crispus* var. *trigranulatus* and *Silene maritima*. The "lows" between the fulls are here uncolonised by plants

Phot. 412. General view of the great apposition shingle beach west of the River Rother (Dungeness lies beyond the picture to the right). Shingle "fulls" (ridges) covered with grass used as sheep pasture in foreground. Beyond is largely bare shingle with patches of scrub.

Phot. 414. Old shingle fulls with vegetation. A tree of elder (*Sambucus nigra*) and a stack of cut gorse (*Ulex europaeus*). Scrub in the distance.

Phot. 415. Flat expanse of shingle with patches of grass. The darker patches are prostrate blackthorns (*Prunus spinosa*).

many south coast fringing beaches but is not recorded from Blakeney Point. *Solanum dulcamara* var. *marinum* is frequent in similar situations on the Sussex fringing beaches.

Dungeness. The great areas of apposition shingle at and near Dungeness on the borders of Kent and Sussex show much later stages of succession, for which there is no opportunity of development on the beaches hitherto described. This vegetation has never been studied at all closely, but the shingle relatively remote from the sea is known to support not only a variety of ruderal plants and grasses, but also scrub of bramble, gorse blackthorn, hawthorn and elder. This varied land vegetation is certainly correlated with greatly increased accumulation of soil, as well as progressive removal from maritime conditions, and there seems no reason why such old shingle areas should not develop climatic climax vegetation if seed parents were available. (Pl. 161 illustrates phases of apposition shingle beach vegetation.)

SHINGLE LICHENS AND MOSSES

The lichen vegetation of shingle is naturally almost confined to stable areas where the stones remain undisturbed for long periods. At Blakeney Point, where the lichens were studied by McLean (1915), the species are comparatively few,[1] while on the Chesil Bank, a provisional list from which was drawn up by Watson (1922), the number of species is larger and the lichen vegetation much more abundant. This difference is probably partly due to the much greater areas of stable shingle and partly to the milder and moister climate of the Dorset coast.

Blakeney Point. All the lichens over the whole Blakeney area, according to McLean, "with the possible exception of those upon the main bank, are subject to submersion at some time or other during the year". *Verrucaria maura* is the species most constantly submerged. It is only found on the edge of shingle where this abuts on the mud on the inner side of the bank, and here "it is submerged for several hours during every high tide, even the lowest neaps". The thallus is intensely black and slow-growing. With *Verrucaria*, Watson (Oliver, 1929) found *Placodium lobulatum* very abundant. At Burnham Overy, a few miles to the west, though not at Blakeney, *Verrucaria* is followed by an overgrowth of *Xanthoria parietina*. To the mud below the high water mark of ordinary spring tides *Collema ceranoides* and *Lecanora badia* are restricted, probably because they depend on a mud substratum: above this level the shingle is comparatively clean.

Among the colonising *Suaeda fruticosa*, on the inner edge of the bank where tidal immersion is regular and frequent, humus-bearing mud is deposited around and over the stones, and stability is high. This shingle

[1] The number recorded was augmented by Watson and by Richards (Oliver, 1929).

bears the following lichens, which cover almost every available spot on the firmly embedded pebbles:

†Lecanora atra	‡Rhizocarpon confervoides
*L. citrina[1]	*Verrucaria maura
*L. badia	†Xanthoria parietina

| * Primary colonisers. | † Saprophytic. | ‡ Ubiquitous. |

The mud slowly overwhelms the stones and reduces the lichens, so that the area will ultimately become a mud flat devoid of lichens.

On the main bank itself the seaward slope is of course destitute of lichens, and so is most of the crest. Here mobility is considerable and where the pebbles are small and there is much sand in the shingle the scour of the wind-driven sand also keeps the stones free from lichens, except in minute depressions where very small thalli may occur. Towards the landward edge of the bank patches of crustose lichens occur opposite the proximal ends of the dormant lateral banks. This is probably because the stable shingle of the lateral resists the inward thrust of mobile shingle from the sea, reducing it to a condition of comparative stability. Since however the whole bank is moving shorewards and actually embedding the proximal ends of the laterals (Fig. 171), this stability cannot be permanent. Probably it is destroyed at intervals of some years, during violent onshore gales. The intervening periods of relative quiescence during which the lichens can develop are therefore limited. The lichens of these patches are young and vigorous—evidently recent colonisers—in contrast with the mature growths on the old laterals. The thalli of these young lichens are remarkably large, as much as 2–3 in. (5–7½ cm.) in diameter, indicating quite rapid growth. The following species occur in such situations:

Lecanora citrina[1]	Rhizocarpon confervoides
*Parmelia fuliginosa	*Squamaria saxicola
P. saxatilis	Xanthoria parietina
Physcia tenella	

The two species marked with an asterisk are confined, at Blakeney, to this community. *Rhizocarpon* is almost ubiquitous above tide marks and *Lecanora citrina*[1] occurs elsewhere only on low shingle among *Suaeda*.

On loose shingle mixed with sand there are:

Lecanora atra	Lecanora galactina
L. citrina[1]	Rhizocarpon confervoides

L. galactina is a decidedly arenicolous species and colonises some of the loosest and sandiest shingle areas.

On the "bound" shingle of the lateral banks there is, according to McLean, a marked distinction between the lichens of the open communi-

---

[1] Probably *Placodium lobulatum* according to Watson (Oliver, 1929).

ties, in which crustose forms occur on the stones, and those of the closed communities dominated by grasses (*Festuca rubra* and *Agropyron pungens*) and characterised by foliose forms. The former has the following species:

Aspicilia gibbosa
Biatorina chalybeia
Buellia colludens
Lecanora atra

Lecanora atroflava
Physcia tenella
Verrucaria microspora
Xanthoria parietina

The latter community is composed of Cladonias and *Cetraria*:

Cetraria aculeata
Cladonia furcata

C. pungens

The dominant *Cetraria* and *Cladonia furcata* were present in astonishing quantity, though quite overshadowed by the grasses. *Peltigera rufescens*, a characteristic species of the grey dunes (see p. 855) occurs also on the "high elbow" of one of the banks.

Lichens of Chesil Bank. With the comparatively poor lichen flora of Blakeney Point on which McLean noted 31 species,[1] most of which have been mentioned in the foregoing account, we may contrast the richer lichen vegetation of the Chesil Bank (Watson, 1922). There is a probably larger number of species, but the more favourable conditions are mainly evidenced by the greater luxuriance of the lichen vegetation, especially on the terrace bordering the Fleet.

Biatorina chalybeia
B. lenticularis
Buellia colludens
B. myriocarpa
B. stellulata
Callopisma cerinum
C. citrinum
Cladonia rangiformis
Evernia prunastri
Lecania erysibe
Lecania porella
L. prosechoides
Lecanora expallens
L. galactina
L. hageni
L. polytropa
L. sophodes
L. umbrina
Lecidea contigua
L. nigroclavata
L. protensa

Parmelia caperata
P. conspersa
P. dubia
P. perlata
P. physodes
P. saxatilis
P. sulcata
Physcia tenella
Porina chlorotica
Ramalina cuspidata
R. farinacea
R. polymorpha
R. scopulorum
R. subfarinacea
Rhizocarpon confervoides
Rinodina demissa
R. exigua
Verrucaria maura
V. nigrescens
Xanthoria parietina

Mosses. Mosses are not common on loose shingle beaches, and they do not colonise shingle until it has been more or less bound and stabilised.

---

[1] Twelve additional species were however recorded by Watson (Oliver, 1929).

On the main bank at Blakeney mosses were found (Richards, 1929) only on the landward face, where species of *Bryum* were prominent:

*Amblystegium serpens
*Brachythecium rutabulum
Bryum pallens
B. pendulum
B. spp.

Ceratodon purpureus
*Eurhynchium praelongum
Tortula ruraliformis
Trichostomum flavovirens

* In the shelter of *Silene maritima.*

On the stable lateral banks the following occurred:

Brachythecium albicans
Bryum spp.
Ceratodon purpureus
Dicranum scoparium var. ortho-
  phyllum

Eurhynchium praelongum
Hypnum cuspidatum var. tectorum
Tortula ruraliformis
Trichostomum flavovirens

Polytrichum piliferum was found on the Yankee bank. It will be noted that this list of species is much like that occurring on the grey dunes (p. 855).

From the Chesil Bank *Bryum capillare, Ceratodon pupureus* and *Hypnum cupressiforme* have been recorded (Watson, 1922).

REFERENCES

HILL, T. G. and HANLEY, J. A. The structure and water content of shingle beaches. *J. Ecol.* 2, 21–38. 1914.

McLEAN, R. C. The ecology of the maritime lichens at Blakeney Point, Norfolk. *J. Ecol.* 3, 129–48. 1915.

OLIVER, F. W. The maritime formations of Blakeney Point in *Types of British Vegetation*, 1911, pp. 354–66.

OLIVER, F. W. The shingle beach as a plant habitat. *New Phytol.* 11, 73–99. 1912.

OLIVER, F. W. Some remarks on Blakeney Point, Norfolk. *J. Ecol.* 1, 4–15. 1913.

OLIVER, F. W. Blakeney Point Report. *Trans. Norf. and Norw. Nat. Soc.* 12, 1929.

OLIVER, F. W. and SALISBURY, E. J. Topography and vegetation of Blakeney Point. *Trans. Norfolk and Norw. Nat. Soc.* 9, 1913 a. (Also issued separately.)

OLIVER, F. W. and SALISBURY, E. J. Vegetation and mobile ground as illustrated by *Suaeda fruticosa* on shingle. *J. Ecol.* 1, 249–72. 1913 b.

RICHARDS, P. W. Notes on the ecology of the bryophytes and lichens at Blakeney Point, Norfolk. *J. Ecol.* 17, 127–40. 1929.

SALISBURY, E. J. The soils of Blakeney Point: a study of soil reaction and succession in relation to the plant covering. *Ann. Bot.* 36, 391–431. 1922.

WATSON, W. List of lichens, etc. from Chesil Beach. *J. Ecol.* 10, 255–6. 1922.

Chapter XLIII

SUBMARITIME VEGETATION

Besides the well-marked maritime vegetation described in the last three chapters there are various plant communities developed near the sea but less directly exposed to maritime influences, and these we may consider together as *submaritime vegetation*.

Because of the gradual dying away of the effects of proximity to the sea as we pass away from the zones directly affected by the tides or by constant sea spray it is impossible to draw a sharp line between maritime and submaritime vegetation. Spray may be carried many miles inland by violent storms, and many non-maritime species are clearly tolerant of its incidence, while some plants which are normally halophytes can equally clearly flourish perfectly well in certain inland habitats. Again, on certain flat coasts salt water may penetrate far up the rivers and drains, rendering the soil water brackish at a considerable distance from the sea. Thus many communities developed near the coast show a mixture of maritime or submaritime and non-maritime plants. It is impossible to analyse these satisfactorily until they have been properly studied and their ecological requirements determined, including the degree of toleration of sea salt.

Tolerance of sea salt

Submaritime species. The *species* which may be called submaritime are more abundant near the sea though not absent from inland stations. Some of these are evidently able to endure considerable amounts of salt, and this capacity may well facilitate their existence in habitats free from the competition of less tolerant species to which the submaritimes might succumb in many inland habitats. Apart from the flora of brackish marshes a number of species occur more commonly in various habitats as the sea is approached, and of these the following are a small selection:

Cerastium tetrandum	Erodium maritimum
Carduus tenuiflorus	Ononis reclinata
Daucus carota	Plantago coronopus
Erodium cicutarium	Urtica pilulifera

(5) BRACKISH WATER AND MARSHES

These are developed especially on flat coasts behind the zone of salt marsh proper where the tides have no direct access and the salt water which penetrates is much diluted by fresh water coming from the land. Brackish marshes may occur along the courses of tidal estuaries at some little distance from the sea, as well as on flat land behind fringing shingle beaches and narrow belts of sand dunes where the water table is brackish. Such land has nearly always been originally reclaimed from salt marsh by

"inning" and drainage and is thus intersected by ditches, and where the
drainage is effective may be made into good pasture land. The drainage
ditches themselves, where the water is brackish, contain aquatic species
characteristic of such situations. And where the drainage is ineffective, so
that the ground is still largely waterlogged, a number of reedswamp and
marsh species occur. The original habitat of such plants is the brackish
water of slow tidal rivers and their adjoining creeks and marshes, where the
fresh water flowing from the land is mixed with salt water brought by the
flood tides.

Besides the characteristic brackish water species a number of fresh
water and marsh plants can tolerate small quantities of salt, and on the
other hand some brackish species can grow in quite fresh water. The
vegetation of the habitats described thus consists of various mixtures of
all three categories. No systematic work has been done on this vegetation
and very little is known about the actual requirements and degree of
toleration of the species concerned.

The following are some of the species met with in these habitats, but it
must be understood that the list is far from being exhaustive, and that a
number of different, mainly unexplored, communities are represented.

Brackish water	*Brackish reedswamps and marshes*
Potamogeton pectinatus	Althaea officinalis
P. filiformis	Blysmus rufus
Ranunculus baudotii	Carex distans
Ruppia maritima	C. divisa
R. rostellata	C. extensa
Zannichellia palustris, etc.	C. punctata
Enteromorpha spp.[1]	C. vulpina
Vaucheria spp.	Eleocharis uniglumis
and many other algae	Oenanthe lachenalii
	Rumex maritimus
	Samolus valerandi
	Scirpus carinatus
	S. maritimus[1]
	S. tabernaemontani
	S. triqueter

Where water from fresh springs enters the upper edge of a salt marsh
there is a mixture of halophytes and freshwater plants, with or without
some of these semihalophytes or "brackish" species (see pp. 839–40). The
salt content of the soil water often varies greatly within a very short
distance.

(6) SPRAY-WASHED ROCKS AND CLIFFS

Well-drained rocks and cliffs above the reach of high tides but constantly
exposed to spray, support a characteristic vegetation of halophytes. Most
of these (e.g. *Limonium vulgare*, *Armeria maritima*, *Plantago maritima*) are

[1] Pl. 162, phot. 416.

abundant also in salt marshes, but the samphire (*Crithmum maritimum*)[1] is quite characteristic of the rocky spray-washed habitats. Other less frequent species of such habitats are beet (*Beta maritima*),[2] fennel (*Foeniculum vulgare*), wild cabbage (*Brassica oleracea*), queen stock (*Matthiola incana*) and sea spleenwort (*Asplenium marinum*).

As examples of definite small halophytic sea cliff communities the following are recorded by Tansley and Adamson (1926, p. 11) from ledges of a fully exposed chalk cliff on the Sussex coast (Cliff End at Cuckmere Haven). All but *Crithmum* and *Beta*, it will be noted, are salt marsh plants.

10 *ft. above the beach*	6 *ft. above the beach*
Glyceria maritima	Beta maritima
Limonium vulgare	Crithmum maritimum
Spergularia marginata	Obione portulacoides

The general vegetation of sea cliffs has never been fully described. It is actually very mixed according to the nature of the rock, the slope, the exposure and the height above the sea. The halophytic species are accompanied by chomophytes in the rock clefts and constituents of grassland and heath communities where the slope is not too precipitous and soil can accumulate; and on the mountainous west coasts of Scotland, and especially of Ireland, many alpine plants descend almost or quite to sea level. On the Irish cliffs, as Praeger has described (many publications summarised in 1934), we have the most extraordinary mixture of species of very various ecological as well as geographical status.

Petch (1933) describes the vegetation of the "high cliffs" up to 1300 ft. (395 m.) of St Kilda, a small isolated island in the Atlantic about 50 miles (80 km.) west of the Outer Hebrides, as consisting of "isolated patches of moorland growing in such cracks and ledges as afford a foothold", but with, in addition, the following curious mixture of species, several of them woodland plants, confined to this habitat:

Angelica sylvestris	f	Rumex acetosa	f
Athyrium filix-femina	f	*Salix herbacea	vr
Dryopteris dilatata	f	*Sedum roseum	a
Lonicera periclymenum	r	Silene maritima	o
Polypodium vulgare	f	Taraxacum palustre	r
Primula vulgaris	o		

* Arctic-alpine species.

The "low cliffs", within reach of the sea spray, similarly bear fragments of surrounding communities, with the addition of the following, all maritime or submaritime species:

Armeria maritima	f	Ligusticum scoticum	lf
Atriplex babingtonii	o	Matricaria maritima	a
Cerastium tetrandrum	a	Silene maritima	f
Cochlearia anglica	a		

Finally the "moorland" which descends to within a few hundred feet of the sea and is therefore affected by the salt spray consists of a very well-

---

[1] Pl. 162, phot. 417.   [2] Pl. 162, phot. 418.

marked community quite comparable with Praeger's *Plantago*-sward (see below) though poorer floristically. All the species are presumably tolerant of salt.

Agrostis tenuis	o	Festuca ovina	a
Armeria maritima	a	Juncus articulatus	o
Carex goodenowii	f	Leontodon autumnalis	f
C. oederi	o	Plantago coronopus	a
Cerastium vulgatum	r	P. lanceolata	f
Euphrasia officinalis	r	Sagina procumbens	f

(7) MARITIME AND SUBMARITIME GRASSLANDS

Grassland close to the sea, though quite out of reach of the tides, is always likely to be more or less affected by spray and usually contains maritime or submaritime species, besides the definite halophytes mentioned in the last section; and this whether it is situated close to sea level or on the tops of comparatively lofty cliffs. Very few ecological observations are available, and determinations of the actual amounts of salt deposited on the sward, together with a study of the autecology of the species present are especially wanted. Nearly all such grassland is more or less grazed, and this of course affects the composition of the turf; but a great deal of such land would certainly be grassland even in the absence of grazing.

Worm's Head. The most elegant observations bearing on the point are contained in McLean's paper (1935) on the maritime grassland of Worm's Head at the extremity of the Gower Peninsula in South Wales. This is developed, like that of the immediately adjoining mainland, on a heavy loam derived from the underlying Carboniferous Limestone by weathering *in situ*, and 2–6 ft. (0·6–1·8 m.) deep. This loam is ferruginous but practically non-calcareous, the calcium carbonate having been leached out during and after the process of weathering. The surface pockets of the limestone contain a dark alkaline rendzina soil (pH 7·2–7·8).

The grassland grazed by sheep contains the following species:

Agrostis tenuis
*‡Armeria maritima
†‡Bellis perennis
  Calluna vulgaris
†Cerastium vulgatum
  Cirsium lanceolatum (occ.)
  Cynosurus cristatus
†Dactylis glomerata
  Erica cinerea
  Festuca ovina
‡Galium verum
  Holcus lanatus
  Leontodon nudicaulis
†‡Lotus corniculatus
  Medicago lupulina (rare)
  Picris hieracioides
*†‡Plantago coronopus

  P. lanceolata
*‡P. maritima
†Poa annua
  P. pratensis
  Potentilla erecta
‡Poterium sanguisorba
  Rubus ulmifolius (rusticanus)
  Rumex acetosa
†‡Sedum acre
  Spiranthes spiralis
‡Thymus serpyllum
  Trifolium repens
  Ulex gallii
  Viola riviniana

Marasmius oreades (fairy rings)

* Maritime and submaritime species.   † "Lair flora."   ‡ Shallow soil over rock.

PLATE 162

Phot. 416. Brackish channel behind the present tidal beach, near Rye. *Scirpus maritimus* in the water, *Enteromorpha* sp. floating and on the bank in the foreground.

Phot. 417. *Crithmum maritimum* on maritime rocks. Kynance Cove, Cornwall. *J. Massart.*

Phot. 418. *Beta maritima* and *Silene maritima* on maritime rocks. Kynance Cove, Cornwall. *J. Massart.*

MARITIME AND SUB-MARITIME VEGETATION

The "lair flora" consists of species specially concentrated in places where sheep habitually lie and thus forms a community dependent on abundant supply of organic nitrogen.

Worm's Head itself is a rocky promontory extending due west into the sea and consisting of three separate eminences. Of these the innermost, on which sheep are grazed and which bears the species enumerated above, is separated from the middle one by a low rocky neck which the sheep do not cross. Thus the last two eminences are entirely free from the effect of the grazing factor. The outer head is too precipitous and rocky for the development of grassland, but the vegetation of the middle head, rising not more than 50 ft. (15 m.) above high water mark, contains the following species, arranged in order of frequency:

Festuca rubra subsp. eu-rubra, var. genuina, sub-var. pruinosa Hack., dominant and
 largely pure.

Dactylis glomerata	Rumex crispus
Holcus lanatus	Plantago lanceolata
*Beta maritima	*Limonium binervosum
*Silene maritima	Cirsium lanceolatum
Rumex acetosa	Spergularia rupicola
Trifolium repens	Leontodon nudicaulis
*Armeria maritima	Galium verum
Sonchus oleraceus	

* Maritime species.

The soil is manured by sea birds (the inner headland both by sheep and by sea birds) whose droppings are full of mussel shells, which also abound in the soil, and is rich in humus (15·1 per cent). Its content in calcium carbonate averages 4·25 per cent, and the pH 7·8. It is colloidal in texture and full of ants.

The dominant fescue forms a thick mattress, through which, in places, a stick may be thrust down 2 ft. (60 cm.) before reaching the soil. The leaves are over a foot (30 cm.) long and the plant flowers freely. The associated species occur as large isolated plants half buried among the dominant grass.

This extreme dominance of the red fescue under conditions of extreme exposure to wind and spray but in the absence of grazing is a very interesting phenomenon which has not been recorded elsewhere (but see below). It will be noted that the number of species present is only half that of the equally exposed but grazed community, largely owing no doubt to the overwhelming dominance of the fescue whose competitive powers are known to be diminished by grazing. The disappearance of the maritime and sub-maritime plantains (*Plantago maritima* and *P. coronopus*) which normally form rosettes is probably due to this cause. On the other hand four maritime species occur (*Beta maritima, Silene maritima, Limonium binervosum* and *Spergularia rupicola*) which are absent from the grazed

57-2

areas. The other species of this Festucetum rubrae absent from the grazed areas are the tall *Sonchus oleraceus* and *Rumex crispus*, which are doubtless intolerant of grazing.

Festucetum on sea stacks. Praeger (1911, p. 40) records "dense deep masses of *Festuca ovina*[1] and other plants" on "some small sea stacks[2] inaccessible to sheep" at Clare Island (Co. Mayo) contrasting "strongly with the closely nibbled grass of the adjoining slopes"; and this seems comparable with McLean's Festucetum on Worm's Head.

Effect of grazing. The great majority of cliff top and cliff slope grasslands are more or less grazed. Sheep reach the most unlikely looking spots and it is hard to say how far the vegetation of such places has been modified, or even in some areas determined as grassland in contrast to heath, by this biotic factor. "Even on the great scarp of Croaghmore" (in Clare Island), writes Praeger, "the sheep have left their mark behind them in the little colonies of *Poa annua*, *Stellaria media*, *Cerastium vulgatum*, etc." But it is probable, as already said, that many such sub-maritime habitats would be dominated by grasses even in the absence of sheep.

***Plantago* sward.** In extremely exposed situations on the west coast of Ireland "where in winter gales the soil becomes soaked with spray" Praeger (1911, 1934) describes constantly recurring communities dominated by *Plantago maritima*, usually in company with *P. coronopus*, and often containing several species of grass. "A mild form" from the south coast of Clare Island, Co. Mayo, at an altitude of 50–100 ft. (15–30 m.) "forms a dense sward ½ in. in height with flower-stems rising to about 2 in.", and has the following composition:

Plantago maritima, P. coronopus   d

Aira praecox	Hypochaeris radicata
Anagallis tenella	Koeleria cristata
Bellis perennis	Lotus corniculatus
Calluna vulgaris	Ophioglossum vulgatum
Carex diversicolor (flacca)	Plantago lanceolata
C. oederi	Polygala depressa
Centaurium umbellatum	Potentilla erecta
Cerastium tetrandrum	Prunella vulgaris
C. vulgatum	Radiola linoides
Cynosurus cristatus	Sagina procumbens
Euphrasia officinalis (agg.)	Succisa pratensis
Festuca ovina	Sieglingia decumbens
Hieracium pilosella	Thymus serpyllum
Hydrocotyle vulgaris	Trifolium repens
Holcus lanatus	Viola riviniana

"An extreme example" consisted of "a smooth shining sheet of *P. maritima* (at least 80 per cent) with mere scraps of other plants. The rosettes

---

[1] Dr Praeger (*in litt.*) is not quite certain that the species was not *F. rubra*.
[2] Precipitous rocky islets.

of *P. coronopus* and *P. maritima* measured $\frac{1}{2}$–$\frac{3}{4}$ in. across." The composition was as follows:

Plantago maritima	d
Aira praecox	P. lanceolata
Anagallis tenella	Potentilla erecta
Festuca ovina	Radiola linoides
Galium saxatile	Sedum anglicum
Jasione montana	
Luzula campestris	Mnium hornum
Plantago coronopus	

The same community occurs on a number of other exposed western islands (Praeger, 1934). Thus on the Great Blasket, a steep and lofty island off the western extremity of the Dingle peninsula in Co. Kerry, entirely exposed to the Atlantic weather, "the south slope is occupied mainly by short grass full of *Plantago maritima* and *P. coronopus* with patches of *Pteridium* in the hollows". On Loop Head (Co. Clare) "*Plantago* sward is well developed, giving way on the low hill tops (about 300 ft.) to dwarf *Calluna*". "Above, over a considerable area, an almost pure *Armeria* sward occupies the ground." On the island of Inishturk the community over a large area is "often composed entirely of *P. maritima* and *P. coronopus* without any other ingredient... as close and smooth as if shaved with a razor". On Achill Island (Co. Mayo) it is well developed in maximum exposure up to 300 or 400 ft. (90–120 m.), as at the extreme point of Achill Head, and forms a dense smooth mat as follows:

Plantago maritima, P. coronopus	d
Agrostis tenuis	Radiola linoides
Aira praecox	Sagina maritima
Armeria maritima	S. procumbens
Cerastium tetrandrum	Spergularia rupicola
Festuca ovina	

Here, it will be seen, other maritime species are present. On Inishkea, Co. Mayo, it is also well developed.

These *Plantago* swards are closely grazed by sheep and rabbits, and their existence, seems to be determined by a combination of extreme exposure to wind, high salt content and close grazing.

Grassland of the Sussex cliff tops. On the flat top of Beachy Head (alt. 500 ft. = c. 150 m.) the turf is very short and is quite a typical sample of chalk grassland without maritime or sub-maritime species. Farther west, where the cliffs are not so high (100–300 ft. = 30–90 m.) and may be supposed to receive more spray during strong onshore gales, the following appear in the grassland of the cliff tops:

Agropyron pungens	Erodium cicutarium
Armeria maritima	Glaucium flavum
Carduus tenuiflorus	Plantago coronopus

and *Daucus carota* becomes locally very abundant. But the maritimes do not form a quantitatively important part of the vegetation. The turf is uniformly short ($\frac{1}{2}$ in.) except in local shelter, and though grazing is widespread the extreme exposure must be an important factor. In sunny summers deficient in rainfall the drought must be very severe, and the exposure to wind maximal at all times (Tansley and Adamson, 1926, pp. 10–11).

REFERENCES

McLean, R. C. An ungrazed grassland on limestone in Wales, with a note on plant "dominions". *J. Ecol.* **23**, 1935.

Petch, C. P. The vegetation of St Kilda. *J. Ecol.* **21**, 1933.

Praeger, R. Lloyd in Clare Island Survey. Part X. Phanerogamia and Pteridophyta. *Proc. Roy. Irish Acad.* **31**, 1911.

Praeger, R. Lloyd. *The Botanist in Ireland.* Dublin, 1934.

Tansley, A. G. and Adamson, R. S. A preliminary survey of the chalk grasslands of the Sussex downs. *J. Ecol.* **14**, 1926.

INDEX

The names of plants in the lists of the various communities are not, as a rule, indexed. Exceptions are made, however, for the more important dominants, for a few large genera whose species occupy markedly different habitats, and for some other plants of vegetational importance; but the selection is necessarily arbitrary.

grassland (*cont.*):
  chalk, 206–7, 491, **525–51**
  limestone, 206–7, 490–1, 552–7
  maritime, 491, 898–902
  neutral, 495, **559–76**
  siliceous, 207–8, 490, 499
  sub-maritime, 898–902
grassland climate, 235
gravel, 80, **97**
  glacial, 26, 125
gravel subsere (Callunetum), 733–4
grazing, 129, 205, 375, 454, 475, 488–9, 496,
  499, 560–1, 723, 753
  aftermath, 567
  by cattle, 129–31
  by rabbits, 133–40
  by sheep, 129–31
  factor, 132–3
  pressure, 261
  regime, 496, 560
"Great Ice Age", 149
Great Ridge Wood (Wiltshire), 174
"greens", 560, 561
grey dunes, 850, 853, 855
grey poplar (*Populus canescens, q.v.*), 259
ground layer, 214, 276
grouse (*Lagopus scoticus*), 4, 751, 757
grouse moor, 124, 128, 751, 765
Grovely Wood (Wiltshire), 174, 176 n., 177
guelder rose (*Viburnum opulus, q.v.*), 265

habitat, **215–16**
Hagley Pool (Oxford), 586, 588
Haines, F. M., on Hindhead Common soil,
  724–6
Hall, A. D. and Russell, E. J., on Kent and
  Surrey sands, 352
halophytes, 819, 845
halophytic vegetation, 99
halosere, 820, 825, 826
Hampshire basin, 107, 111, 125
Hampshire uplands, 107, 171, 180 n.
hares, 142
Harley, J. L., on beech mycorrhiza, 368, 387
  on two upland ashwoods, 433–6
Harris, G. T., on species from Wistman's
  Wood, 301
Hastings beds, 19, 113
hawthorn (*Crataegus monogyna*), **260–1**, 262,
  306, 353, 372, 373, 377, 379, 383, 384, 391,
  396, 397, 429, 432, 475, 476, 480, 657
  large fruited (*C. oxyacanthoides*), 260, 306,
  480
hawthorn scrub, 295, 372–5, 376, 377, 380, 382,
  383, 441, 480–4
  sere, 372, 373, 375, 382, 383
hawthorn-mercury sere, 384
hay crop, 179, 205
Hay Meads (Oxford), 568–70
hazel (*Corylus avellana, q.v.*), 154, 181, 203,
  259–60, 267, 306, 311, 382, 383, 429, 432,
  442, 445, 474
  coppice, 245, 250, 259, 267
  scrub, 260, 474
heat, 30, 78, 79

heath, 199–200, 202, 203, 351, 353, 354, 356,
  357, *and see* Callunetum
  and heather moor, 743–7
  formation, **723–72**
    distribution, 723–4, 763–5
    status, 765–6
  grasses, 353
  invasion by, 490, 500
  "pasture", 489
  plants, 490
heather, *see Calluna*
"heather moor", 128, 674, 743–7, 753–7
heathland, 176, 224
Hedera helix, **266**, 279, 293, 298, 300, 315, 331,
  336, 341, 343, 344, 368, 369, 382, 384, 392,
  429, 435, 437, 461, 473, 474
helophytes, 235
hemicryptophyta rosulata, 371
  scaposa, 371, 407
hemicryptophytes, 235, 369, 371, 487
Hercynian continent, 15, 16
Herefordshire plain, 14, 17 (Fig. 5), 121, 175
Hertfordshire sessile oakwoods, 304–11, 350–1
Heslop Harrison, J. W., on salt marsh, 825, 830,
  831
"high moor", 674
"Highland Line", 2 (Fig. 1), 9 (Fig. 4), 17
  (Fig. 5)
Highland oakwoods, 343–9
"Highland zone", 194
Highlands, *see* Scottish Highlands
Hill, T. G. and Hanley, J. A., on the water con-
  tent of shingle beach, 879
"hill pasture", 489, 499 n.
Hippophaë rhamnoides, 266, 864
Hippophaëtum, 864
historical factors, 216
Historical Period, the, 150 (Fig. 38), **171–93**
Hochmoor, 201, 673, 674
Holcus lanatus, 316, 318, 345, 496, 521, **538**,
  547, 550, 554, 564, 570, 572, 575, 666, 696,
  888
  mollis, 281, 285, 286, 301, 307, 308, 315, 318,
  325, 327, 345, 346, 353, 389, 392, 406, 412,
  413, 417–19, 452, 506, 507, 562
Holland, percentage of woodland, 196
holly (*Ilex aquifolium, q.v.*), **257**, 276, 306, 328,
  330, 336, 374, 391, 410, 412, 442
Honckenya peploides, 847, 884, 886, 889
honeysuckle (*Lonicera periclymenum, q.v.*), **266**,
  279, 353, 429
hooks (of shingle beaches), 869, 870–3 (Fig.
  168), 874 (Fig. 169), 875 (Fig. 170), 876
  (Fig. 171), 878
hop, *see Humulus lupulus*
Hope Simpson, J. F., on chalk grasses, 535
  on chalk bryophytes, 540–1
  on chalk spoil heaps, 546–8
Hopkinson, J. W., on Sherwood Forest, 352–4
  on Sherwood Forest Callunetum, 729
Horwood, A. R., on relict *Nardus*, 518
hornbeam (*Carpinus betulus*), **256–7**, 305, 311,
  374
  coppice, 305–7
hornblende, 81